AF831399

GIScience for the Sustainable Management of Water Resources

GIScience for the Sustainable Management of Water Resources

Edited by
Gowhar Meraj
Shruti Kanga
Majid Farooq
Suraj Kumar Singh
Sudhanshu

APPLE ACADEMIC PRESS

First edition published 2023

Apple Academic Press Inc.
1265 Goldenrod Circle, NE,
Palm Bay, FL 32905 USA

760 Laurentian Drive, Unit 19,
Burlington, ON L7N 0A4, CANADA

CRC Press
6000 Broken Sound Parkway NW,
Suite 300, Boca Raton, FL 33487-2742 USA

4 Park Square, Milton Park,
Abingdon, Oxon, OX14 4RN UK

© 2023 by Apple Academic Press, Inc.

Apple Academic Press exclusively co-publishes with CRC Press, an imprint of Taylor & Francis Group, LLC

Reasonable efforts have been made to publish reliable data and information, but the authors, editors, and publisher cannot assume responsibility for the validity of all materials or the consequences of their use. The authors, editors, and publishers have attempted to trace the copyright holders of all material reproduced in this publication and apologize to copyright holders if permission to publish in this form has not been obtained. If any copyright material has not been acknowledged, please write and let us know so we may rectify in any future reprint.

Except as permitted under U.S. Copyright Law, no part of this book may be reprinted, reproduced, transmitted, or utilized in any form by any electronic, mechanical, or other means, now known or hereafter invented, including photocopying, microfilming, and recording, or in any information storage or retrieval system, without written permission from the publishers.

For permission to photocopy or use material electronically from this work, access www.copyright.com or contact the Copyright Clearance Center, Inc. (CCC), 222 Rosewood Drive, Danvers, MA 01923, 978-750-8400. For works that are not available on CCC please contact mpkbookspermissions@tandf.co.uk

Trademark notice: Product or corporate names may be trademarks or registered trademarks and are used only for identification and explanation without intent to infringe.

Library and Archives Canada Cataloguing in Publication

Title: GIScience for the sustainable management of water resources / edited by Gowhar Meraj [and four others].
Names: Meraj, Gowhar, editor.
Description: First edition. | Includes bibliographical references and index.
Identifiers: Canadiana (print) 20220285772 | Canadiana (ebook) 20220285845 | ISBN 9781774910481 (hardcover) | ISBN 9781774910498 (softcover) | ISBN 9781003284512 (ebook)
Subjects: LCSH: Water-supply—Management—Geographic information systems.
Classification: LCC TD345 .G57 2023 | DDC 628.10285—dc23

Library of Congress Cataloging-in-Publication Data

CIP data on file with US Library of Congress

ISBN: 978-1-77491-048-1 (hbk)
ISBN: 978-1-77491-049-8 (pbk)
ISBN: 978-1-00328-451-2 (ebk)

About the Editors

Gowhar Meraj (lead and corresponding editor)

Dr. Gowhar Meraj is currently working as a Young Scientist Fellow under the Department of Science and Technology, Government of India's Scheme for Young Scientists and Technologists (SYST) at the Department of Ecology, Environment and Remote sensing, Government of Jammu and Kashmir, India. Previously, he has worked as a Programme Officer on the ENVIS Scheme of the Ministry of Environment, Forests, and Climate Change (MoEF&CC) for five years. He has also worked as a consultant with World Bank Group, New Delhi, on its South Asia Water Initiative Program. He has more than 10 years of experience in research and teaching in various fields related to satellite remote sensing and geographic information science (GIS), such as hazard vulnerability assessments, hydrological modeling, flood assessment, watershed management, natural resource management, and ecosystem services and modeling. He has more than 50 publications in various international and national journals. He is also serving as academic editor in PeerJ-Life and Environment Journal and International Journal of Ecology, Hindawi. He is on reviewer boards of various SCI journals such as *Information Fusion, Natural Hazards, Arabian Journal of Geosciences, Geocarto International,* etc. He earned MSc, MPhil, and PhD degrees in Environmental Sciences with a special research focus on spatial information sciences. He has also holds a postgraduate diploma in Remote Sensing and GIS.

Google scholar profile: https://scholar.google.com/citations?user=tLe6BeEAAAAJ&hl=en&oi=ao

Shruti Kanga

Dr. Shruti Kanga is currently working as Associate Professor and Coordinator at the Centre for Climate Change and Water Resources, Suresh Gyan Vihar University, Jaipur, India. She has previously worked as Assistant Professor in the Centre for Land Resource Management, Central University of Jharkhand (CUJ), Ranchi. She also served as a research associate in the Ministry of Environment, Forest, and Climate Change (MoEF) and GoI-funded project on forest fires and has teaching experience at different universities, including Central University of HNB Garhwal and the University of Jammu. She has attended and organized around 35 national and international conferences. She was also a course coordinator of several modules of EDUSAT-based distance learning program organized by the Indian Institute of Remote Sensing (IIRS), Indian Space Research Organization (ISRO), Government of India. She has 12 years of teaching and research experience. She did her PhD in Technology (Geoinformatics) at the Department of Remote Sensing, Birla Institute of Technology, Ranchi, India. She has worked in the area of forest fires risk modeling and management, tourism, resource management, etc.

Google scholar profile: https://scholar.google.co.in/citations?user=RSkmve8AAAAJ&hl=en&oi=ao

Majid Farooq

Dr. Majid Farooq is currently Scientist-D at the Department of Ecology, Environment and Remote Sensing, Government of Jammu and Kashmir, India. He has more than 15 years of experience in research, teaching, and consultancy related to remote sensing and GIS, such as climate change vulnerability assessments, flood modeling, ecosystem assessment, watershed management, natural resource management, and ecosystem services and modeling. He has more than 35 publications in various international and national journals. In addition, he has also served as a reviewer for various SCI journals, such as *Environmental Monitoring and Assessment* (Springer), *International Journal of Remote Sensing* (Taylor and Francis), etc. He has done his MTech in remote sensing and GIS from SRM University, Chennai, India.

Google Scholar profile: https://scholar.google.co.in/citations?user=6X2n7iMAAAAJ&hl=en

Suraj Kumar Singh

Dr. Suraj Kumar Singh is currently working as Associate Professor and Coordinator at the Centre for Sustainable Development, Suresh Gyan Vihar University, Jaipur, India. Previously, he worked as an Assistant Professor in the Centre for Land Resource Management, Central University of Jharkhand

(CUJ), Ranchi. He has also served as research associate at the National Remote Sensing Centre (NRSC) and at the Indian Space Research Organization (ISRO) and on GoI-sponsored projects, such as the National Wasteland Mapping and National Urban Information System for the northern parts of India. He has worked in the area of waterlogging and flood hazards, geospatial applications in water resources, disaster management, hydrogeomorphology, urban planning, and wasteland mapping. He was also a course coordinator of different modules of the EDUSAT-based distance learning program organized by the Indian Institute of Remote Sensing (IIRS), Indian Space Research Organization (ISRO), Government of India. He has 12 years of teaching and research experience. He had the responsibility of an NSS program officer at the Central University of Jharkhand and coordinator of Centre for Sustainable development, Suresh Gyan Vihar University, Jaipur, India. He earned his PhD degree in Technology (Geoinformatics) at the Department of Remote Sensing, Birla Institute of Technology, Ranchi, India.

Google Scholar profile: https://scholar.google.co.in/citations?user=wXlG45AAAAAJ&hl=en

Sudhanshu

Dr. Sudhanshu is the chief mentor at the Suresh Gyan Vihar University, Jaipur, Rajasthan, India. He is an expert geologist with over 30 years of experience in research and teaching.

Contents

Contributors

Ifeka Adolphus
Federal University of Technology, Akure, Ondo State, Nigeria; E-mail: adolphusifeka@yahoo.com

Ummar Ahad
Centre of Central Asian Studies, University of Kashmir, Srinagar, Jammu and Kashmir, India

Faheem Ahamad
Department of Environmental Science, Keral Verma Subharti College of Sciences (KVSCOS), Swami Vivekanand Subharti University, Meerut 250005, Uttar Pradesh, India;
E-mail: faheem.ahamad170390@gmail.com

Mukeem Ahmad
Department of Geology, Aligarh Muslim University, Aligarh, UP, India

Uzair Ahmed
Department of Geography and Geomatics, University of Peshawar, Pakistan;
E-mail: uzairahmedsparc770@gmail.com

Syed Ahmad Ali
Department of Geology, Aligarh Muslim University, Aligarh, UP, India

Umair Ali
Department of Earth Sciences, University of Kashmir, Srinagar, Jammu and Kashmir, India;
E-mail: umairgeo121@gmail.com

Sher Singh Baghel
Gateway College of Architecture and Design, Sonipat, Haryana131001, India

N. Balakrishnan
Water Technology Centre, TNAU, Coimbatore, India

Ekta Baranwal
Department of Civil Engineering, Jamia Millia Islamia, New Delhi 110025, India;
E-mail: ektabaranwal01@gmail.com

Vikrant Bhade
Department of Geography, Savitribai Phule Pune University, Pune, India

Rakesh Bhutiani
Limnology and Ecological Modelling Laboratory Department of Zoology and Environmental Science, Gurukula Kangri Vishwavidyalaya, Haridwar 249404, Uttarakhand, India

Sneha Rani Deb
Department of Geography, Gauhati University, Guwahati, Assam, India

P. Dhanya
Tamil Nadu Agricultural University, Coimbatore, Tamil Nadu, India; E-mail: dhanya.eptri@gmail.com

Ugwuegbu Ikechukwu Emmanuel
Nigerian Meteorological Agency, Abuja, Nigeria

Vinay Gangare
SISL Infotech, New Delhi, India

Abiy Gebremichael
Southern Agricultural Research Institute, Bonga Agricultural Research Center, Bonga, Ethiopia;
E-mail: gebremichaelaby@gmail.com

V. Geethalakshmi
Tamil Nadu Agricultural University, Coimbatore, Tamil Nadu, India

Himanshu Govil
Department of Applied Geology, National Institute of Technology, Raipur, India

Subhanil Guha
Department of Applied Geology, National Institute of Technology, Raipur, India;
E-mail: himgeo@gmail.com

Shazia Gulzar
The Fazal Haq College, Mardan, Pakistan

N. Jayaraju
Department of Geology, Yogi Vemana University, Kadapa 516005, Andhra Pradesh, India

Senthilraja Kandasamy
Directorate of Crop management, Tamil Nadu Agricultural University, Coimbatore, India

Avinash Kandekar
Department of Geography, Savitribai Phule Pune University, Pune, India

S. Karthigayini
Department of Agricultural Engineering and Soil Science, Rajarata University of Sri Lanka;
E-mail: ammusacthivel@gmail.com

Mona Khakhar
CEPT University, Ahmedabad, India; E-mail: mona.khakhar@gmail.com

Fazli Amin Khalil
Department of Electronics, University of Peshawar, Peshawar, Pakistan

Tanzeel Khatoon
Department of Geology, Aligarh Muslim University, Aligarh, UP, India

Binay Kumar
Centre for Development of Advanced Computing, Pune, India

C. Navin Kumar
Water Technology Centre, Tamil Nadu Agricultural University, Coimbatore 641003, Tamil Nadu, India;
E-mail: navindpi@gmail.com

P. Kumaresan
Water Technology Centre, TNAU, Coimbatore, India; E-mail: acfkumagri@gmail.com

B. Lakshmanna
Department of Geology, Yogi Vemana University, Kadapa 516005, Andhra Pradesh, India

Abaas Ahmad Mir
ICSSR, New Delhi, India

Alka Mishra
Department of Rural Technology and Social Development, Guru Ghasidas University, Bilaspur (C. G.)

Rajashree Naik
Department of Environmental Science, School of Earth Sciences, Central University of Rajasthan, Bandarsindri-305817, Ajmer (Raj) India

Kumarasen Palaniappan
Water Technology Centre, Tamil Nadu Agricultural University, Coimbatore, India

Mayank Pandey
Department of Rural Technology and Social Development, Guru Ghasidas University, Bilaspur, India; E-mail: makepandey1995@gmail.com

S. Panneerselvam
Water Technology Centre, TNAU, Coimbatore, India

Sudarshan Singh Parmar
School of Electronics Devi Ahilya University, Indore, MP, India

Smit Patel
CEPT University, Ahmedabad, India

S. Pazhanivelan
Department of GIS & RS, TNAU, Coimbatore, India

S. Arul Prasad
Water Technology Centre, Tamil Nadu Agricultural University, Coimbatore 641003, Tamil Nadu, India

Dhanya Praveen
Directorate of Crop management, Tamil Nadu Agricultural University, Coimbatore, India

Nadia Qayum
Department of Meteorology, COMSATS University, Islamabad, Pakistan

Rimsha Qureshi
Department of Geography and Geomatics, University of Peshawar, Pakistan

K. P. Ragunath
Department of GIS & RS, TNAU, Coimbatore, India

M. Rajasekhar
Department of Geology, Yogi Vemana University, Kadapa, Andhra Pradesh 516005, India

G. Sudarsana Raju
Department of Geology, Yogi Vemana University, Kadapa, Andhra Pradesh 516005, India; E-mail: gsraju05@gmail.com

Mugilan Raman
Department of RS & GIS, Tamil Nadu Agricultural University, Coimbatore, India

Mukesh Ruhela
Department of Environmental Engineering (SITE), Swami Vivekanand Subharti University, Meerut-250005, Uttar Pradesh, India

S. Sahana
Department of Biosystems Technology, Eastern University of Sri Lanka, Sri Lanka

Abid Sarwar
Directorate of Soil Survey, Agriculture Department, Govt. of KP, Khyber Pakhtunkhwa, Pakistan

Priyanka Sengupta
Department of Geography, Gauhati University, Guwahati, Assam, India;
E-mail: senguptapriyanka741@gmail.com

Vinothkanna Shanmugam
Department of Geography, Madurai Kamaraj University, Madurai, India

Laxmi Kant Sharma
Department of Environmental Science, School of Earth Sciences, Central University of Rajasthan, Bandarsindri-305817, Ajmer, India; E-mail: laxmikant_evs@curaj.ac.in

Vineesha Singh
Department of Earth Sciences, Barkatullah University, Bhopal, MP, India; E-mail: Vgangare@gmail.com

G. Sreenivasulu
Department of Geology, Yogi Vemana University, Kadapa 516005, Andhra Pradesh, India;
E-mail: seenu9441@gmail.com

Y. Sreenivasulu
Department of English, SSL, Vellore Institute of Technology, Vellore, Tamil Nadu 632014, India

S. L. Swami
College of Agriculture and Research Station, Korba, IGKV (C.G.), India

Pawan Thapa
Department of Geomatics Engineering, Kathmandu University, Dhulikhel, Nepal;
E-mail: pawan.thapa@ku.edu.np

N. Thavaprakaash
Water Technology Centre, Tamil Nadu Agricultural University, Coimbatore 641003, Tamil Nadu, India

B. Upendra
ESSO-National Centre for Earth Science Studies, Thiruvananthapuram 695011, Kerala, India

Geethalakshmi Vellingiri
Directorate of Crop management, Tamil Nadu Agricultural University, Coimbatore, India

Guhan Velusamy
Agro Climate Research Centre, Tamil Nadu Agricultural University, Coimbatore, India;
E-mail: guhanthiran@gmail.com

Anjana Vyas
CEPT University, Ahmedabad, India

Swapnil Vyas
Department of Geography, Savitribai Phule Pune University, Pune, India

Editorial Advisory Board

Prof. M. S. Nathawat
School of Sciences
Indira Gandhi National Open University
New Delhi 110068, India

Dr. (Mrs) Tapati Banerjee
National Atlas & Thematic Mapping Organisation
Department of Science and Technology
Government of India
Kolkata 700064, India

Prof. Milap Punia
Centre for the Study of Regional Development
Jawaharlal Nehru University
New Delhi 110067, India

Prof. Rajendra Prasad
Department of Physics
Indian Institute of Technology (BHU)
Varanasi 221005, India

Dr. Devendra Pradhan
Indian Meteorological Department
Ministry of Earth Sciences
Government of India
New Delhi 110003, India

Prof. Manoj K. Pandit
Department of Geology
University of Rajasthan
Jaipur 302004, India

Dr. Snehmani
Snow and Avalanche Study Establishment
RDC, DRDO
Chandigarh 160017, India

Prof. Shakeel Ahmed
Department of Geography
Jamia Millia Islamia
New Delhi 110025, India

Prof. Suresh Prasad Singh
Himalayan University
Itanagar 791111, India

Mr. Peeyush Gupta
NMCG, Ministry of Jal Shakti
Government of India
New Delhi, India

Abbreviations

ACZ	agro climatic zone
AGNPS	agricultural non-point source pollution
AMC	Ahmedabad Municipal Corporation
ANSWERS	Areal Nonpoint Source Watershed Environment Response Simulation
API	air pollution index
AQI	air quality indices
AR	artificial recharge
ARS	artificial recharge structures
ASTER	Advanced Spaceborne Thermal Emission and Reflection Radiation
BIS	Bureau of Indian Standard
BOD	biological oxygen demand
CB	contour bund
CD	check dam
CDOM	color dissolved organic matter
CI	consistency index
CPCB	Central Pollution Control Board
CPHEEO	Central Public Health and Environmental Engineering Organisation
CR	consistency ratio
CREAMS	Chemical Runoff and Erosion from Agricultural Management Systems
CZCS	Coastal Zone Color Scanner
DD	drainage density
DEM	digital elevation model
DO	dissolved oxygen
DOM	dissolved organic matter
DOS	department of space
DSSAT	decision support system for agro technology transfer
EC	electrical conductivity
EIA	effective impervious area
EQI	environmental quality index
ESA	European Space Agency

ETM	enhanced thematic mapper
FP	farm pond
GIS	geographic information system
GM	geomorphology
GMC	Guwahati Municipal Corporation
GPS	global positioning system
HSG	hydrological soil groups
ICMR	Indian Council of Medical Research
IDF	intensity duration frequency
IMD	India Meteorological Department
ISC	impervious surface change
LASCAM	large-scale catchment model
LDI	landscape development index
LDN	land degradation neutrality
LISEM	Limburg Soil Erosion Model
LULC	land use land cover
LID	low-impact development
MAR	managed aquifer recharge
MERIS	medium resolution imaging spectrometer
MF	mesic forest
MGWB	model ground water bill
MMF	Morgan-Morgan-Finney
MODIS	Moderate Resolution Imaging Spectroradiometer
MoEFCC	Ministry of Environment, Forest and Climate Change
MOS	modular optical scanner
MUSLE	modified universal soil loss equation
NAPCC	national action plan on climate change
NDVI	normalize differential vegetation map
NDWI	normalized difference water index
NRCD	national river conservation directorate
NRCP	National River Conservation Programme
NRDWP	National Rural Drinking Water Programme
NWFL	national water framework law
NWM	national water mission
NWP	national water policy
OM	organic matter
PCM	pairwise comparison matrix
PE	potential evaporation
PT	percolation tank

QI	quality indices
RCMs	regional climate models
RCP4.5	Representative Concentration Pathway 4.5
RMF	rich mesic forest
RPI	river pollution index
RS	remote sensing
RT	radiative transfer
RUSLE	revised universal soil loss equation
SAR	synthetic aperture radar
SEDD	sediment delivery distributed
SLEMSA	Soil Loss Estimation for South Africa
SOI	Survey of India
SRTM	Shuttle Radar Topographic Mission
SWAT	soil and water assessment tool
SWC	soil water conservation
SWM	stormwater management
TDS	total dissolved solids
TM	Thematic Mapper
WEPP	water erosion prediction project
WOA	weighted overlay analysis
WQI	water quality index
WRI	World Resource Institute
WRM	water resource management
YAP	Yamuna Action Plan

Introduction

Water is one of the most critical resources of nature that sustains life both in natural and artificial ecosystems. The uneven spatiotemporal distribution of water resources is one of the vital factors responsible for various anthropogenic pressures (e.g., pollution) that we are facing in the 21st century. Due to our imperfect knowledge about the distribution of water resources, it has become incumbent that spatial information techniques are used for understanding the root causes behind the degradation of our water resources. Satellite remote sensing provides essential data to map water resources, hydrology flux measurement, monitoring drought, and flood inundation. The geographic information system (GIS) provides the best tools for modeling and assessing water resources for drought flood risk management.

Even in the absence of a complete understanding of the processes and relationships involved or sufficient data, the construction of flow charts and mind maps can help develop an appreciation of the issues and help build consensus among the various stakeholders. GIS has influenced the development and implementation of a hydrological model at several different levels. GIS has been used to address water supply, water quality, and stormwater management problems and allows users to run more traditional lumped models more efficiently and include at least some degree of spatial effects by partitioning the entire watershed into smaller subwatershed. GIS has been used to transform what was originally a specific model into a spatially distributed model. GIS is used for a variety of input models and comparing the model output with field data to improve the scientific basis of policy and the critical water quality management plan. Management of water resources is crucial as we look for ways to build environmentally and socially sustainable societies and lifestyles.

In some cases, we need to find new methods to alter water resources supply and demand. In other cases, we need to find an effective strategy that is faster and more effective in identifying pollution sources. GIS can contribute to solving the problems in each of these things. The GIs technology can help guide the implementation of water resources policies and promote a more efficient allocation of natural resources and the fair and the community as we strive to achieve the above goals. This volume contains chapters from eminent researchers and experts.

Sustainable exploitation of water resources requires planning and control methods that allow incorporating a large number of spatial and temporal variables. Because of its features, GIS seems to be the most suitable tool to aid in managing water resources available. GIS and modeling can make an essential contribution to integrated water resources management: indeed, given the scarcity of public health and environmental data, some form of modeling tends to be a prerequisite.

The primary focus of this book is to fill the gap in the available literature on the subject by bringing the concepts, theories, and experiences of specialists and professionals in this field. The editors have worked hard to bring the best literature in this field in a book form for helping students, researchers, and policymakers develop a complete understanding of the vulnerabilities and solutions to the whole environmental system. This publication is ideally designed for urban planners, environmentalists, policymakers, ecologists, researchers, academicians, students, and professionals in remote sensing, civil engineering, social science, computer science, and information technology.

The book is based on clear conceptual understanding, internationally acclaimed authorship, the latest research on the subject, and an understanding of hydrology spatially while providing a holistic treatise on water science and its management. We hope the book shall do service to humanity.

We acknowledge the help of all the reviewers who tirelessly read chapters and sent suggestions to authors that greatly enhanced their quality and prospective reach. Special thanks to Dr. Neelu Gera, Dr. Muhammad Muslim, and Dr. Muzamil Amin for their valuable suggestions during the book's proofreading. Finally, we thank our families who supported us in thick and thin at each stage of our lives; nothing would have been possible without their help and support.

Editors
Gowhar Meraj
Shruti Kanga
Majid Farooq
Suraj Kumar Singh
Sudhanshu

Preface

Chapter 1 introduces low-impact development (LID) measures to increase infiltration capacity in urban areas. This study describes the alarming issues concerning the rate at which the urban population is growing worldwide. This chapter focuses on a flooding case study in Ahmedabad city's west zone, which is entirely built up and incorporates an assessment of the stormwater network's connectivity. This study includes identifying areas that flood when it rains in the monsoon season and checking the network capacity. LID structures suitable for locations with such characteristics are determined and the impact of the LID structures on water removal is examined. Spatial data like stormwater drainage network, ground level, and catch basins' invert level, digital elevation, and contour maps are all used in this study. An efficiency of 67% of the drainage network was found to carry stormwater. The chapter concludes that flooding occurred primarily due to network and local features such as topography/slope and land use land cover (LULC).

Chapter 2 focuses on remote sensing and GIS-based satellite images playing a critical role in monitoring groundwater resources and water spread in tanks in coastal Tamil Nadu, India. This chapter discusses a case study on water resources and crop area diversification using remote sensing and GIS and Tools for Natural Information Management in the African Great Lakes (TNIAMP). The study used Sentinel 1A, and RISAT, and assessment was done for optical data during the cropping season. Sentinel 1A SAR data were used to pinpoint and aid relief efforts after the recent floods in Tamil Nadu. The study's main objective is to monitor water resources to keep track of water availability changes, particularly water dispersion and sustainable crop intensification and diversification.

Chapter 3 discusses how protecting groundwater resources is significant today because of falling of water tables and precipitation occurring in arid and semiarid regions. This chapter is a case study that aimed at locating locations where artificial recharge could be applied and adapted, with an area of increasing groundwater in mind, in the district of Kadapa in India. The study concludes that with groundwater recharge maps that identify recharge areas are categorized into low, moderate, high, and very high potential. From the present study, it is found that 1.18% (1.99 km^2) of the study area is located in the high and high groundwater recharge zone has fallen to 54.14% (91.06 km^2). Meanwhile, a substantial 44.36% (74.61 km^2) of the land area and a smaller percentage of 0.32% (0.55 km^2) of the whole place falls under the ARS with reduced government funding.

Chapter 4 discusses a study to lay down the water and soil conservation measures in Solani Watershed, India. The chapter describes the role of GIScience in providing watershed researchers a new way to observe their landscape. The study carried out a morphometrical analysis of the Solani watershed in the Upper Ganga drainage basin to determine the priority subwatersheds (SWS) for soil conservation. Stream order ranged up to fourth order was found to have a dendritic to parallel drainage pattern. In the watershed, the drainage density is between 0.93 and 1.07 km/km^2. The analysis found that SWS-3, located in the Solani watershed, was the most vulnerable SWS in the watershed.

Chapter 5 uses a case study to understand and assess the erosion in Imo State, Nigeria. It demonstrates how it is essential to comprehend the causes of gully and flood erosion to understand the flooding and less fertility in the region. Flood and gully erosion vulnerability has been modeled using Landsat satellite data (e.g., Landsat TM, ETM+, and OLI) for 1986, 2000, and 2015. The study concludes that 62.9% of Imo State was in a low-risk area, 17.2% was in the medium-risk area, and 19.9% was in the high-risk area due to the fuzzy overlay. About 62 of the 214 regions that fall into the low-risk category are rural towns and villages, whereas 64 of the 214 regions that fall into the medium-vulnerable category are also rural towns and villages, and 52 of the 214 regions that fall into the highly vulnerable category are also rural towns and villages. An estimate of soil loss is between 0 and 123,922 ha/ton/year. The map that indicates erosion risk was categorized into four risk levels: low, moderate, high, and very high. Approximately 87.3% of the Imo state is located in the very low-risk area, and approximately 8.1% is in the low-risk area.

Chapter 6 discusses GIScience's role as an effective method for identifying possible groundwater locations based on satellite-based remote sensing data. Further, they used the multifaceted influencing factors (MIF) approach for developing the groundwater potential zonation in any area. The chapter attempts to find out if Matta Tehsil, Swat, Pakistan, has any potential groundwater zones. The geospatial analysis has been performed using ArcGIS 10.4 and multifaceted influencing factors (MIF) techniques. This study suggests that very high GWPZ cover 8.86 km^2, high, 115.27 km^2, 335.53 km^2, and high, respectively, each of which has potential for groundwater, respectively, low or promising potential.

Chapter 7 discusses the assessment of the hydrological characteristics of a river basin for its management strategies. The fundamental components of a fluvial landscape are drainage basins. This chapter explores the Parambikulam Aliyar basin's morphometric characteristics through GIScience to assess its hydrological behavior. The chapter studies the basin's morphometric characteristics, including stream order, stream length, drainage density, bifurcation ratio, and other details. The chapter clarifies the traditional approach's primary challenge; the difficulty of comprehensively covering the entire area and doing ground-level inspections of stream networks is catered well by GIScience. Using digital elevation models (DEMs), the hydrological flow can be detected from higher to lower elevations. It helps in creating an automated stream module with additional thematic layers using the geoprocessing tools. Geoprocessing can be used to supplement morphometric analysis results.

Chapter 8 investigates the strategies and methods used in applying web-based technology to exchange geospatial data on the Internet. Many blocks in Rajasthan, Madhya Pradesh, and Gujarat have been labeled as groundwater-depleted zones in the CGWB's annual report. This study has tried to offer a concrete case study of groundwater depletion at the block level. A distributed GIS system or the web comprises dynamic client/server configurations where client and server roles are switched depending on who needs to perform specific GIS analysis tasks. Geospatial information can be accessed via the web interface, and with web-enabled GIS, users have access to geospatial data. Different geospatial datasets were obtained by identifying various features and objects that can be extracted from multiple satellites. This web-based interface makes it possible to identify and then start up a process for each relevant piece of spatially referenced data and monitor the process's output so that the client can access it.

Chapter 9 addresses erosion in a Himalayan watershed by identifying areas vulnerable to soil erosion so that protective measures can be put in place to help reduce the long-term damage caused by soil erosion in the Himalayan watershed of Lolab Kashmir. To identify areas with increased erodibility, subsurface linear, shape, and relief aspects were calculated and assigned a ranking based on their contribution to erodibility. The four classifications of the area (viz. high, medium, low, and very high) were based on soil conservation and management. The study concludes that most watersheds in this study area are at risk for soil erosion because of steep slopes, deep valleys, and high topography, leading to increased soil erosion in the upper loose soil layer covered by structured rock. The parameters of morphometry and other terrain characteristics are imperative in determining soil erosion and developing soil and water conservation policies, and limiting landslides in the region.

Chapter 10 discusses the role of geospatial approaches in discovering and quantifying COVID-19 viral diversity and abundance even when there are no field measurements. India had the most extended COVID-19 lockdown period, beginning March 24, 2020, and ending on May 30, 2020. In the absence of onsite inspections, ongoing scientific research in current studies seeks to examine India's largest inland saline wetland, Sambhar Salt Lake, to discuss its COVID-19 situation using remote sensing technology. The chapter correlates substrates with the algal sheets found in the literature.

Chapter 11 discusses the enormous potential for land degradation and climate change impact assessments in geospatial technologies. This chapter focused on investigating the long-term trends in moisture and aridity conditions and decadal project changes in peanut yield under RCP 4.5, using the DSSAT model. The study has calculated the moisture index using precipitation and potential evaporation data to investigate the temporal and spatial variations in the studied areas' aridity. The spatiotemporal assessment demonstrated a better grasp of experimental hydro-climatic conditions and crop yield changes with climate change. The study has shown that coastal aridity is likely to increase due to climate change. The PNUTGRO crop simulation model results have shown that peanut yields will decline at the end of the 21st century.

Chapter 12 assesses the effect of urban growth on landscape dynamics and natural resources by applying GIScience and using various datasets, including LANDSAT- TM 4-5, LANDSAT-8, between 1990 and 2020 Bilaspur, India. The authors discuss how people's constant migration from rural areas to urban areas to get employment, provide education, or maintain living standards change the urban centers' demography and infrastructure. The chapter describes how sustainable management of land resources is necessary to understand the complex interaction between landscape patterns and natural processes influencing many ecosystems like food, water quality, wood, climate regulation, pollination of the crop. Large-scale changes in the landscape in the last two centuries have been superimposed on long-term dynamics associated with climate variability.

Chapter 13 investigates the feasibility of constructing a major commercial port in Dugarajapatnam, Tupilipalem, Nellore district of Andhra Pradesh, India, and addresses a set of lip dynamics on-and-off. The study evaluates the relationship between coastal morphodynamics and the distribution of physicochemical parameters. Multivariate statistics were used to minimize data and derive a limited number of latent factors. The study shows the lagoon mouth opening and closing with monsoon and pre-monsoon uses Landsat 8 OLI/TIRS multitemporal satellite images dated October 27, 2014 (Monsoon) and August 27 of 2015 (premonsoon). Parameters such as pH, temperature, electrical conductivity, salinity, dissolved oxygen, and organic matter have been estimated. The study's observations included dominance Avicennia marina spp. in the mangrove swamps. At the Tupilipalem lagoon shore, higher salinity has been found during the lagoon mouth closing period. During the monsoon, the silica concentration has been observed to get increased from premonsoon to postmonsoon. Data revealed that lagoon mouth dynamics change detection occurs at irregular intervals.

Chapter 14 discusses how soil loss is a significant factor in depleting a critical watershed located in the mountains. Using Revised Universal Soil Loss Equation analysis, which integrates remote sensing data and GIS techniques, the soil erosion rate is assessed in the Charnawati Watershed, Nepal. The erosion rate in the low zone area is 58.55%. Approximately 18% area is classified as being on the verge of a high to very severe intensity rating, and the erosion rate is approximately 18%. Such information will help sustainable watershed and ecological management planning, especially around catchment areas supporting conservation areas.

Chapter 15 describes river Yamuna's state of the environment using satellite imagery, surveying, and identifying the factors causing the river's degradation. Four decades of satellite images have shown that the river Yamuna has drastically changed in size and flow direction over its entire length. Due to the continued lack of an adequate understanding of water resources management, India faces a continuing decline in the overall number of water bodies and a further reduction in water resources' dimensions and quality. This chapter looks at the study of water resources, their sources of degradation, and how they can be managed appropriately in India using the river Yamuna as a case study.

Chapter 16 focuses on the rates of deglaciation and recession over the last three decades, in and around Kangchenjunga National Park in Sikkim, India, and Kangchenjunga Conservation Area in Nepal. Global glacier receding is one of the most reliable indicators of climate change. Approximately 30% of the study area has snow accumulation areas, exposing ablation areas, ablation areas covered by debris, and scarp faces. Over 30 years, a 43.5-km^2 decrease was measured in the glaciers. Finally, the chapter addresses the impact of climate change on the glacier resources of the North Eastern Himalayas.

Chapter 17 reviews the literature concerning the role of remote sensing and GIS in water resources management. The chapter describes how GIS integrates the spatial analytical functionality for spatially distributed data and benefits the users for updating, managing, and presenting spatial data. Further, it perfectly elucidates how the multitemporal satellite data can measure different aspects of land degradation, forest cover, and land-use change, making it a perfect tool for solving multiple soil and water conservation problems.

Chapter 18 reviews the calculation methodologies regarding various environmental quality indices (water-quality indices or air-quality indices) in India. The overall role of EQI (air and water quality index is to quantify a source's quality by evaluating either one or the other filtration method and each constituent's concentrations in a sample. EQI has been used to convey the air and water pollution level to humans in India. The study demonstrates how human health and the environment are inextricably linked, and the quality of the environment depends on different parameters, including water, air, and soil. However, since a small portion of the population is aware of the pollution level in a particular part of the environment, this chapter addresses the need for mass awareness.

Chapter 19 discusses groundwater conservation's role in the sustenance of human life on the planet Earth. Further, the chapter aims to lay out a foundation for groundwater protection and groundwater contamination control. It discusses the occurrence of groundwater, its storage forms, characteristics of groundwater, and groundwater's influence on discharge. The chapter concludes that every human being must care for groundwater because of its effects on people's lives.

Chapter 20 addresses the Ramsar Wetland Deepor Beel's deteriorating condition, which forms part of the Brahmaputra Delta. The chapter reviews the research that has been done on the ecology of Deepor Beel, the potential impact that urban development may have on it, and various solutions proposed to address this problem. The chapter also discusses the threats and management plans for this wetland. Recent research endeavors that due to decreased absorption capacity of the Beel, the city now experiences artificial waterlogging during floods. The wetland's self-purifying capability is weakened because of pollution. Moving the dumping site to an alternative location will help protect the swamp. It is possible to utilize ecotourism's benefits to conserve a wetland through sound management of the ecosystem, monetary services, and better well-being for the community from "responsible tourism" in a natural setting.

Chapter 21 discusses the role of atmospheric boundary layers' science in assessing the water resources management possibilities and GIScience's role in its evaluation. The atmospheric boundary layer is divided into microscale, mesoscale, and gamma scale. Carbon fluxes occurring within and outside significantly impact crop growth and canopy. Landforms, vegetation, water bodies, soil, and buildings such as trees and other structures have also affected the microclimate. Studies have shown that the atmospheric boundary layer needs to be well researched out. It shall help crop production as the latter is directly affected by the increasing relative humidity, lowering canopy temperature, raising soil temperature, increasing the amount of photosynthetically active radiation, and raising leaf gas exchange.

PART I

Integrating tools of GIScience for Planning Sustainable Development at Landscape Level

CHAPTER 1

Managing Urban Stormwater for Sustainable Urban Development: A Geospatial Approach

MONA KHAKHAR*, SMIT PATEL, and ANJANA VYAS

CEPT University, Ahmedabad, India

Corresponding author. E-mail: mona.khakhar@gmail.com

ABSTRACT

Across the globe, urban population is increasing at an alarming rate. The increase in urban population causes the increase in impervious areas due to the changed land use land cover (LULC). This in turn significantly influences the hydrology of the area causing flooding in urban areas. In this research, an attempt has been made to adopt low impact development (LID) measures to increase infiltration in the study area. The present study analyzes flooding in the west zone of Ahmedabad city with 100% urban land use and served by the stormwater network connection. The study comprises verifying the network capacity followed by identification of the areas that get flooded during the monsoon season since past few decades. The appropriate LID structures for such locations are identified and their impact to reduce the flooding is analyzed. The study uses spatial data such as stormwater drainage network, ground level and invert level of catch basins, digital elevation, and contour map. The network efficiency to carry the stormwater was done using Bentley CivilStorm and was found to be 67%. It was derived that the flooding occurred primarily due to two reasons—insufficient capacity of network, and other local features existing in the area such as natural topography/ slope, LULC, and vicinity to existing network. The analysis indicated that

GIScience for the Sustainable Management of Water Resources. Gowhar Meraj, Shruti Kanga, Majid Farooq, Suraj Kumar Singh, & Sudhanshu (Eds.)
© 2023 Apple Academic Press, Inc. Co-published with CRC Press (Taylor & Francis)

after implementing the LID structures such as vegetated swale, infiltration trench, and bioretention in the identified 11 sites, 7.73 million L of water may be infiltrated into the ground out of 11.34 million L of flooded water. This would in turn retain the quality and quantity of water in urban areas. The study finally demonstrates that the flooded water can be reduced by 68% by using these practices in west zone of Ahmedabad.

1.1 INTRODUCTION

It is estimated by the United Nations that almost 68% of world's population will be living in urban areas due to better opportunities to earn a living (UN-DESA, 2018). With this trend in the world, the urban population in India is projected to add around 34% urban population by 2050 to the existing 31% of urban population in 2020 (UN DESA, 2018). The process of increase in urban areas is directly related to the rise in the impervious areas as observed in various studies across the world (Zhang and Wu, 2020) (Sofia et al., 2019; Sovan and Sahoo, 2017; Su et al., 2018; Wang et al., 2017; Coles, 2004; Sahoo and Sreeja, 2016). Dramatic increase in urban population results into significant changes in the land cover within the watersheds that can adversely affect the stream hydrology and modification of stream channels. The entire phenomenon may induce higher and earlier hydrograph peak flow rates, increased peak runoff volumes, and reduction in time of concentration (Farahmand et al., 2007; Leopold, 1968). The increase in the peak flows create pressure on the existing infrastructure and may further pose threat to lives of people, properties, and environment (Kourtis et al., 2020).

Besides, with the increasing urban population and the increasing LULC changes to house these individuals, cities are at an increased risk of the consequences of climate change (Kanga et al., 2017a; Avashia and Garg, 2020). The impact of change in LULC result into increasing instances of floods in urban areas as reported in India (Meraj et al., 2015, 2016; Kanga et al., 2020a; Avashia and Garg, 2020; Zope, 2016). Further, according to the study by Ali et al. (2014) the rainfall maxima at 1–3 day durations and 100 year return period may increase significantly as per the projected climate in future for majority of urban areas in India. As per recent study by Swerts (2014), half of India's population will be urban in less than 30 years resulting in large scale changes in the land cover. Similar observation is noted by Wang et al. (2017) with Indian subcontinent displaying 2274.62 km^2 of area as impervious surface change (ISC) with the rise in population for the decade from 2000 to 2010. As per other studies done in India, the effective imperviousness that may significantly influence the hydrology has increased within the range of 32.73–96.09% for seven different watersheds in Guwahati from 2006 to 2011 also demonstrated that for urban catchment of the Oshiwara River in Mumbai, the area in high flood hazard zone increased by 64.29% with the total flood hazard area increasing by 22.27% from 1966 to 2009 (Sahoo and Sreeja, 2016; Zope, 2016). Similar study was done for the city of Ahmedabad indicating the increase in the effective impervious area (EIA) in the range of 4.7–9.45% from the year 2000 to 2014 (Sankalp and Sahoo, 2017). The corresponding runoff has increased by 14% for the increased EIA as estimated by Sovan and Sahoo for the duration of the study. This is also evident as per many recent instances of floods in Ahmedabad during the monsoon rainfall events as experienced by the city dwellers since 2010 (NDTV, 2010; Indian Express, 2016; Times of India, 2020). Considering the percentage decadal population growth rate of 22.31 for Ahmedabad as compared to 19.17 for Gujarat and 17.64 for India from 2001 to 2011, it is important that the instances of urban flooding are reduced in the city of Ahmedabad. Thus, it is of utmost importance to maneuver the stormwater management (SWM) strategies in a holistic way that would reduce the stress on urban infrastructure and generate the opportunity to device resilient solutions to adapt to climate changes (Kanga et al., 2017a, b; Kanga et al. 2020b). Of the many possible options for the SWM, the use of low impact development (LID) controls is one of the successful measures that aim to return the developed areas/watersheds to the predevelopment hydrologic condition (Damodaram et al., 2010).

This research is divided in two stages. The first stage of the research assesses the existing stormwater network efficiency to carry the stormwater during the rainfall events in the city and the related local flooding. The second stage of the research involves analysis of the topography, land use, and vicinity to the existing stormwater network for the study area to derive the areas prone to flooding due to these multiple causative factors. Further, the research proposes the LID measures for few identified sites experiencing flooding within the study area. Finally, the study proposes LID measures for all the flooding locations and estimates the total volume of the runoff that can be attenuated due to the LID measures if they are implemented and thereby the benefit of recharging the groundwater for the study area.

1.2 STUDY AREA

The existing stormwater drainage network of the west zone Ahmedabad is considered in this study. West zone covers 36 km^2 area of Ahmedabad city.

Elevation of this area is varies between 41 meter and 57 meter from mean sea level. The annual rainfall is distributed during the monsoon from July to October. Average annual rainfall is around 800 mm (30 inches) in the study area, but several rainfall events with the intensity of 40 to 50 mm per hour or greater causes flooding in many areas. The west zone although has separate stormwater drainage network experiences flooding every year during monsoon at locations such as Vijay cross roads, Sabarmati road and several others. Figure 1.1 represents the study area.

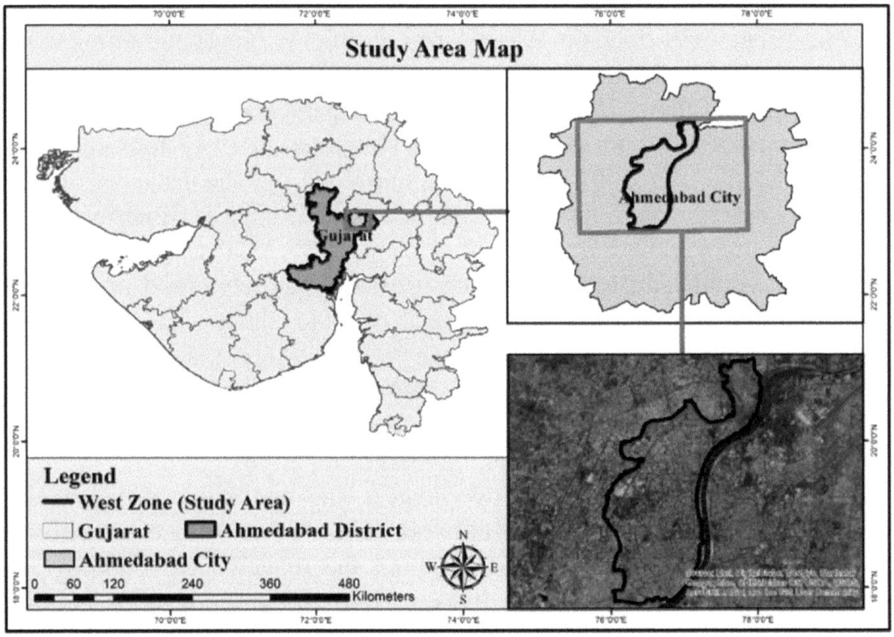

FIGURE 1.1 Location map of the study area.

1.3 METHODOLOGY

The flow diagram of the methodology used for this study is depicted in Figure 1.2.

The methodology involved acquiring various database from sources that are mentioned in Table 1.1. The first stage of analysis required digitization of the network to assess the network efficiency. It involved stormwater network components such as pipe diameter, lengths, manhole/catch basins and their individual catchment areas, ground level, invert level, and material of pipe.

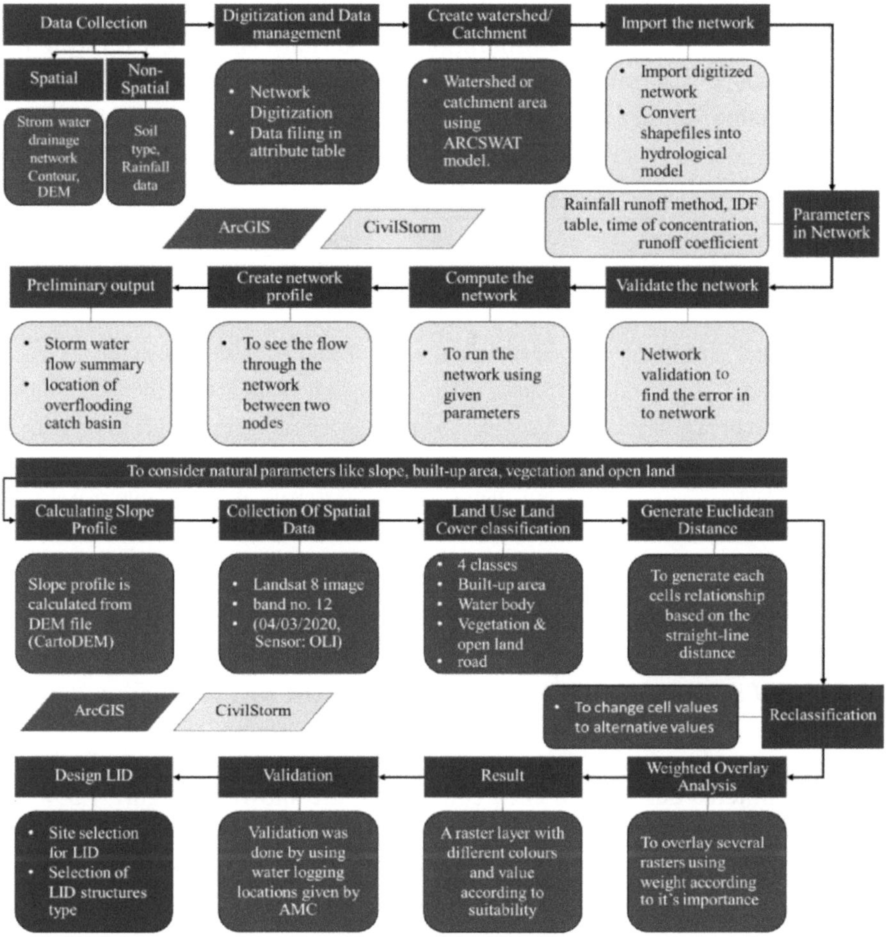

FIGURE 1.2 The flow diagram of the methodology.

Catchment areas were derived for each of the catch basin or manhole using Digital Elevation Model (DEM) using Soil and Water Assessment Tool (SWAT), that is, ArcSWAT in ArcGIS 10.5 (Texas A and M University, 2020). The SWAT is a tool that can be used to simulate the quality and quantity of surface and groundwater and predict the environmental impact of land use, land management practices, and climate. Network performance is analyzed using Bentley CivilStorm to derive the network efficiency. Further, the locations prone to flooding are also derived using LULC, slope and vicinity of the network parameters employing weighted sum tool in ArcGIS 10.5 (Rashid et al., 2013; Singh et al., 2017a, b; Meraj et al., 2018a, b). These locations were then validated with the sites experiencing frequent flooding as per Ahmedabad Municipal Corporation (AMC) database. The LID structures were proposed at all such locations and the runoff attenuation and infiltration was estimated for all such measures.

TABLE 1.1 Data for the Study Area and its Sources.

Data	Format	Source
Stormwater drainage network	PDF	AMC Central Zone Office
Ground level and invert level	Tabular	AMC West Zone Office
DEM file (cell size 30 m)	Spatial data	CartoDEM, Bhuvan portal
Contour map	Spatial data	AMC West Zone Office
Rainfall data of last 50 years	Maximum hourly intensity per year	Meteorological Centre, Ahmedabad, Indian Metrological Department
Soil type data	Soil map	Narmada, Water Resource, Water Supply and Kalpsar Department

1.4 RESULTS AND DISCUSSION

1.4.1 DERIVATION OF THE CATCHMENT FOR NETWORK

The catchment and subcatchment areas were obtained for the network area. The catchments derived for each catch basin are shown in Figure 1.3.

The digitized network is imported to Bentley CivilStorm. The property of the network components such as length, diameter, ground level, invert level, ground elevation for catch basin are assigned to the shapefile using model builder tool in CivilStorm. Subsequently, the hydrological model is created for further calculation of runoff volume with selected parameters and method.

1.4.2 ESTIMATION OF RAINFALL RUNOFF

The rainfall runoff was estimated using the rational method as recommended by Central Public Health and Environmental Engineering Organisation (CPHEEO), (CPHEEO, 2019). This method is used for stormwater drainage studies for Ahmedabad city by other researcher also (Rawal, 2018; Gajjar and Dholakia, 2014). The estimation of rainfall runoff using rational method is shown as below:

$$Q_p = 10\,C\,I\,A$$

Where, Q_p: Peak flow in m^3/hr
C: Coefficient of runoff
I: Average rainfall intensity in mm/h
A: Area of catchment in hectares

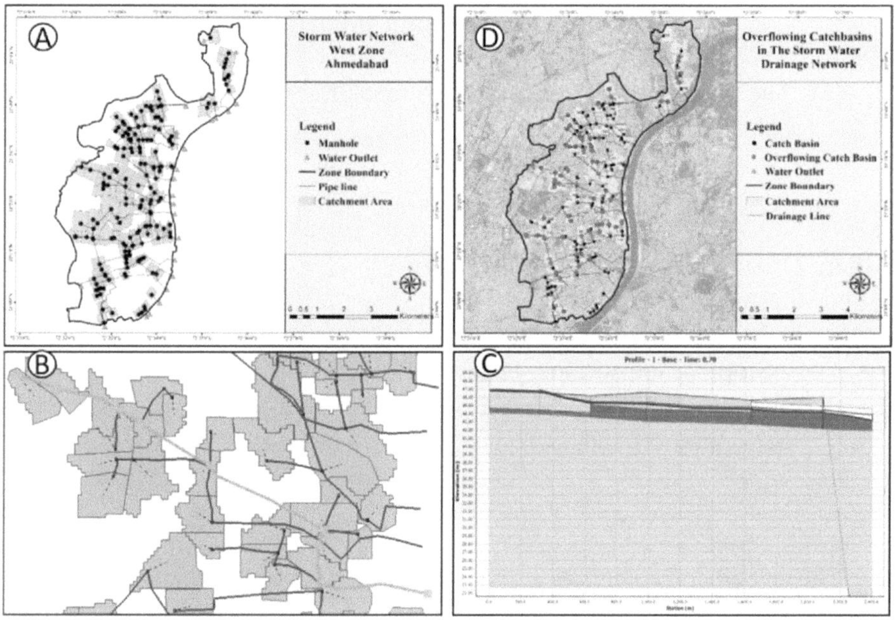

FIGURE 1.3 (a) Stormwater network and catchments for the catch basins, (b) network profile path, (c) sample analysis of network profile, and (d) overflowing catch basin locations.

For this study area, it was important to avoid frequent flooding. With the recent past experiencing frequent events of flooding and the limited scope to update the infrastructural facilities for stormwater, the design return period of 2 years, that is, 44 mm/h for the SWM strategy is adopted. Considering the mixed land use for the study area, the average runoff coefficient of 0.8 is considered as per CPHEEO (CPHEEO, 2019). The rainfall intensity was arrived using the intensity duration frequency (IDF) relationship on the basis of maximum hourly rainfall data from Meteorological Centre, Ahmedabad, India Meteorological Department (IMD). The rainfall intensity in the rational method is the average rainfall intensity over a given duration equal to the time of concentration (t_c) for the drainage area. Time of concentration (t_c) is defined as the time taken by the rainfall runoff to travel from farthest point to catchment outlet. Time of concentration is given by following equation. The sample calculations for the time of concentration (t_c) and velocity are shown in Table 1.2.

$$t_c = t_o + t_f,$$

where t_o: Time for the surface flow to reach the first inlet

t_f: Time to flow through the storm drainage system to the point of consideration

(a) The time for the surface flow to reach the first inlet, that is, (t_o) is as below:

$$t_o = \frac{0.994(1.1 - C)L^{0.5}}{S^{0.333}},$$

where t_o: Time of surface flow (minutes)

C: Runoff coefficient

L: Length of flow on the surface (m)

S: Slope of the surface, in percentage (%)

(b) Time to flow through the storm drainage system to the point of consideration, that is, (t_f) is given as below

$$t_f = \frac{L_{drain}}{V}.$$

The velocity of water flow in m/s is calculated from the Manning's equation.

$$V = \frac{1}{n} R^{0.67} S^{0.5},$$

where, V: Velocity of Flow, m/sec

t_f: Time of travel, minutes

n: Manning's coefficient

R: Hydraulic radius (m)

S: Longitudinal slope

1.4.3 ESTIMATION OF NETWORK EFFICIENCY

The network was run in the Bentley CivilStorm with all the input parameters mentioned above. Figure 1.3 (b) and (c) represents the network profile path and sample analysis of network profile respectively in the CivilStorm. The network results for the west zone show that there is total inflow volume of

353 million L, total outflow volume is 236 million L, and overflow volume is 60 million L. The network summary indicates a gap of 56 million L of rain water that may be attributed to infiltration into the ground or the losses in the network. Thus, it is estimated that the network efficiency is around 67% in the study area. Figure 1.3 (d) shows the overflowing catch basins in the map. Out of 172 catch basins in the network, 70 catch basins overflow during storm events.

TABLE 1.2 Sample Calculation for Time of Concentration and Velocity of Flow.

Sr. no.	Pipe length (m)	Dia (mm)	Hydraulic radius (mm)	Slope	Velocity (mm/s)	t_0 (min)	t_f (min)	t_c (min)
1	33.00	1800	900.00	0.004	463.91	3.44	0.07	3.51
2	43.60	1050	525.00	0.002	228.60	5.48	0.19	5.67
3	80.70	600	300.00	0.004	222.21	4.25	0.36	4.62
4	91.80	600	300.00	0.003	192.44	5.45	0.48	5.93
5	96.00	450	225.00	0.001	91.63	7.69	1.05	8.74
6	99.00	900	450.00	0.001	145.78	6.84	0.68	7.52
7	99.30	600	300.00	0.001	111.10	8.28	0.89	9.17
8	106.20	600	300.00	0.001	111.10	8.04	0.96	9.00
9	117.30	450	225.00	0.001	91.63	7.94	1.28	9.22

1.4.4 IDENTIFICATION OF FLOODING LOCATIONS

The overflow from the 70 catch basins shown in Figure 1.3 (d) may cause flooding in the nearby areas according the topography of the area. Besides, the flooding may also be caused due to land use when the stormwater cannot reach/access the network. Therefore, the flooding may be attributed to two reasons in the study area—due to insufficient capacity of network and natural topography condition of study area (Sharma and Kanga, 2020; Kanga et al., 2020b).

In order to derive all possible locations of flooding, it is necessary to thoroughly study the LULC and topography of the study area (Altaf et al., 2013; Bhatt et al., 2017). Hence, the parameters such as slope, built-up area, water body, open land, distance from the catch basin, and conduit were considered for the weighted overlay in ArcGIS 10.5 (SMR Eslami, 2011). It may be noted that the locations more prone to flooding during the storm event may also be considered favorable for proposing the LID measures.

1.4.5 SLOPE CALCULATION

The stormwater tends to flow through the areas that have greater slopes and will get stagnated resulting into inundation/flooding at places with flat slopes. Apparently, these may also be the sites/locations where LIDs may be proposed. For slope generation DEM file with 30 m resolution from CartoDEM was used (Altaf et al., 2014). The slope percentage varies from 0 to 17% in the study area. Figure 1.4 (a) depicts the variation of slope in the study area indicating most of the areas with less than 5% slope with scattered pockets having higher degree of slope.

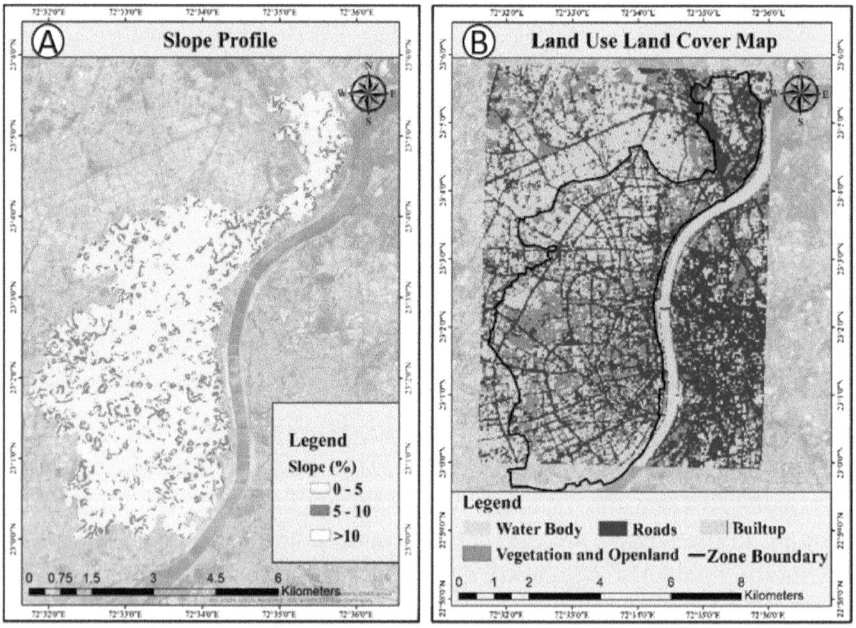

FIGURE 1.4 (a) Slope variation for the study area and (b) LULC for the study area.

1.4.6 LULC CLASSIFICATION

For weighted overlay analysis, LULC is the first important parameter comprising built-up areas, roads, water body, vegetation, and open land that will impact the stormwater movement. Besides, the best location for LID is near or within vegetation area and open land. LID structures should not be constructed in built-up areas. For the present study, the land use classification was done using Landsat 8 satellite image for March 4, 2020 in following four classes:

1. Water body
2. Built-up area
3. Roads and
4. Vegetation and open land

The land use classification was done using ERDAS IMAGINE software. Classified image is shown in Figure 1.4 (b) indicating the area under water-bodies as 2.96 km^2, roads with 25.59 km^2, built-up areas with 35.30 km^2, and the vegetation and open land with 16.59 km^2.

1.4.7 EUCLIDEAN DISTANCE

Euclidean distance is a common tool, mostly used in multicriteria analysis. This tool calculates the Euclidean distance to the closest source for each cell. In this study, the two feature class layers namely conduit and catch basin are used as source to estimate the Euclidean distance for each cell as shown in Figure 1.5 (a) and (b). The distance of 0–20 m and 20–50 m is defined from the conduit feature as in Figure 1.5 (a) class while 0–50 m, and 50–100 m distance is defined for the catch basin point feature class as in Figure 1.5 (b).

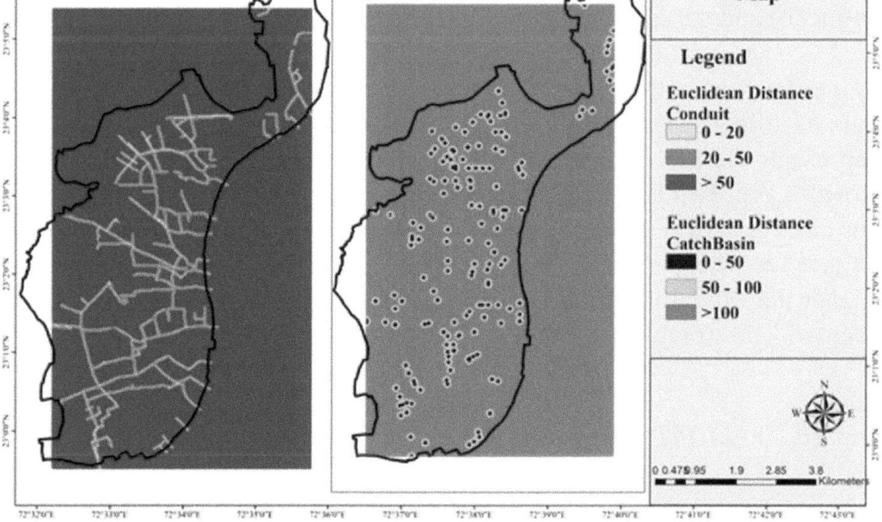

FIGURE 1.5 Euclidean distance map for the study area.

1.4.8 RECLASSIFICATION

Reclassify tool in ArcGIS is used for reclassification of the range or value for the LULC, slope, network conduit, and catch basin parameters. This tool changes the values or range values into raster file. To run this tool input raster must have valid statistics. For inputs with range of values, there should be no overlap. The reclassified values for the four raster layers viz. slope, LULC, conduit, and catch basin are shown in Table 1.3.

TABLE 1.3 Reclassified Values.

Slope		LULC		Conduit		Catch basin	
Old values	New values	Old values	New values	Old values	New values	Old values	New values
0–5	1	Water body	1	0–20	1	0–50	1
5–10	2	Vegetation and open land	2	20–50	2	50–100	2
> 10	3	Roads	3	> 50	3	>100	3
-	-	Built-up	4	-	-	-	-

1.4.9 WEIGHTED OVERLAY ANALYSIS

Weighted overlay tool is used to overlay several raster files or layers using measurement weights and scale according to its importance. The weights/influence and the scale values for each of the four layers—slope, catch basin, LULC, and conduit are assigned. The weights/influence value and the scale values for each of the parameter were adapted from SMR Eslami (2011). The sum of influence values of all the parameters should be 100. The field value represents the new reclassified values. Scale values are assigned from 1 to 9 with 1 representing lowest favorable and 9 representing highest favorable for placing the LID. Final result of weighted overlay analysis can be seen in Figure 1.6. The map shows the values for locations varying from 3 to 9. The higher the value for a location signifies the favorable location for the placement of LID structures or the location most prone to flooding and vice versa.

1.4.10 VALIDATION OF FLOODING LOCATIONS

The locations most prone to flooding/suitable locations for LID is overlapped with locations of the overflowing catch basins as shown in Figure 1.7 (a). It shows that majority of the overflowing catch basins coincide with area that is highest prone to flooding, that is, value 9. This confirms that the cause of flooding in the study area is due to network efficiency as well as due to factors such as LULC, slope, and the vicinity to the network. Further, the database for the locations experiencing frequent flooding was obtained from the AMC. Upon placing those locations on map as shown in Figure 1.7 (b) shows that out of 15 locations shared by AMC, 11 of the locations coincide with findings of this study.

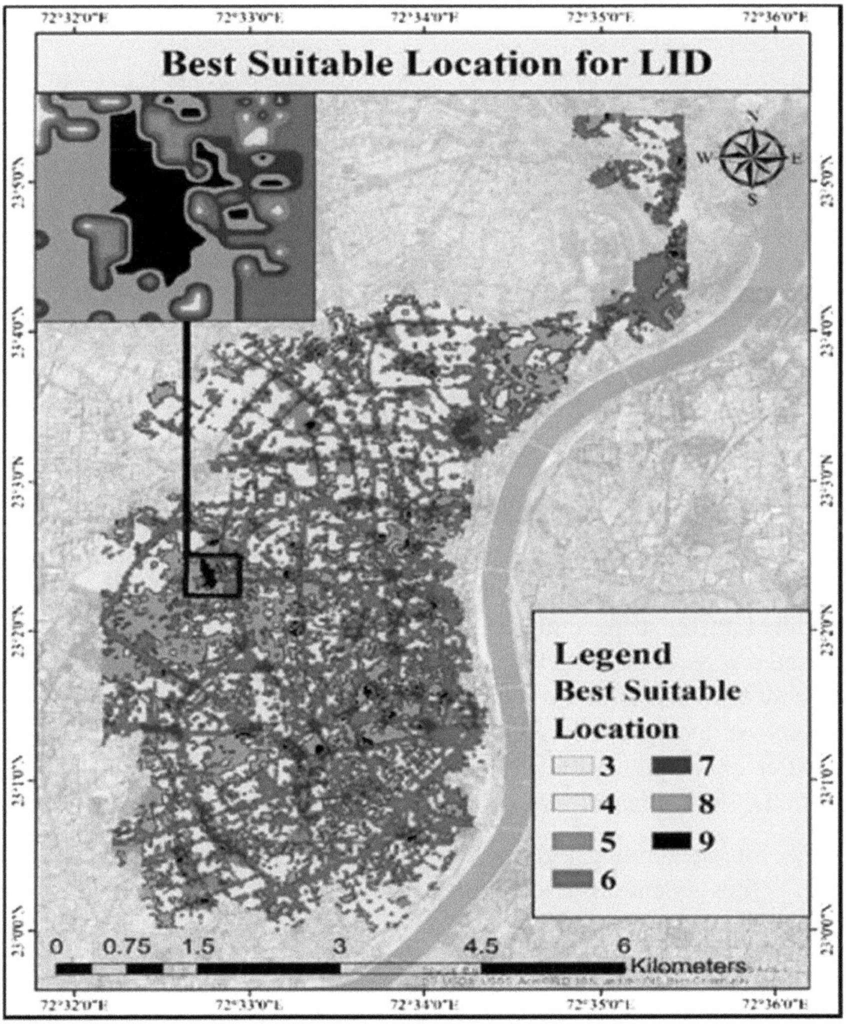

FIGURE 1.6 Suitable locations for LID/locations prone to flooding.

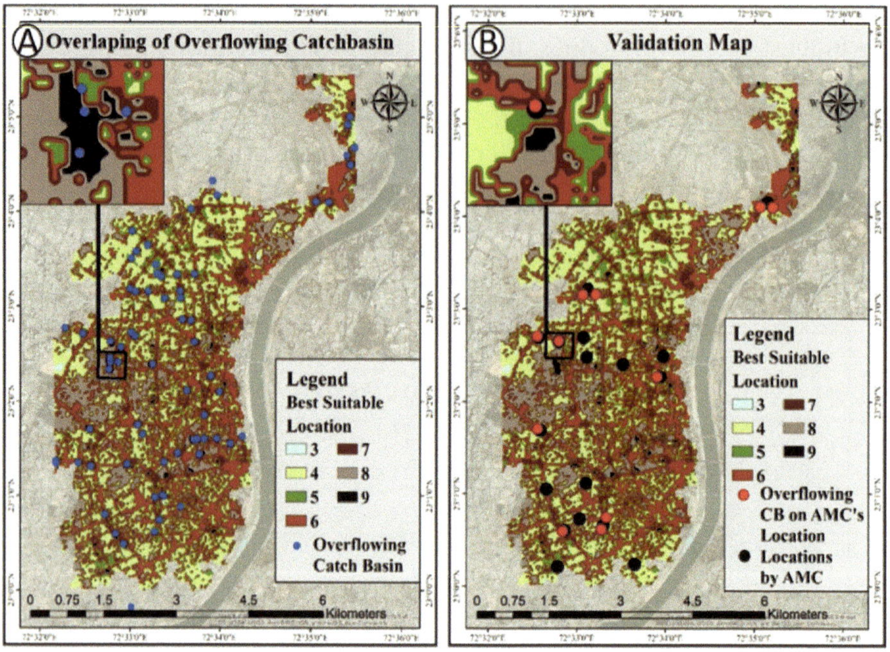

FIGURE 1.7 Weighted overlay for favorable location for LID measures.

1.4.11 SITTING OF LID STRUCTURES

LID measures entail diversified solutions to reserve and retain natural drainage flow (Pandey et al., 2010, 2013). Many types of LID structures are available in the world such as rooftop, vegetated swale, bioretention, infiltration trenches, vegetated buffer strips, and others. The selection of type of LID structure to be used at a particular site/location depends on presence of favorable aspects at the identified sites in the study area. LID practices are widely used for two reasons—reducing the flooding condition and retaining quality of water.

After observing study area and overflowing volume of the 11 flooding locations, 3 site categories were identified for the placement of LID structures:

1. Sites generating high volume of runoff and have large space availability.
2. Sites generating low volume of runoff and have large space availability.
3. Sites generating high volume of runoff and have very low space availability.

These site categories helped to establish the decision for selection of the LID structure. Besides, the physical parameters required for various LID structures were considered using guidelines from CASQA (2003). The combination of the physical parameters required to construct the LID along with the possible site category aided the decision to select the appropriate LID measure for all the 11 locations in the study area. On the basis of availability of land area, slope and the infiltration capacity of the soil, various LID measures were selected for all the 11 locations as shown in Table 1.4.

TABLE 1.4 LID structures for Flooding Locations.

No.	Name of the area	Area availability	Flood (million L)	volume	LID type
1	Khadyani Chali	No space	2 million L	No space	N.A.
2	Near Keshav Nagar	0.49 acres	1.9 million L	Low volume and small area	Bioretention
3	Near A.E.C Bridge	0.1 acres	1.1 million L	Low volume and small area	Bioretention
4	A.E.C Bridge to 132 ft R.R	0.88 acres	1.3 million L	Low volume and large area	Infiltration trench
5	Near Vijay Cross Road	4.1 acres	15 thousand L	Low volume and large area	Infiltration trench
6	Helmet Circle	8.5 acres	2.1 million L	High volume and large area	Vegetated swale
7	Commerce 6 Road to Shreeji Bag	0.65 acres	11 thousand L	Low volume and small area	Bioretention
8	Near Meghani House	No space	1.3 million L	No space	N.A.
9	Near Shantivan Char Rasta	No space	0.35 million L	No space	N.A.
10	Panjarapol Char Rasta	8.1 acres	0.7 million L	High volume and large area	Vegetated swale
11	Near Aayojan Nagar	0.31 acres	0.6 million L	Low volume and small area	Bioretention

The time for infiltration of stormwater runoff was estimated for the each of the above LID structures. It was estimated that the attenuated runoff volume will be gradually infiltrated into the ground within the time range of 6 h–24 h. Thus, the provision of LID structures at all the 11 locations contributes to 11.4 million L of water that can be recharged into the ground. This will add fresh groundwater and reduce the stress on the groundwater in the city area.

1.5 CONCLUSION

The network efficiency using Bentley CivilStorm was found to be 67% for the study area. It was derived that the flooding occurred primarily due to two reasons—insufficient capacity of network, and other local features existing in the area such as natural topography/slope, LULC, and vicinity to existing network. Further, the weighted overlay tool was used for all the parameters to derive the locations that are most prone to flooding. The result demonstrated that 73% of the locations were same as identified by AMC for frequent flooding. These locations were proposed to design the LID structures. The various LID structures were proposed based on the category of site with respect to the stormwater runoff volume. These categories comprised sites that have large space availability and generate high volume of stormwater runoff, sites with low volume of runoff and large space availability, and the ones with high volume of runoff and low space availability. The analysis indicated that after implementing the LID structures such as vegetated swale, infiltration trench, and bioretention in the identified 11 sites, 7.73 million L of water can be infiltrated into the ground out of 11.4 million L of flooded water. This would in turn retain the quality and quantity of water in urban areas (Singh and Pandey, 2014). The study finally demonstrates that the flooding in the study area may be reduced by 68% by using these practices.

KEYWORDS

- **stormwater network**
- **land use land cover (LULC)**
- **urban flooding**
- **satellite remote sensing**
- **GIS**
- **ArcSWAT**

REFERENCES

Ali, H.; Mishra, V.; Pai, D. S. Observed and Projected Urban Extreme Rainfall Events in India. *J. Geophysical Res. Atmospheres* **2014**.

Altaf, F.; Gowhar, M.; Romshoo, S. A. Morphometric Analysis to Infer Hydrological Behaviour of Lidder Watershed, Western Himalaya, India. *Geogr. J.* **2013,** *2013.*

Altaf, S.; Gowhar, M.; Romshoo, S. A. Morphometry and Land Cover Based Multi-Criteria Analysis for Assessing the Soil Erosion Susceptibility of the Western Himalayan Watershed. *Environ. Monitoring Assess.* **2014,** *186*(12), 8391–8412.

Avashia, V.; Garg, A. Implications of Land Use Transitions and Climate Change on Local Flooding in Urban Areas: An Assessment of 42 Indian Cities. Land Use Policy, 2020.

Bhatt, C. M.; et al. Satellite-Based Assessment of the Catastrophic Jhelum Floods of September 2014, Jammu & Kashmir, India. *Geomat. Nat. Hazards Risk* **2017,** *8*(2), 309–327.

CASQA, California Stormwater Quality Association. California Stormwater BMP Handbook—New Development and Redevelopment. Menlo Park, CA, USA: California Stormwater Quality Association, 2003.

Coles, J. F. The Effects of Urbanization on the Biological, Physical, and Chemical Characteristics of Coastal New England Streams (No. 1695). US Geological Survey, 2004.

CPHEEO. *Manual on Storm Water Drainage Systems*, Vol. 1; Part A. Ministry of Housing and Urban Affairs, Government of India, 2019.

Damodaram, C.; et al. Simulation of Combined Best Management Practices and Low Impact Development for Sustainable Stormwater Management. *J. Am. Water Res. Assoc.* **2010,** 907–918.

Farahmand, T.; Fleming, S. W.; Quilty, E. J. Detection and Visualization of Storm Hydrograph Changes Under Urbanization: An Impulse Response Approach. *J. Environ. Manag.* **2007,** 93–100.

Harshil Gajjar, H.; Dholakia, M. B. Storm Water Network Design of Jodhpur Tekra. **2014,** 744–747.

Indian Express. After Rain, Many Areas Face Water-Logging in Ahmedabad. 25 August 2016. 10 August 2020. https://indianexpress.com/article/cities/ahmedabad/after-rain-many-areas-face-water-logging-in-ahmedabad-2994961/.

Kanga, S.; Sheikh, A. J.; Godara, U. GIscience for Groundwater Prospect Zonation. *J. Crit. Rev.* **2020b,** *7*(18), 697–709.

Kanga, S.; Singh S. K.; Sudhanshu. Delineation of Urban Built-Up and Change Detection Analysis using Multi-Temporal Satellite Images. *Int. J. Recent Res. Aspects* **2017a,** *4*(3), 1–9.

Kanga, S.; Kumar, S.; Singh, S. K. Climate Induced Variation in Forest Fire Using Remote Sensing and GIS in Bilaspur District of Himachal Pradesh. *Int. J. Eng. Comput. Sci.* **2017b,** *6*(6), 21695–21702.

Kanga, S.; et al. Modeling the Spatial Pattern of Sediment Flow in Lower Hugli Estuary, West Bengal, India by Quantifying Suspended Sediment Concentration (SSC) and Depth Conditions Using Geoinformatics. *Appl. Comput. Geosci.* **2020,** 100043.

Kourtis, I. M.; Tsihrintzis, V. A.; Baltas, E. A Robust Approach for Comparing Conventional and Sustainable Flood Mitigation Measures in Urban Basins. *J. Environ. Manag.* **2020.**

Leopold, L. B. Hydrology for Urban Land Planning: A Guidebook on the Hydrologic Effects of Urban Land Use. US Department of the Interior, Geological Survey, 1968.

Meraj, G.; Yousuf, A. R.; Romshoo, S. A. Impacts of the Geo-Environmental Setting on the Flood Vulnerability at Watershed Scale in the Jhelum Basin; MPhil Dissertation, 2013.

Meraj, G.; Romshoo, S. A.; Yousuf, A. R.; Altaf, S.; Altaf, F. Assessing the Influence of Watershed Characteristics on the Flood Vulnerability of Jhelum Basin in Kashmir Himalaya: Reply to Comment by Shah 2015. *Nat. Hazards* **2015b,** *78*(1), 1–5.

Meraj, G.; et al. Assessing the Influence of Watershed Characteristics on the Flood Vulnerability of Jhelum Basin in Kashmir Himalaya. *Nat. Hazards* **2015,** *77*(1), 153–175.
Meraj, G.; Romshoo, S. A.; Altaf, S. Inferring Land Surface Processes from Watershed Characterization. In *Geostatistical and Geospatial Approaches for the Characterization of Natural Resources in the Environment.* Springer: Cham, 2016b; pp 741–744.
Meraj, G.; et al. An Integrated Geoinformatics and Hydrological Modelling-Based Approach for Effective Flood Management in the Jhelum Basin, NW Himalaya. *Multidisciplinary Digital Publishing Institute Proceedings* **2018,** *7*(1), 8.
Meraj, G.; et al. Geoinformatics Based Approach for Estimating the Sediment Yield of the Mountainous Watersheds in Kashmir Himalaya, India. *Geocarto Int.* **2018a,** *33*(10), 1114–1138.
NDTV. Heavy Rains Cause Flooding in Ahmedabad, 8 August 2010, 10 August 2020. https://www.ndtv.com/cities/heavy-rains-cause-flooding-in-ahmedabad-426469.
Pandey, A. C.; Singh, S. K.; Nathawat, M. S. Waterlogging and Flood Hazards Vulnerability and Risk Assessment in Indo Gangetic plain. *Nat. Hazards* **2010,** *55*(2), 273–289.
Pandey, A. C.; Singh, S. K.; Nathawat, M. S.; Saha, D. Assessment of Surface and Subsurface Waterlogging, Water Level Fluctuations, and Lithological Variations for Evaluating Groundwater Resources in Ganga Plains. *Int. J. Digital Earth* **2013,** *6*(3), 276–296.
Rashid, I.; et al. Assessing the Impact of Anthropogenic Activities on Manasbal Lake in Kashmir Himalayas. *Int. J. Environ. Sci.* **2013,** *3*(6), 2036–2047.
Rawal, S. K. Storm Water Drainage Solutions for Urban. *Int. J. Innov. Res. Sci.* **2018,** 3706–3711.
Sahoo, S. N.; Sreeja, P. Determination of Effective Impervious area for an Urban Indian Catchment. *J. Hydrol. Eng.* **2016**.
Sankalp, S.; Sahoo, S. N. Identification of Urbanization Growth Trend for Ahmedabad City in India Using Remote Sensing and GIS, 2017.
Sharma, A.; Kanga, S. Surface Runoff Estimation of Sind River Basin Using SCS-CN Method and GIS Technology, 2020.
Singh, S. K.; Pandey, A. C. Geomorphology and the Controls of Geohydrology on Waterlogging in Gangetic Plains, North Bihar, India. *Environ. Earth Sci.* **2014,** *71*(4), 1561–1579.
Singh, S. K.; Kanga, S. Role of Geoinformatics in Site Suitability Analysis of Infrastructures Using PRA Approach. *Am. Int. J. Res. Sci. Technol. Eng. Math.* **2017a,** *1*(18), 81–85.
Singh, S. K.; Mishra, S. K.; Kanga, S. Delineation of Groundwater Potential Zone Using Geospatial Techniques for Shimla City, Himachal Pradesh (India). *Int. J. Sci. Res. Dev.* **2017b,** *5*(4), 225–234.
SMR Eslami, HR Fazeli, Mehraban, A. GIS-Based Model to Evaluate Low Impact Development Suitability Location. *Geospat. World* **2011**.
Sofia, G.; et al. On the Linkage Between Runoff Generation, Land Drainage, Soil Properties, and Temporal Patterns of Precipitation in Agricultural Floodplains. *Adv. Water Res.* 2019, 120–138.
Su, M., et al. The Influence of Landscape Pattern on the Risk of Urban Water-Logging and Flood Disaster. *Ecol. Indic.* **2018,** 133–140.
Swerts, E.; Pumain, D.; Denis, E. The Future of India's Urbanization. *Futures* **2014,** 43–52.
Texas A & M University. The Soil & Water Assessment Tool. n.d. November 19, 2020, Texas, 2020.
Times of India. Ahmedabad: 20 mm of Rain in an Hour Causes Waterlogging. 26 July 2020. 10 August 2020. https://timesofindia.indiatimes.com/city/ahmedabad/ahmedabad-20mm-of-rain-in-an-hour-causes-waterlogging/articleshow/77176917.cms.

UN DESA. 2018 Revision of the World Urbanization Prospects, 2018.

Wang, P.; Huang, C.; Brown de Colstoun, E. C. Mapping 2000–2010 Impervious Surface Change in India Using Global Land Survey Landsat Data. *Remote Sens.* 2017.

Zhang, Q.; Wu, Z. Identifying Dominant Factors of Waterlogging Events in Metropolitan Coastal. n.d.

Zope, P. E.; Eldho, T. I.; Jothiprakash, V. Impacts of Land Use–Land Cover Change and Urbanization on Flooding: A Case Study of Oshiwara River Basin in Mumbai, India. *Catena* **2016,** 142–154.

CHAPTER 2

Remote Sensing and GIS-Based Water Resource Monitoring for Sustainable Crop Intensification and Diversification

S. PANNEERSELVAM[1], S. PAZHANIVELAN[2], K. P. RAGUNATH[2], P. KUMARESAN[1*], and N. BALAKRISHNAN[1]

[1]*Water Technology Centre, TNAU, Coimbatore, India*

[2]*Department of GIS & RS, TNAU, Coimbatore, India*

[*]*Corresponding author. E-mail: acfkumagri@gmail.com*

ABSTRACT

Remote sensing is being used successfully as an alternative that provides spatially and temporally consistent information required for efficient management. Remote sensing information from LISS IV and SAR data through RISAT (India), Cosmo Skymed (Italy), as well as Sentinel 1A (Europe) is often used to estimate the crop area as well as yield with a reliability of 87–93% in Tamil Nadu. TNIAMP used remote sensing data as a method to investigate the effect on water supplies, crop sector, and diversification. For practical assessment, Synthetic Aperture Radar (SAR) data from Sentinel 1A and RISAT were utilized along with optical data during the cropping season. The ability has been demonstrated in rice, cotton, maize, groundnut, and coconut. The satellite technology was effective in assessing the impact of water shortages in the agriculture ecosystem. The recent floods of Tamil Nadu were precisely mapped using Sentinel 1A SAR data which helped in forwarding relief measures. Near real-time information, using satellites, on water spread and water availability duration are generated across the

GIScience for the Sustainable Management of Water Resources. Gowhar Meraj, Shruti Kanga, Majid Farooq, Suraj Kumar Singh, & Sudhanshu (Eds.)
© 2023 Apple Academic Press, Inc. Co-published with CRC Press (Taylor & Francis)

system tanks. This study aims to monitor water resources, especially for water spreads and sustainable crop intensification and diversification. This chapter highlights the remote sensing and GIS techniques and presents a brief review on applying these two emerging techniques for groundwater resource management and development.

2.1 INTRODUCTION

In recent years, it has become increasingly important to map natural resources like forests and waterways using satellite imagery (Kanga et al., 2017a, 2017b, 2021; Rather et al., 2018; Hassanin et al., 2020). Forest and water resources are intensively exploited, and their sustainable management must always be monitored at regular intervals. The extent and rate of degradation and disappearance of waterbodies, which play a critical role in the ecosystems and climate variations, is analyzed and assessed in the spatial–temporal domain (Singh et al., 2017b; Kanga et al., 2020a, 2020b). Surface hydrological indicators are a promising scientific tool for assessing and managing water resources (Magesh et al., 2011; Thomas et al., 2012).

Geospatial tools will provide significant assistance in determining the scale of environmental impact to implement relevant conservation measures. Multiple satellite data with varying spatial, spectral, and temporal characteristics have been used worldwide to generate various thematic maps of LULC emphasizing essential water bodies (Rashid et al., 2013; Singh et al., 2017a). Different abstraction methods have been established for these features, and each method has its advantages and disadvantages. Satellite data visualization effectively delineates large and small water bodies but is resource-intensive and lengthy when working with high-resolution data. Unsupervised classification, an approach that uses interactive self-organizing data analysis, yields results with very low accuracy (Meraj et al., 2013, 2018a, 2018b). There is a spectral overlap between bodies of water and other groups, however, this simple and immediate approach produces excellent results (Farooq and Muslim, 2014; Nathawat et al., 2010; Kumar et al., 2018; Joy et al., 2019).

The supervised classification provides more precise and reliable outputs than unsupervised methods. A considerable number of spectral training data are also required for the supervised technique. It is not a process that is automated. The space features of the objects are also disregarded (Huang et al., 2018).

Information on water spread areas and changes during the cropping season in these tanks can be used to make policy decisions on irrigation management. The conventional method of assessing the spread is tedious, and hence remote sensing-based analysis was carried out in all the PWD tanks of Phase I and Phase II subbasins of the TNIAM Project. Sentinel 1A satellite data with 12 days intervals were downloaded and processed for water spread under each PWD tanks during the September and November months of 2019 and January 2020, and the area in hectares was estimated. During these 3 months in all the tanks, the water spread area was subjected to change detection to determine the variation in the spread during the major cropping season Rabi in Tamil Nadu. A total of 3409 PWD tanks pertaining to TNIAMP consisting of 1852 PWD tanks in Phase I and 1557 PWD tanks in Phase II subbasins were digitized in ArcGIS software, and all the available attributes were added for use in extracting the water spread area. Subbasin-wise results on water spread derived using the digitized PWD tanks, and the satellite-derived water spread is presented in Tables 2.1 and 2.2. The maps of individual subbasins under Phase I and Phase II with the digitized PWD tanks are illustrated in Figures 2.3 and 2.4. However, water-intensive or water scarcity is a major problem for future climate change in Cauvery delta region farmers preparing for crop expansion and amplification (Abraham et al., 2014).

Diversifying of crop area essentially facilitates the transition from growing a single crop (food/nonfood) to cultivating varying amounts in much the same piece of land. It is a viable strategy to achieve food and nutritional security goals, revenue growth, judicious use of freshwater resources, increased external input production, and improved long-term agricultural development (Nyamai et al., 2012). Crop diversification, on the other hand, is an ecological imperative in Tamil Nadu, India, primarily to prevent further destruction of the already grossly misused natural resources. Reduced or increased usage efficiency of exogenous variables (e.g., fertilizers, irrigation water, pesticides) increases farm incomes by lowering cultivation costs while increasing profitability and soil and environmental health sustainability. The state has guided an unprecedented increase in the area and production of food grains, especially cereal-based, intensive rice–rice monocropping systems, particularly in the Cauvery delta region. This paper presents a method for extracting the surface water from the a Sentinel 1A (Europe) using a basic model (Sheffield et al., 2018). The main goal of this research paper is to explain the issues and solutions that arise when dealing with large amounts of water spread data in a geospatial setting and determining the best locations for the appropriate crop.

2.2 METHODOLOGY

The study of the water spread area, crop diversification, and intensification of land use/land cover of the study area has been analyzed. In the water bodies, different techniques were followed using Sentinel 1A satellite images. According to the pixel value of the original images supervised, classification is performed to generate the land used/land cover classes. The following classes, that is, settlement, cropland, water body, agricultural land, and crop diversification are identified for the land use/land cover map. The attributes are given to all digitized features against their real alignment on the earth surface. The values of water level were mapped in pre-monsoon and post-monsoon phases and the isopleths were created for the overlay analysis. Every type of crop depends on the different characteristics of attribute and the final crop suitability zones are identified using the weighted overlay analysis techniques. Then the layers are transferred to the maps with proper annotation through the GIS software.

In order to determine the water spread area changes over the period of last 25 years, necessary geospatial data were acquired, processed, and analyzed. The open source software ArcGIS were used for the geospatial processes (Fig. 2.1). Using the GIS and remote sensing techniques, the thematic maps, such as location map, base map, and drainage map were prepared. The surface water bodies of the year 2019–2020 were digitized using Survey of India toposheet and the present water bodies were digitized.

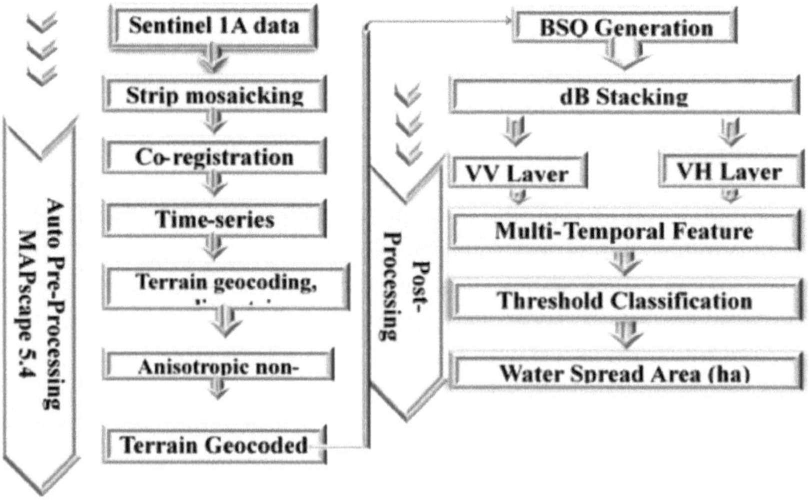

FIGURE 2.1 Methodology.

2.2.1 STUDY AREA

Specific subcatchments in Tamil Nadu are studied, with a total area of 393,388 ha occurring in Tamil Nadu, India. Tamil Nadu, India's latitude is 11.127123, and its longitude is 78.656891. Tamil Nadu, India is situated at 11°7′ 37.6428″ N and 78° 39′ 24.8076″ E in the India region (Fig. 2.2).

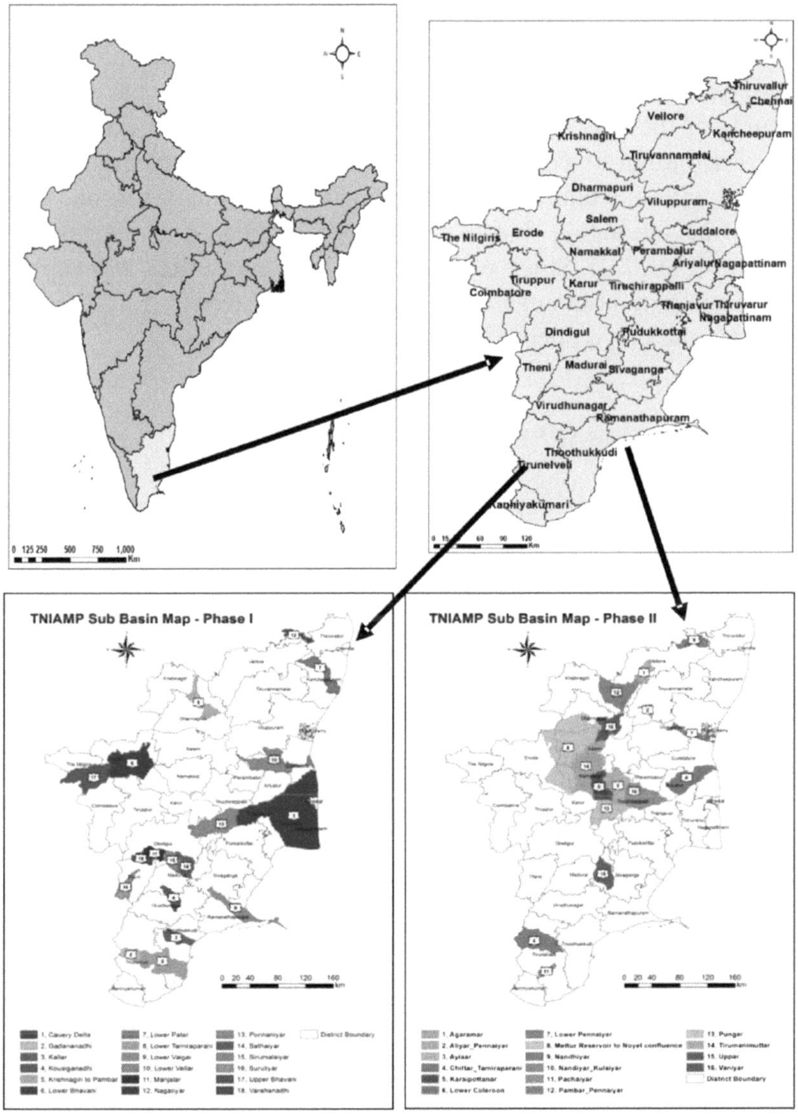

FIGURE 2.2 Study area.

2.3 RESULT AND DISCUSSION

The results of the study of spatiotemporal satellite-derived water distribution estimations of 34 subbasins, as well as a model system for the processing of surface water bodies from Tamil Nadu's Sentinel 1A SAR satellite, are presented in this chapter. With the aid of figures, the discussion is based on the satellite-derived water spread location (Fig. 2.3a–f). Surface water layers were detected from the Sentinel 1A SAR image via the model for water body extraction. The focus of the investigation is on the surface water (Table 2.1). The analysis of the spread of water in the tanks revealed that relative to the beginning of the season in September to November, there is still a decrease in the distribution of water in most of the basins due to more tank usage, compared with November and the ending of the January season, indicating increased accumulation due to rain in the middle of use for agricultural purposes (Table 2.2). Since water is consumed more than it is replenished, the subbasin with both negative and positive differences from September to November and November to January is the one to discuss.

2.4 RICE AREA MAPPING AND STATISTICS

With the advent of high spatial and temporal resolution, synthetic aperture radar (SAR) satellite sensors, crop area mapping with higher precision have become a reality. During the 2017 and 2018 cropping season, Sentinel 1A SAR satellite data from the European Space Agency (ESA) was acquired at 12-day intervals (Table 2.3). As influenced by the growth parameters of the rice crop and the exposed soil surface/water, the backscattering coefficient (σ^{o}), coherence and multitemporal characteristics were extracted.

Rice area maps were developed using time series C band synthetic aperture radar data of Sentinel 1A for the Cauvery delta districts, viz., Ariyalur, Cuddalore, Nagapattinam, Thanjavur, Thiruvarur, and Tiruchirappalli districts. For the six districts covering 68 blocks, that is, 6, 13, 17, 11, 14, 10, and 14 blocks for the Ariyalur, Cuddalore, Nagapattinam, Thanjavur, Thiruvarur, and Tiruchirappalli districts, maps, and rice area statistics at district and block level were extracted.

Cuddalore district registered a rice area of 99,177 and 70,995 ha during 2017 and 2018. Kattumannarkoil, Keerapalayam, Kumaratchi, Kurinjipadi, Mel Bhuvanagiri, and Parangipettai were the major blocks contributing for rice area. Among the 13 blocks, Keerapalayam recorded the largest rice area of 13812 ha during 2017 followed by Mel Bhuvanagiri, Kurinjipadi, and

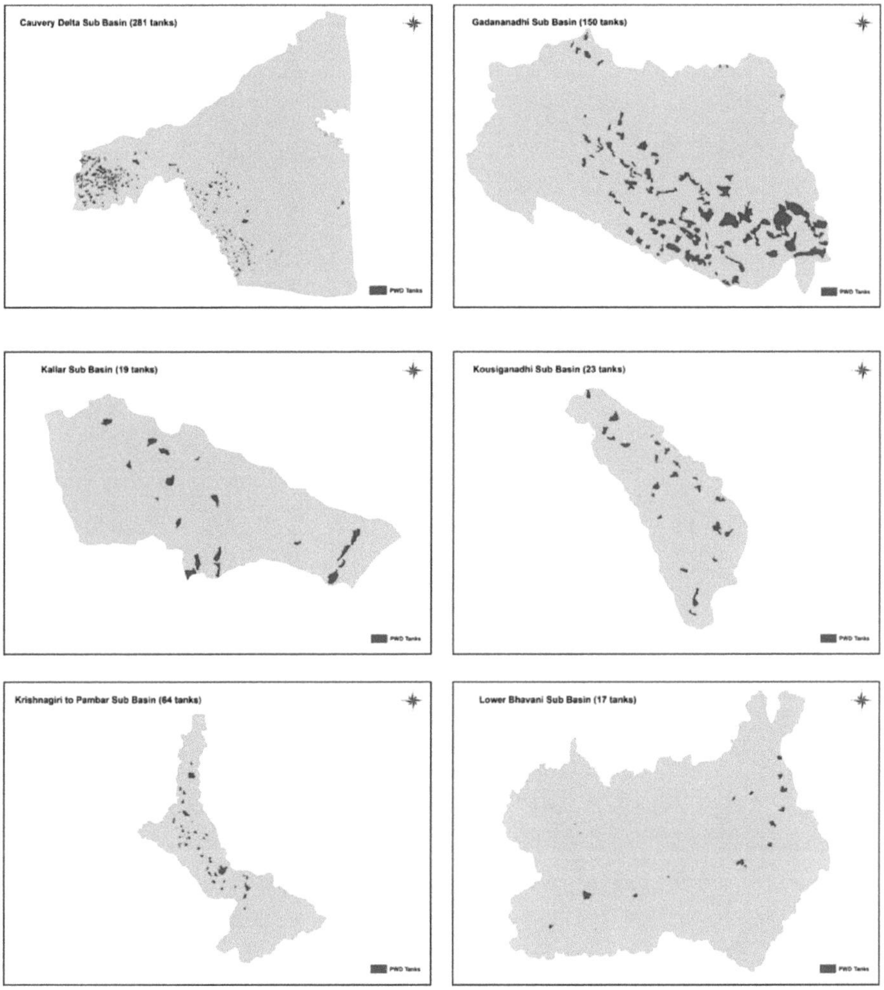
Cauvery Delta Sub Basin (281 tanks)
Gadananadhi Sub Basin (150 tanks)
Kallar Sub Basin (19 tanks)
Kousiganadhi Sub Basin (23 tanks)
Krishnagiri to Pambar Sub Basin (64 tanks)
Lower Bhavani Sub Basin (17 tanks)

(a)

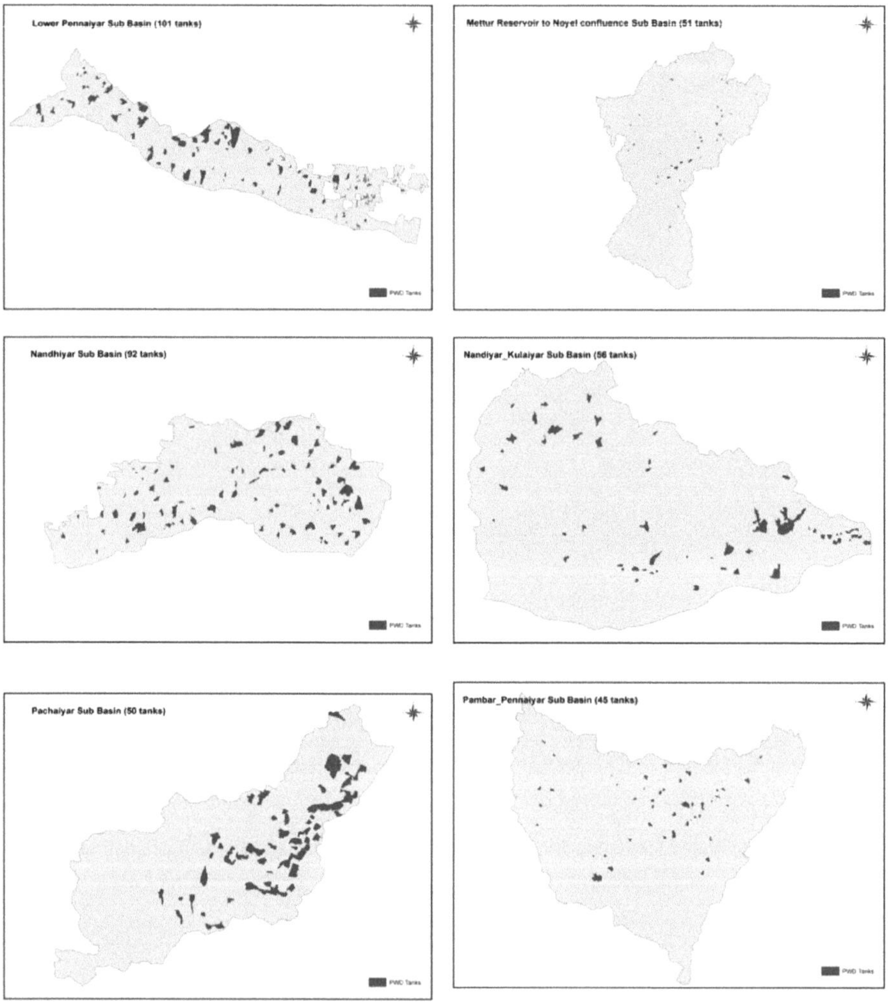
Lower Pennaiyar Sub Basin (101 tanks)
Mettur Reservoir to Noyel confluence Sub Basin (51 tanks)
Nandhiyar Sub Basin (92 tanks)
Nandiyar_Kulaiyar Sub Basin (56 tanks)
Pachaiyar Sub Basin (50 tanks)
Pambar_Pennaiyar Sub Basin (45 tanks)

(b)

Agaramaru Sub Basin (9 tanks)
Aliyar_Pennaiyar Sub Basin (12 tanks)
Ayiaar Sub Basin (29 tanks)
Chittar_Tamiraparani Sub Basin (406 tanks)
Karaipottanar Sub Basin (34 tanks)
Lower Coleroon Sub Basin (13 tanks)

(c)

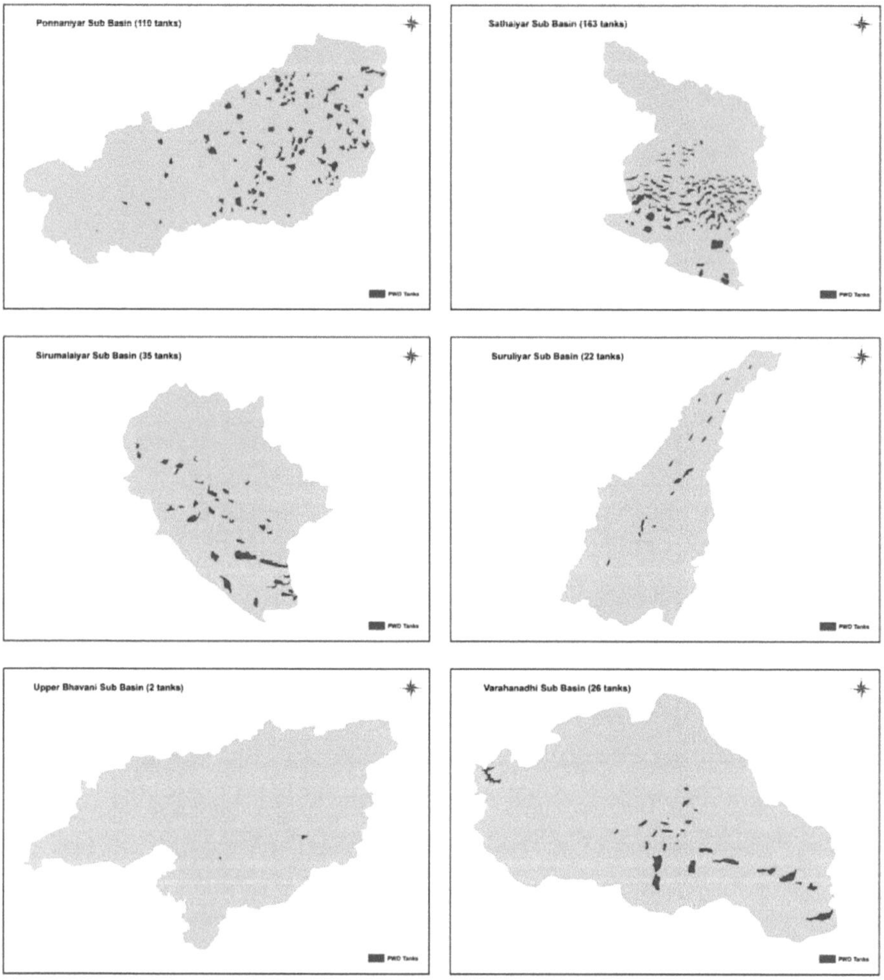
Ponnaniyar Sub Basin (110 tanks)
Sathaiyar Sub Basin (163 tanks)
Sirumalaiyar Sub Basin (35 tanks)
Suruliyar Sub Basin (22 tanks)
Upper Bhavani Sub Basin (2 tanks)
Varahanadhi Sub Basin (26 tanks)

(d)

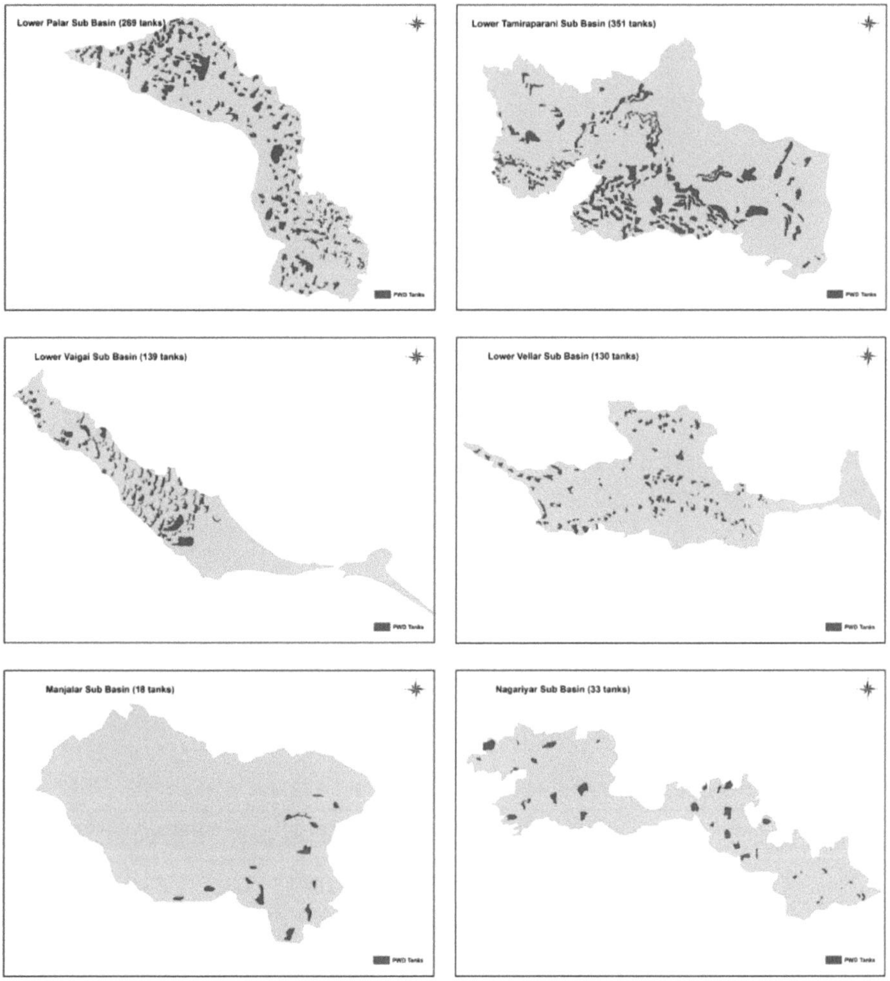
Lower Palar Sub Basin (269 tanks)
Lower Tamiraparani Sub Basin (351 tanks)
Lower Vaigai Sub Basin (139 tanks)
Lower Vellar Sub Basin (130 tanks)
Manjalar Sub Basin (18 tanks)
Nagariyar Sub Basin (33 tanks)

(e)

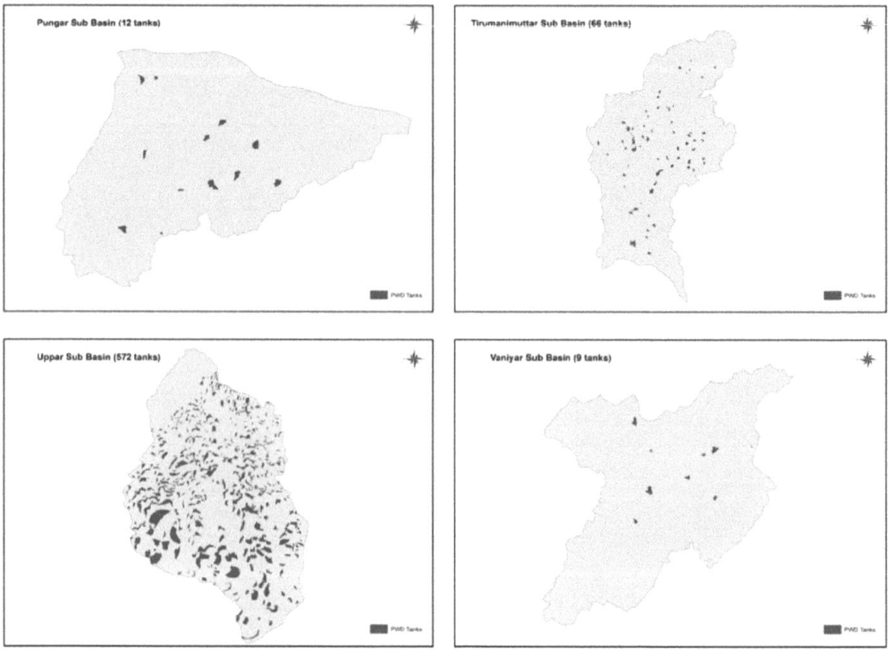

(f)

FIGURE 2.3 (a-f) Tamil Nadu river basins.

TABLE 2.1 Details of Water Spread in the PWD Tanks of Phase I Subbasins—1852 Tanks.

Subbasins	No. of tanks	Water spread area (ha)			Change in spread (ha)		
		Sep. 2019	Nov. 2019	Jan. 2020	Sep.–Nov.	Nov.–Jan.	Sep.–Jan.
Cauvery Delta	281	1346	1542	1480	196	**−62**	134
Gadananadhi	150	784	519	510	**−265**	**−9**	**−274**
Kallar	19	292	225	220	**−67**	**−5**	**−72**
Kousiganadhi	23	182	38	72	**−144**	34	**−110**
Krishnagiri to Pambar	64	596	622	663	26	41	67
Lower Bhavani	17	81	177	197	96	20	116
Lower Palar	269	2279	3818	5337	1539	1519	3058
Lower Tamiraparani	351	8985	4405	5635	**−4580**	1230	**−3350**
Lower Vaigai	139	689	305	607	**−384**	302	**−82**
Lower Vellar	130	224	221	395	**−3**	174	171
Manjalar	18	66	108	126	42	18	60
Nagariyar	33	20	181	207	161	26	187
Ponnaniyar	110	1158	574	721	**−584**	147	**−437**
Sathaiyar	163	599	547	518	**−52**	**−29**	**−81**
Sirumalaiyar	35	72	156	161	84	5	89

TABLE 2.3 *(Continued)*

Subbasins	No. of tanks	Water spread area (ha)			Change in spread (ha)		
		Sep. 2019	Nov. 2019	Jan. 2020	Sep.–Nov.	Nov.–Jan.	Sep.–Jan.
Suruliyar	22	60	42	42	**−18**	0	**−18**
Upper Bhavani	2	22	23	21	1	**−2**	**−1**
Varahanadhi	26	22	145	111	123	**−34**	89

TABLE 2.2 Details of Water Spread in the PWD Tanks of Phase II Subbasins—1557 Tanks.

Subbasins	No. of tanks	Water spread area (ha)			Change in spread (ha)		
		Sep. 2019	Nov. 2019	Jan. 2020	Sep.–Nov.	Nov.–Jan.	Sep.–Jan.
Agaramaru	9	3	2	15	**−1**	13	12
Aliyar_Pennaiyar	12	35	27	66	**−8**	39	31
Ayiaar	29	513	235	240	**−278**	5	**−273**
Chittar_Tamiraparani	406	2951	1721	2050	**−1230**	329	**−901**
Karaipottanar	34	92	93	128	1	35	36
Lower Coleroon	13	2154	2029	2114	**−125**	85	**−40**
Lower Pennaiyar	101	237	296	414	59	118	177
Mettur Reservoir to Noyyal confluence	51	46	130	139	84	9	93
Nandhiyar	92	92	510	654	418	144	562
Nandiyar_Kulaiyar	56	458	487	492	29	5	34
Pachaiyar	50	769	215	529	**−554**	314	**−240**
Pambar_Pennaiyar	45	204	188	266	**−16**	78	62
Pungar	12	145	14	21	**−131**	7	**−124**
Tirumanimuttar	66	166	253	260	87	7	94
Uppar	572	1372	879	1119	**−493**	240	**−253**
Vaniyar	9	2	1	84	**−1**	83	82

Kattumannarkoil with an area of 11,471, 10,308 and 10,142 ha, respectively. Kumaratchi, Kammapuram, Vridhachalam, and Parangipettai blocks had contributed a rice area of 10,098, 9912, 9865, and 9066 ha, respectively during 2017 season. Nallur and Cuddalore were the next best blocks to register rice area of 5153 and 4772 ha, respectively. The other blocks, viz., Annagramam, Panruti, and Mangalore accounted for 4566 ha cumulatively (Fig. 2.4).

TABLE 2.3 Rice Area Statistics in Selected Subbasins—Area (ha).

Subbasin	Rice	Late rice	Total area (ha)
Cauvery Delta	344,706	0	344,706
Lower Coleroon	35,941	16	35,957

FIGURE 2.4 Rice area.

2.5 CROP DIVERSIFICATION

Crop diversification enables farmers to have a wider range of crops for a given area, expending resources on multiple crops while also reducing the risk. People often refer to crop diversification in India as a transition from traditionally grown crops that have a lower return to crops that are more remunerative. Cropping diversification is also occurring because of government policies that try to make some crops more profitable, market reforms that allow for higher profits, infrastructure development that supports them, government subsidies, and higher profitability production of crops (Flachsbarth et al., 2015). In dryland areas, crop diversification and the planting of a significant number of crops are used to minimize the likelihood of crop failure due to frequent droughts. Crop replacement and crop change also are occurring in the region that is experiencing some soil issues (Fig. 2.5). Apart from the rice crop, SAR-based temporal signatures were created for crop identification and delineation in order to track the crop.

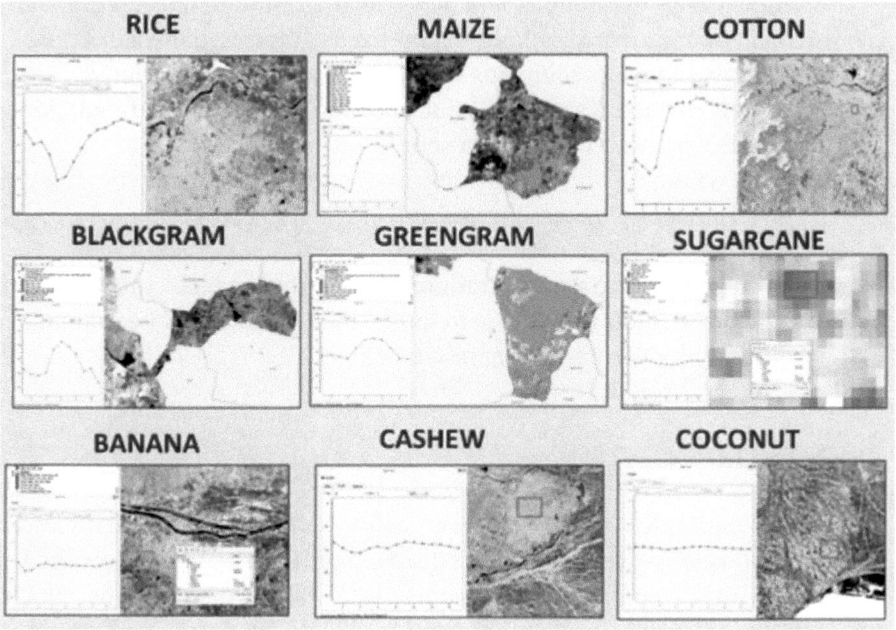

FIGURE 2.5 Crop diversification.

2.6 CONCLUSION

The results of the work showed that there is negative and positive reduction of surface water bodies within the span of two decades. The local community has encroached on several water bodies that used to bring storm water from one water source to the next in a river basin. The growing population and developments in the field of real estate, acted as major driving forces which converted agricultural land into residential plots and there by exerting pressure on the nearby surface water bodies. The study area has undergone continuous monsoon failure and the water resources were abandoned without proper conservation. The paper concludes that in this major issue regarding surface water bodies, In terms of innovative policy formulation, regulation, and execution, the government shares equal responsibility with private entities. The following measures should be taken to ensure water sources for current and future generations.

1. Water resource protection committees including members from the native community, governmental parties, teachers in the nearby

educational institutions and an official from the concerned government department should be set up for monitoring the water bodies.
2. Removal of encroachments should be enforced very strictly.
3. For all surface water bodies, a boundary should be defined. Along the border, strong protection should be installed.
4. Plots within this boundary must no longer be cleared by the relevant government agencies. Further infringements can be avoided in this way.
5. Even though the rainfall failure occurs, the surface water resource structures, such as ponds, streams, lakes, etc. should be maintained properly by regular conservative actions.

ACKNOWLEDGMENTS

The author expresses his appreciation to the book's editors and reviewers for their informative reviews and critical remarks, which have resulted in substantial improvements to the manuscript's early versions. The authors are grateful to TNIAMP scheme funded by World Bank. They are thankful to Mr. G. R. Mugilan Raman, Senior Research Fellow and Mr. S. Aravinth Senguttuvan, Junior Research Fellow, Department of RS & GIS, Tamil Nadu Agricultural University, Coimbatore, for their valuable suggestions and help during this work.

KEYWORDS

- **water resource management**
- **water spread area**
- **remote sensing and GIS**
- **crop diversification**
- **intensification**

REFERENCES

Abraham, B.; Araya, H.; Berhe, T.; Edwards, S.; Gujja, B.; Khadka, R. B.; Uphoff, N. The System of Crop Intensification: Reports from the Field on Improving Agricultural

Production, Food Security, and Resilience to Climate Change for Multiple Crops. *Agric. Food Sec.* **2014,** *3* (1), 1–12.

Araya, H.; Edwards, S.; Asmelash, A.; Legasse, H.; Zibelo, G. H.; Mohammed, E.; Misgina, S. SCI: Planting with Space. *Farm. Matters* **2013,** *29* (1), 35–37.

Behera, D.; Chaudhury, A. K.; Vutukutu, V. K.; Gupta, A.; Machiraju, S.; Shah, P. *Enhancing Agricultural Livelihoods through Community Institutions in Bihar, India. South Asia Livelihoods Learning Note, Series 3, Note 1*, 2013.

Cafer, A. M.; Rikoon, J. S. Adoption of New Technologies by Smallholder Farmers: The Contributions of Extension, Research Institutes, Cooperatives, and Access to Cash for Improving TEF Production in Ethiopia. *Agric. Human Values* **2018,** *35* (3), 685–699.

Dash, T. K.; Pal, A. *Growing Crops with SRI Principles*; SRI Secretariat and Sir Dorabji Tata Trust: Bhubaneswar, 2011. http://sdtt-sri.org/wp-content/themes/SDTT-SRI/Document/output.pdf.

Farooq, M.; Muslim, M. Dynamics and Forecasting of Population Growth and Urban Expansion in Srinagar City—A Geospatial Approach. *Int. Arch. Photogramm., Remote Sens. Spat. Inf. Sci.* **2014,** *40* (8), 709.

Flachsbarth, I.; Willaarts, B.; Xie, H.; Pitois, G.; Mueller, N. D.; Ringler, C.; Garrido, A. The Role of Latin America's Land and Water Resources for Global Food Security: Environmental Trade-offs of Future Food Production Pathways. *PLoS One* **2015,** *10* (1), e0116733.

Hassanin, M.; Kanga, S.; Farooq, M.; Singh, S. K. Mapping of Trees Outside Forest (ToF) from Sentinel-2 MSI Satellite Data Using Object-based Image Analysis. *Guj. Agric. Univ. Res. J.* **2020,** 45, *207*–213.

Huang, C.; Chen, Y.; Zhang, S.; Wu, J. Detecting, Extracting, and Monitoring Surface Water from Space Using Optical Sensors: A Review. *Rev. Geophys.* **2018,** *56* (2), 333–360.

Joy, J.; Kanga, S.; Singh, S. K. Kerala Flood 2018: Flood Mapping by Participatory GIS Approach, Meloor Panchayat. *Int. J. Emerg. Technol.* **2019,** *10* (1), 197–205.

Kanga, S.; Singh, S. K.; Sudhanshu. Delineation of Urban Built-Up and Change Detection Analysis Using Multi-temporal Satellite Images. *Int. J. Rec. Res. Aspects* **2017a,** *4* (3), 1–9.

Kanga, S.; Kumar, S.; Singh, S. K. Climate Induced Variation in Forest Fire Using Remote Sensing and GIS in Bilaspur District of Himachal Pradesh. *Int. J. Eng. Comput. Sci.* **2017b,** *6* (6), 21695–21702.

Kanga, S.; Rather, M. A.; Farooq, M.; Singh, S. K. GIS Based Forest Fire Vulnerability Assessment and Its Validation Using Field and MODIS Data: A Case Study of Bhaderwah Forest Division, Jammu and Kashmir (India). *Indian Forest.* **2021,** *147* (2), 120–136.

Kumar, S.; et al. *Delineation of Shoreline Change along Chilika Lagoon (Odisha), East Coast of India using Geospatial Technique*, 2018.

Magesh, N. S.; Chandrasekar, N.; Soundranayagam, J. P. Morphometric Evaluation of Papanasam and Manimuthar Watersheds, Parts of Western Ghats, Tirunelveli District, Tamil Nadu, India: A GIS Approach. *Environ. Earth Sci.* **2011,** *64* (2), 373–381.

Meraj, G.; Yousuf, A. R.; Romshoo, S. A. *Impacts of the Geo-environmental Setting on the Flood Vulnerability at Watershed Scale in the Jhelum Basin.* M. Phil. Dissertation, 2013.

Meraj, G.; et al. An Integrated Geoinformatics and Hydrological Modelling-based Approach for Effective Flood Management in the Jhelum Basin, NW Himalaya. *Multidiscip. Dig. Publ. Inst. Proc.* **2018a,** *7* (1), 8.

Meraj, G.; et al. Geoinformatics Based Approach for Estimating the Sediment Yield of the Mountainous Watersheds in Kashmir Himalaya, India. *Geocarto Int.* **2018b,** *33* (10), 1114–1138.

Nathawat, M. S.; et al. Monitoring & Analysis of Wastelands and Its Dynamics Using Multiresolution and Temporal Satellite Data in Part of Indian State of Bihar. *Int. J. Geomat. Geosci.* **2010,** *1* (3), 297–307.

Nyamai, M.; Mati, B. M.; Home, P. G.; Odongo, B.; Wanjogu, R.; Thuranira, E. G. Improving Land and Water Productivity in Basin Rice Cultivation in Kenya through System of Rice Intensification (SRI). *Agric. Eng. Int.: CIGR J.* **2012,** *14* (2), 1–9.

Rashid, I.; et al. Assessing the Impact of Anthropogenic Activities on Manasbal Lake in Kashmir Himalayas. *Int. J. Environ. Sci.* **2013,** *3* (6), 2036–2047.

Rather, M. A.; et al. Remote Sensing and GIS Based Forest Fire Vulnerability Assessment in Dachigam National Park, North Western Himalaya. *Asian J. Appl. Sci.* **2018,** *11* (2), 98–114.

Sheffield, J.; Wood, E. F.; Pan, M.; Beck, H.; Coccia, G.; Serrat Capdevila, A.; Verbist, K. Satellite Remote Sensing for Water Resources Management: Potential for Supporting Sustainable Development in Data Poor Regions. *Water Resour. Res.* **2018,** *54* (12), 9724–9758.

Singh, S. K.; Kanga, S. Role of Geoinformatics in Site Suitability Analysis of Infrastructures Using PRA Approach. *Am. Int. J. Res. Sci., Technol., Eng. Math.* **2017a,** *1* (18), 81–85.

Thomas, J.; Joseph, S.; Thrivikramji, K. P.; Abe, G.; Kannan, N. Morphometrical Analysis of Two Tropical Mountain River Basins of Contrasting Environmental Settings, the Southern Western Ghats, India. *Environ. Earth Sci.* **2012,** *66* (8), 2353–2366.

CHAPTER 3

Using Analytical Hierarchical Process (AHP)-Based Geospatial Approach for Identification of Suitable Sites for Artificial Recharge Structures for Sustainable Water Resource Management

M. RAJASEKHAR[1], G. SUDARSANA RAJU[1*], and Y. SREENIVASULU[2]

[1]*Department of Geology, Yogi Vemana University, Kadapa 516005, Andhra Pradesh, India*

[2]*Department of English, SSL, Vellore Institute of Technology, Vellore 632014, Tamil Nadu, India*

[*]*Corresponding author. E-mail: gsraju05@gmail.com*

ABSTRACT

The sustainable development of the groundwater resources is of utmost importance these days due to a decrease in the water table and low precipitation, especially in arid and semiarid areas. The present research is an effort to identify suitable areas for applying and adapting artificial recharge structures (ARS) to increase groundwater through the decision-making and geospatial approach in Kadapa district, South India. The main components for preparing ARS maps for the identification of suitable charging structures are geospatial and decision-making. These weights are focused on its effect on future groundwater recharge in the field analysis with expert opinions on ARS.

GIScience for the Sustainable Management of Water Resources. Gowhar Meraj, Shruti Kanga, Majid Farooq, Suraj Kumar Singh, & Sudhanshu (Eds.)
© 2023 Apple Academic Press, Inc. Co-published with CRC Press (Taylor & Francis)

Further, the weights are standardized with the analytical hierarchical process AHP approach to identify the ARS to improve sustainable development policies. In the study, recharge areas are divided into different types, such as low, moderate, high, and very high groundwater recharge potential. The ARS of the present study shows that 1.18% (1.99 km^2) of the area is found to be very high, and the high zone for artificial groundwater recharge fell to 54.14% (91.06 km^2). On the other hand, 44.36% (74.61 km^2) and 0.32% (0.55 km^2) of the whole area fall under moderate and poor individual ARS. Hence, the current modified method helps to quickly identify a location for SWC structures and manage groundwater resources for sustainable growth in the arid and semiarid regions.

3.1 INTRODUCTION

According to the world base survey, India is the world's largest groundwater user in which 60% of the water being used for farming and 85% for drinking (World Bank, 2010). Nowadays, groundwater depletion resulted in deepening water levels and degradation of shallow groundwater drainage systems (CGWB, 2015, 2016). Simultaneously, despite the demand controls, artificial recharge, and rainwater harvesting are feasible apart from demand control (Senthilkumar et al., 2019). Both financial stability and the development of rural areas in India are standout among the essential effects for the advancement of the Indian economy (Kumbhar et al., 2013). In such locations, water source is a released hydrological unit, from which the runoff takes place, and after the precipitation, it crosses a single place in a vast stream, such as a river, lake, or pond (Chowdary et al., 2009). It includes evaluating the different thematic layers about prospects, problems, and collecting forms of educational activity combined with auditory financial influences that can be connected to the field without endangering the environment (Kumar et al., 2008; Chowdhury et al., 2018).

Therefore, it refers to resource management that ensures adequate land use, land shelter against all forms of erosion, soil fertility improvement and conservation, water conservation and management (Kadam et al., 2012; Meraj et al., 2021). Several strategies for obtaining immediate or conventional soil water conservation (SWC) and their possible existence, internally that hydrogeological and geospatial research is used regularly and appears to be common (Emmanuel et al., 2008; Kadam et al., 2017; Kudnar and Rajasekhar, 2020).

The geohydrological studies include some spatial layers of geology that affect the structures of the SWC. However, the potential of vast quantities of water in surface water sources is always impractical due to the level alleviation in these alluvial systems. This unlocks the prospect of exploiting the subsurface potential of water (Bouwer, 2002), for instance, Managed Aquifer Recharge (MAR) (Dillon, 2005) where infiltration wells access surplus water, simultaneously aquitards regularly, and overlay springs in coastal alluvial areas by making surface-reviving approaches inadmissible. ASR-wells will invade spring water and focus on demand (Pyne, 2005). Underground storage was called water banking (McClurg, 2001). Water finance, despite everything, can prompt quantity and quality issues, such as surface water maintenance, supplements, and contamination charges. An adequate amount of such problems become worsened later due to the expanding effects of serious farming (Bouwer, 2002).

To investigate the suitability of SWC sites, the Kadapa district has been selected as characteristic of semiarid areas in southern India for the screening of moderate rainfall. In the selected area, accessible groundwater must be collected and misused especially for drinking water, irrigation, and domestic purposes (Rajasekhar et al., 2018a, 2018b). This allows soil degradation in reasonable areas and maintains soil and groundwater management. Taking this into account, the study area is chosen with the aim of (1) understanding the area's runoff and recharge potential, (2) identifying possible soil water storage sites for rainwater runoff and soil erosion prevention, and (3) validating the derived SWC site with field survey to assess research accuracy.

3.2 STUDY AREA

This study was conducted in areas between latitudes entangling 14° 18′ 310″ N–14° 29′ 50″ N and longitudes 78° 21′ 10″ to 78° 33′ 40″ E. The research area encompasses the western part of Kadapa District, South India and is bordered on Veerapanayani Palle, Vemula, Pulivendula, and Pendlimarri (Figure 3.1a). It lies in the Indian toposheet carrying nos. 57J07 and 57J11, with a gross geographic area of 168.21 km^2 and the central area of this sample is dominated by pediment and pediplain complex with gentle slope of B soils. Further, some areas in the southern regions have a hilly complex with a higher slope moderate to very steep (Ramachandra et al., 2019). Furthermore, the area is drained by streams which are tributaries of Papaghni River and the streams are mostly ephemeral and the average rainfall in the research area is 550.33 mm/year.

3.3 METHODOLOGY

To delineate the suitable sites for artificial recharge structures, the decision using ArcGIS and sites of groundwater channel junctions with adequate surface runoff potential were determined. ArcGIS has been processed with input layers via weighted overlay analysis (WOA) and the output map was achieved using WOA toolbox analytical hierarchical process (AHP) standardization in ArcGIS environment. In the study, eight spatial layers such as geology, drainage density (DD), land use/land cover (LULC), hydrological soil groups (HSG), slope, geomorphology (GM), rainfall and runoff were analyzed using decision-making techniques comprising geometric mean and standardized weight calculation to delineate artificial recharge structures (Kaliraj et al., 2014; Rajasekhar et al., 2018). The detailed methodology of the study is shown in Figure 3.1b.

Using Landsat8 satellite, the LULC layer was prepared by image classification in Erdas Imagine software (Kumar et al., 2014). Simultaneously, drainage layer was prepared with the help of Survey of India (SOI) topographical sheets and the soil layer was prepared using the district soil resource map by Groundwater Department, Kadapa. Further, the slope map was derived from NRSC Bhuvan's Cartosat DEM data and the rainfall data from the groundwater department, Anantapur were used for rain gauge stations from 1998 to 2018. Geomorphology and geology layers were the interpretations of FCC satellite imagery to identify various geology and geomorphological units. Field tests were subsequently carried out in significant areas to verify the veracity of remote sensing data and also to integrate the field information in the map. Further, the geological map of the study area was created by interpreting satellite data based on the geological map prepared by the Geological Survey of India (GSI, 2002) and digitizing the satellite imagery. Runoff map was prepared using LULC and HSG curve intersection using ArcGIS software (Rajasekhar et al., 2020). An aggregate of eight thematic layers, such as geology, DD, LULC, HSG, slope, GM, rainfall, and runoff with their inter-relationship is inferred in the Kadapa District to incite AR methodology (Figure 3.2).

3.3.1 ASSIGNING THE WEIGHTS BY AHP

The AHP simplifies the PCM parameters and depends on the assigning of experts concerning the significance of the influencing factors (Meraj et al., 2018). AHP is one of the most common decision-making approaches

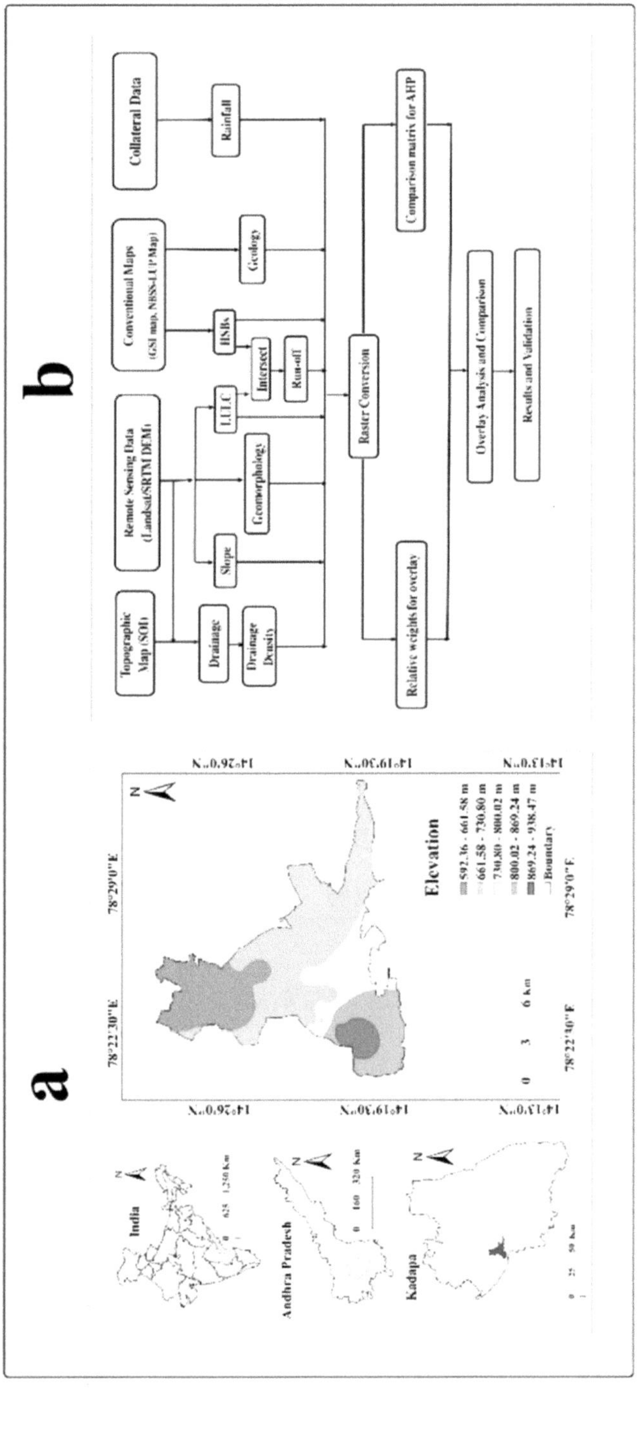

FIGURE 3.1 (a) Location map: Vempalli, and (b) methodology.

(Saaty, 1980, 1999, 2008; Kadam et al., 2017; Shailaja and Kadam, 2018). This process can also analyze proportionally various data sets through the PCM analysis of each parameter and flexibility in the situation of intentions (Kumar et al., 2016; Rajasekhar, 2019a; Shailaja and Kadam, 2018). Saaty (1977) initially carried out this systematic exercise to show sustainability and the consequences in a multilateral decision-making system. The matrix structure of AHP offers the possibility to quantify and synthesize different features of the multilateral decision-making process hierarchically (Singh et al., 2017; Kanga et al., 2020).

In this analysis, AHP is used to estimate the weight of all related parameters concerning groundwater recharge capacity in the research area (Table 3.4). AHP implies an assessment of the significance between the parameters, the normalization, and calculating the consistency ratio (CR). Depending on the hierarchical order constructed, it is decided that the influencing factors' priority recognizes the significance of the features at various levels of the hierarchy (Mustafa et al., 2011; Das, 2019; Kaliraj et al., 2016). To obtain the qualitative data, the factors that influence each level are compared in pairs. The Saaty scale from 1 to 9 (Table 3.1) is used to measure the factor's relative importance (Saaty, 2008).

TABLE 3.1 AHP Scale and Random Index Values.

Scale	1	2	3	4	5	6	7	8	9
Importance	Equal	Weak	Moderate	Moderate plus	Strong	Strong plus	Very strong	Very, very strong	Extreme
RCI	0	0	0.58	0.89	1.12	1.24	1.32	1.41	1.45

3.3.2 PAIRWISE COMPARISON MATRIX

This prepares a diagonal matrix by placing values in the upper triangle and using their reciprocal values to fill the triangular matrix. The expert's judgment was used for PCM (Zolekar and Bhagat, 2015), and the relative weights were normalized too. The normalized principal eigenvector is obtained by averaging the order to verify the consistency of the priority vector. The key own value is acquired by applying the products to each part of the own vector and summing the columns of Tables 3.2 and 3.3. This fundamental eigenvalue is used to degree the consistency of ideas through an analysis of the consistency ratio (CR) (Saaty, 1980). Further, a CR value lower than 0.1

can be considered as opinions with less uncertainty in weight determination (Kadam et al., 2018; Rajasekhar, 2018d, 2019b). Also the Consistency Index (CI) is calculated using the below equations and the comparison matrix to compare all parameters for calculating CR.

$$\text{CI} = \frac{\lambda max - n}{n-1} \tag{3.1}$$

where, λmax = consistency vector
n= number of criteria or factors

The CR has been calculated based on Equation (3.2).

$$\text{CR}= \frac{CI}{RCI} \tag{3.2}$$

where
RCI = random consistency index, provided by Saaty (2008) (Table 3.1).

TABLE 3.2 PCM for the AHP Approach.

Theme	Geology	LULC	DD	Soils	Slope	GM	Rainfall	Runoff
Geology	1.00	0.50	0.50	4.00	4.00	0.50	1.00	0.20
LULC	2.00	1.00	0.25	4.00	1.00	0.50	0.50	3.00
DD	2.00	4.00	1.00	5.00	2.00	0.33	3.00	0.25
Soils	0.20	0.25	0.20	1.00	0.33	0.25	0.20	0.33
Slope	0.20	1.00	0.50	3.00	1.00	0.33	2.00	1.00
GM	2.00	0.50	3.00	4.00	3.00	1.00	3.00	2.00
Rainfall	1.00	2.00	0.33	5.00	0.50	0.33	1.00	0.03
Runoff	5.00	0.33	4.00	3.00	1.00	0.50	5.00	1.00

TABLE 3.3 Normalized Matrix for AR Normalized Eight Calculation.

Theme	Geology	LULC	Soils	DD	GM	Slope	Rainfall	Runoff	% Weight	CR
Geology	0.075	0.052	0.051	0.138	0.312	0.133	0.064	0.026	11	0.060
LULC	0.149	0.104	0.026	0.138	0.078	0.133	0.032	0.384	13	
Soils	0.149	0.418	0.102	0.172	0.156	0.089	0.191	0.032	16	
DD	0.015	0.026	0.020	0.034	0.026	0.067	0.013	0.043	3	
GM	0.015	0.104	0.051	0.103	0.078	0.089	0.127	0.128	9	
Slope	0.149	0.052	0.307	0.138	0.234	0.267	0.191	0.256	20	
Rainfall	0.075	0.209	0.034	0.172	0.039	0.089	0.064	0.004	9	
Runoff	0.373	0.034	0.409	0.103	0.078	0.133	0.318	0.128	20	

TABLE 3.4 Criteria, Subcriteria and AHP Weights for ARS.

Theme	Weight	Normalized weight	Class
Geology	11	0.11	Dolomite
			Quartzite/Conglomerate
			Quartzite
			Shale
			Waterbodies
LULC	13	0.13	Agriculture, crop land
			Agriculture, fallow land
			Barren/Uncultivable/Wasteland
			Forest land
			Waterbodies
HSG	16	0.16	Group B
			Group C
DD	3	0.03	0–0.6 km/km^2
			0.6–1.2 km/km^2
			1.2–1.7 km/km^2
			1.7–2.3 km/km^2
			2.3–3.6 km/km^2
GM	9	0.09	Anthropogenic terrains
			Pediments pediplains
			Structural hills
			Waterbodies
Slope	20	0.20	Plain
			Very gentle
			Gentle
			Moderate
			Strong
			Moderate to steep
			Very steep
Rainfall	9	0.09	496.05–533.09 mm/year
			533.09–548.89 mm/year
			548.89–567.41 mm/year
			567.41–591.38 mm/year
			591.38–634.96 mm/year
Runoff	20	0.20	Very low
			Low
			Moderate
			High
			Very high

3.4 RESULT AND DISCUSSION

A detailed procedure for groundwater resource identification of a particular area needs exhaustive investigation of various thematic layers, such as geology, DD, LULC, HSG, slope, GM, rainfall, and runoff to identify ARS. In this study, typical thematic layers were digitized from geospatial and field data which specifics were prepared from RS information, field overview, and criterion in weighted overlay analysis of various layers mentioned in Table 3.4 and composed with the groundwater potential outcomes (Raju et al., 2019; Balaji et al., 2019; Veeraswamy et al., 2020).

3.4.1 GEOLOGY

The geological formation of this area incorporates dolomites, quartzites, conglomerates, and shales. It is observed that the study region is commonly composed of quartzites overlain by shales in the major parts of the region. The shales are in a low degree of infiltration due to their porosity and permeability nature, while the dolomites formations have a high degree of infiltration due to their permeable nature (GSI, 2002). The quartzites formation of the present study has moderate recharge zones as their nature of the impact, which indicates moderate groundwater recharges zones (Figure 3.2a).

3.4.2 LAND USE/LAND COVER

Land use depicts how a piece of land is utilized for irrigation, home, or industry, identifies with human movement or financial capacity related to a particular real estate. Land use is interpretable by satellite imageries (Rajasekhar et al., 2017). Different sorts of land use examples are recognized in the present study which incorporates forest, cropland, fallow land, barren land, and water bodies, among which, cropland and water bodies are very good groundwater recharge zones with the highest priority (Figure 3.2b). Further, sandy regions have good recharge areas in the southeastern and northwestern areas of the research region, though the water bodies and built-up areas have poor ARS (Table 3.4).

3.4.3 HYDROLOGICAL SOIL GROUPS

Soils are most important in the artificial recharge which are classified into four types based on infiltration rate. Soils that are heavily containing clay, have not allowed infiltration and high runoff rate. On the other hand, the sandy soil allows high infiltration and low runoff rates (Rajasekhar et al., 2020). The SCS assemblies are decided on the important features or centered on the slightest infiltration rate. The CN for valuation reliant on the identifications of pixels, intersecting the LULC, and soil layers (Table 3.5). The HSGs soil group of the study area has been classified as B and C classes which are instigated in the present study (Fig 3.2c). The HSG B soil is in the central part of the region and an area of 69.64 km^2 (41.40%) whereas HSG C occurs in the south-western and northern parts of the study region and an area of 98.57 km^2 (58.60%). The HSG B soils have a higher infiltration rate as compared with the HSG C group of soils based on their porosity and permeability nature of the soils.

TABLE 3.5 USDA-SCS Soil Classification.

HSG type	Soils classification	Percentage
HSG B	Moderate infiltration rates and fine to coarse textures	41.40
HSG C	Low infiltration and tolerably fine textured soils	58.60

3.4.4 DRAINAGE DENSITY

DD is an inverse permeability function which is a significant influence in identifying GWPZs. The study's drainage map is used to prepare the DD map (Figure 3.3d). High DD values favor runoff resulting in low artificial recharge. The drainage density of the region has a straight relation with the landscape, the geomorphological patterns, the subsurface geology, and the land use in the extent, and stream order has enhanced the condition of suitable localities of protection measures regarding groundwater storage capacities (Horton, 1932, 1945; Nathawat et al., 2010; Kumar et al., 2018; Joy et al., 2019). The DD classes in the study area are classified from 0 to 3.6 km/km^2, respectively (Figure 3.2d). Deriving DD and its relation to surface runoff and rock permeability help in identifying the ARS (Table 3.4).

3.4.5 GEOMORPHOLOGY

Geomorphology is a landform analysis that involves its classification, forms, and physical processes that helped determine potential groundwater areas. The main geomorphological units (Figure 3.3a) of the studied area are divided mainly into four types, that is, structural hills and valleys, anthropogenic terrains, pediment–peneplain complex, and water bodies. The maximum elevation is 760 m above MSL is near Reddivaripalli and minimum elevation is 220 m above MSL near Krishnamgaripalli. The average elevation of the area is about 210 m. On the other hand, the northerly flowing Madduleru River forms the main drainage in the western part of the area. Besides, several seasonal streams are draining into some large tanks situated near Kothakota, Pamdurthi, Kondagattupalli, Mangalamadaka, Siddaramapuram, and Bandamadapalli. The hills ranges display the subparallel drainage patterns, while the plains and isolated hills display the subdendritic and subparallel drainage patterns. Further, subdendritic and subradiating drainage patterns with channel filled are most prevalent in the area.

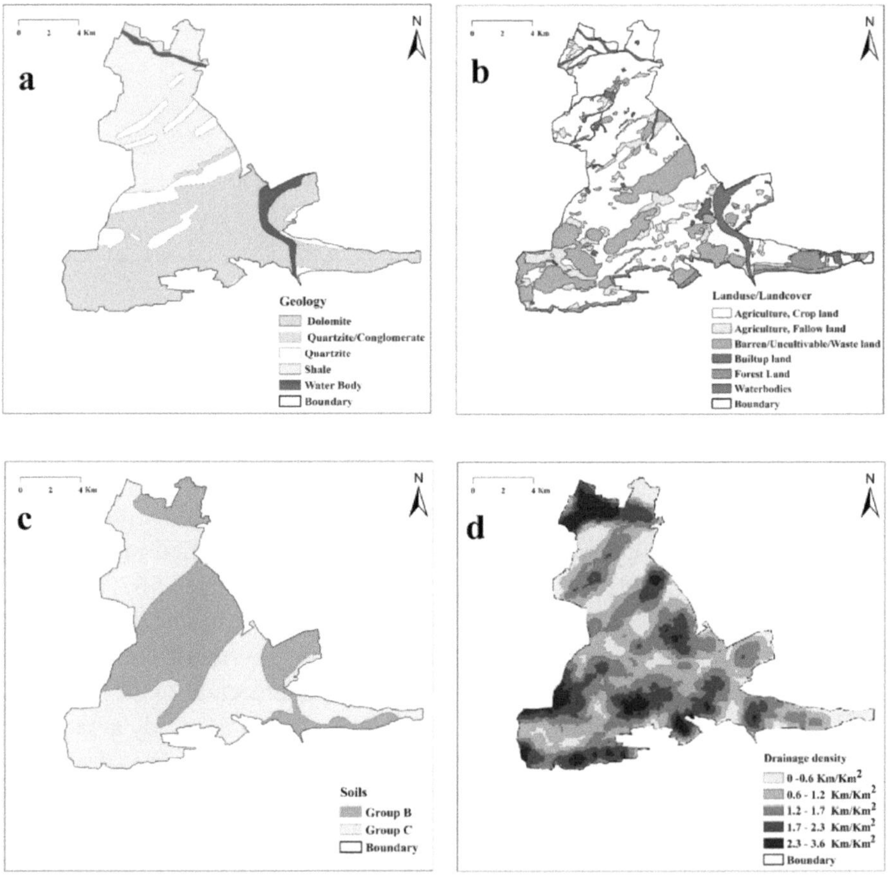

FIGURE 3.2 (a) Geology, (b) land use/land cover, (c) soils, and (d) drainage density.

3.4.6 SLOPE

The slope monitors subsurface groundwater penetration and recharge forms. It also shows the reasonableness of ground water prospecting (GWP) detection. In the gentle slope area, the surface runoff was slightly allowing more water to permeate, while the high slope zone allows high runoff allowing less rainwater dwelling time, thus moderately less infiltration. Subsequently, the slope map was classified into four categories of slope (Kadam et al., 2017) and its obtained measures are 0–5, 5–15, 15–35, and >35 (Figure 3.3b). The foremost portion of the region shows that low-to-moderate slope class represents low-to-moderate surface runoff and infiltration which is suitable for ARS sites (Table 3.6).

TABLE 3.6 Slope Classes.

Slope	Significance
Plain	Low runoff
Very gentle	Low runoff
Gentle	Low runoff
Moderate	Medium runoff
Strong	Medium runoff
Moderate to steep	High runoff
Very steep	High runoff

3.4.7 RAINFALL

The rainfall availability was reflected as the main cause of recharge (Rajasekhar et al., 2019a). It significantly affects the groundwater recharge prospective and the proficiency of AHP (Kadam et al., 2017). Monthly rainfall map for a period of two decades (i.e., 1998–2018) was observed from the C. M Dashboard Govt of A. P. in a regular interval. The rainfall map was reclassified into five main categories: 496.05–533.09, 533.09–548.89, 548.89–567.41, 567.41–591.38, and 591.38–634.96 mm/year (Figure 3.3c). In Figure 3.3g, the normal yearly rainfall in the high elevated regions is the greater the runoff compared with low elevated zones (Raju et al., 2018, 2019). The intersection of LULC, HSGs, and rainfall intensity results in the runoff potentiality of the study area (Figure 3.3c). The rainfall accessibility was reflected as the main cause of recharge. The rainfall significantly affects the groundwater recharge potential and the proficiency of the AHP (Rajasekhar et al., 2019b; Rahmati et al., 2015). Thus, rainwater harvesting approach is the key parameter in the surface water development process with the help of the rainfall parameter.

3.4.8 RUNOFF POTENTIAL

The integration of land use/land cover, hydrological soil groups, and rainfall information with evaluated curve number values are analyzed to get the result of the runoff potential layer through the ArcGIS environment (Figure 3.3d). The surface water runoff is the utmost important factor for water availability in the agriculture field. The runoff layer of the research area is classified into various categories, such as very low, low, moderate, high, and very high (Kadam et al., 2016). The surface water runoff in the central part and south-eastern regions of the present study, a moderate-to-high and the very low to low. However, the most very high runoff potential is from the southeastern side and the lower elevated ranges in the study region.

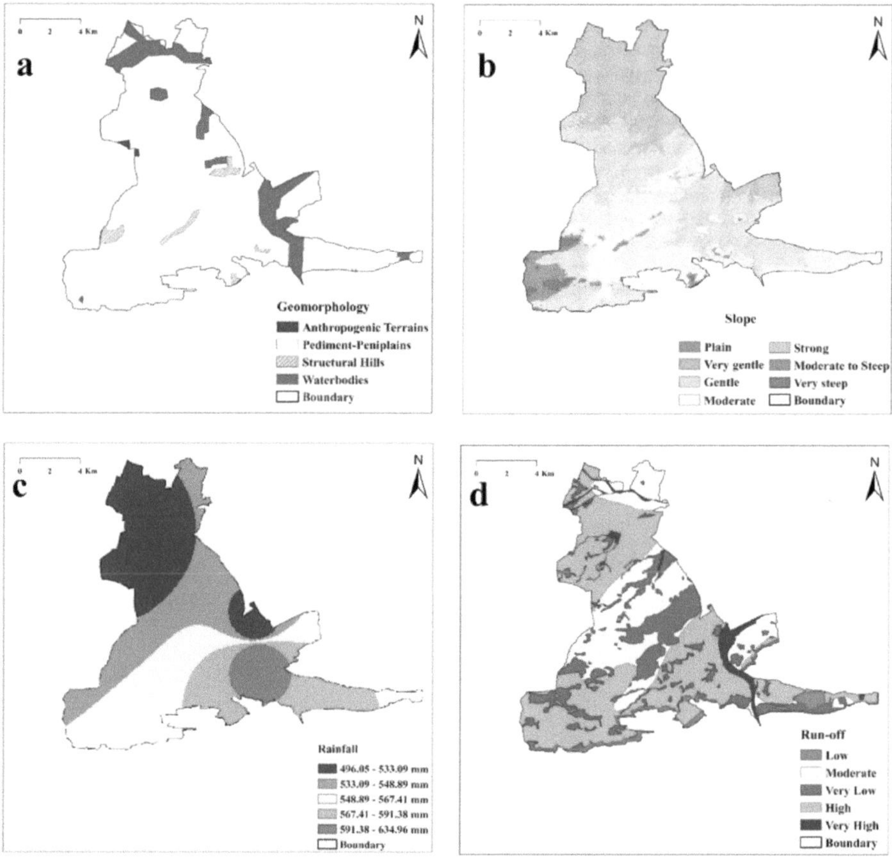

FIGURE 3.3 (a) Geomorphology, (b) slope, (c) rainfall, and (d) runoff.

3.4.9 ARTIFICIAL RECHARGE

Artificial recharge of the present study was identified through a weighted overlay method of various thematic maps, such as geology, DD, LULC, HSG, slope, GM, rainfall, and runoff in ArcGIS software. The weights assigned for the above eight thematic layers (Shailaja and Kadam, 2018; Shankar and Mohan, 2005) (Table 3.7) are normalized weights for various features (Table 3.2). Artificial recharge map was prepared based on the decision-making approaches in the ArcGIS environment (Figure 3.4). As per the quintile process, the AR assessment has been reclassified into various categories, that is, low, moderate, high, and very high categories (CGWB, 2006, 2007, 2013a, 2013b). The AR results of the present study reveals that 1.18% (1.99 km^2) of the region was found as very good and 54.14% (91.06 km^2) of the fallen in a good zone for artificial recharge of the groundwater.

On the other hand, 44.36% (74.61 km^2), and 0.32% (0.55 km^2) of the complete zone goes under individually moderate and poor ARS. (Table 3.7). Subsequently, the very high ARS are located in the southeastern part of the area as a result of high slope, high DD, and geology with low rainfall under low AR (Figure 3.4). It is obvious from the AR map that northwestern and south-western region of the present study is most reasonable for AR. Thus, distinctive SWC sites have recommended in the study by various topographical and hydrogeological limits and the possibility of structures in the present study (Kadam et al., 2012; Kaliraj et al., 2013; Rajaveni et al., 2015).

Artificial recharge specifies the need for the improvement of groundwater resources in the study region which implies to diversify the various existing structures that are identified through field surveys. Taking into account the field surveys and artificial recharge layer (Figure 3.4), there is high need of constructing of various recharge structures for sustainable development (Kumar et al., 2016). Hence, the proposed and existing appropriate ARS to develop the study region are principally check dam (CD), contour bund (CB), farm pond (FP), and percolation tank (PT). Further, the study area signifies the proposed and existing recharge structures that are identified around 75 recharge structures to obtain the green growth of the study area (Figure 3.4).

3.4.10 VALIDATION

It can be observed from the artificial recharge map that high recharge potential areas are located in the central and southwestern part, and a cross-validation analysis was performed in this part to ensure groundwater recharge potential as per CGWB (2016–17) field data. The defined recharge zone is also a buried pediment shallow in this class. And, the degree of fluctuation in the sample area ranges from 12.65 to 26.98 mbgl.

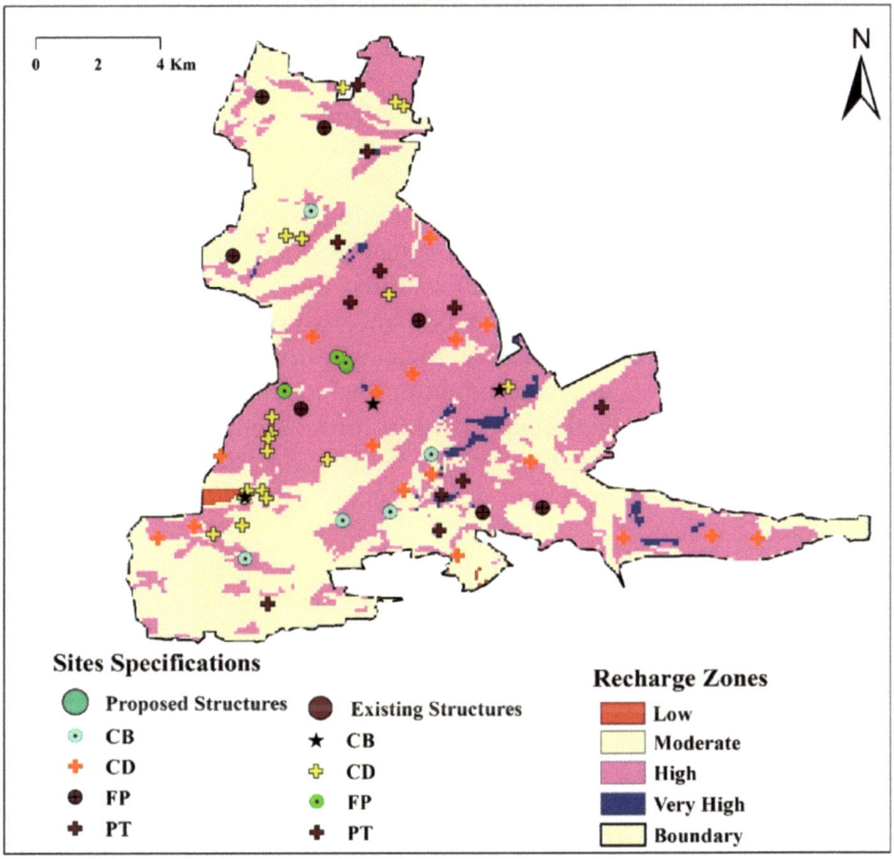

FIGURE 3.4 Artificial recharge zones.

TABLE 3.7 Artificial Recharge.

Classes	Area	
	%	km²
Poor	0.32	0.55
Moderate	44.36	74.61
Good	54.14	91.06
Very good	1.18	1.99

3.5 CONCLUSION

The process of site suitability analysis for identifying the artificial recharge structures using geospatial and decision-making approaches has an added benefit over conventional assessment methodologies. This may be due to proficiency about WOA integrative for pertinent parameters, such as geology, DD, LULC, HSG, slope, GM, rainfall, and runoff which give smaller suitability components as a composite layer. Subsequently, the utilization of the SCS-CN method for runoff mapping also provided an exceptional study of area runoff, giving suitable consequences of provincial runoff complexity using the integration of LULC and HSGs parameters. The composite layer obtained from the multiplication of the layer's weightage was further averaged into four categories, that is, very high, high, moderate, and low. Simultaneously, the suitability of CDs, PTs, CBs, FPs were consigned to zones of proximity to villages. Each of the structures was assessed based on the specific construction needs as well as the ecological supportability limit. From the study area, 39 locations were found suitable for CDs, 13 PTs, 8 CBs, and 15 FPs. Thus, southeastern areas were found unsatisfactory for groundwater recharge structures.

Considering the result analysis, the present research will endorse suitable enhancement, improvement, and conservation of groundwater assets in the study region. Moreover, integrated approaches to RS, GIS, and AHP proved gainful in separating rational groundwater ARS from other regular techniques. Hence, the study encourages adequate conservation and management of groundwater assets for sustainable green growth.

KEYWORDS

- **groundwater**
- **water conservation**
- **AHP**
- **remote sensing**
- **GIS**

REFERENCES

Bouwer, H. Integrated Water Management for the 21st Century: Problems and Solutions. *J. Irrigat. Drain. Eng.* **2002,** *128* (4), 193–202. DOI:10.1061/(ASCE)0733-9437(2002)128:4 (193).

CGWB. *Manual on Artificial Recharge of Groundwater*; Central Groundwater Board: New Delhi, 2002.

CGWB. *Dynamic Groundwater Resource of Chhattisgarh*; Central Groundwater Board: Raipur, 2006.

CGWB. *Manual on Artificial Recharge of Groundwater*; Central Groundwater Board: New Delhi, 2007.

CGWB. *Master Plan for Artificial Recharge to Groundwater in India*; Central Groundwater Board: Faridabad, 2013a.

CGWB. *Manual on Artificial Recharge of Groundwater*; Central Groundwater Board: New Delhi, 2013b.

Chowdary, V. M.; Ramakrishna, D.; Srivastava, Y. K.; Chandran, V.; Jeyaram, A. Integrated Water Resource Development Plan for Sustainable Management of Mayurakshi Watershed, India Using Remote Sensing and GIS. *Water Resour. Manage.* **2009,** *23*, 1581–1602.

Chowdhury, A.; Jha, M. K.; Chowdary, V. M. Delineation of Groundwater Recharge Zones and Identification of Artificial Recharge Sites in West Medinipur District, West Bengal, Using RS, GIS, and MCDM Techniques. *Environ. Earth Sci.* **2010,** *59* (6), 1209–1222.

Chowdhury, S. J.; Rahaman, M. F.; Arefin, R.; Ali, M. S.; Mazumder, Q. H. Delineation of Groundwater Potential Zones of Atrai–Sib River Basin in North-West Bangladesh Using Remote Sensing and GIS Techniques. *Sustain. Water Resour. Manage.* **2018**. DOI:10.1007/s40899-018-0240-x.

Das, S. Comparison among Influencing Factor, Frequency Ratio, and Analytical Hierarchy Process Techniques for Groundwater Potential Zonation in Vaitarna Basin, Maharashtra, India. *Groundwater Sustain. Dev.* **2019,** *8*, 617–629.

Dillon, P. Future Management of Aquifer Recharge. *Hydrogeol. J.* **2005,** *13* (1), 313–316. DOI:10.1007/s10040-004-0413-6.

Etikala, B.; Golla, V.; Li, P.; Renati, S. Deciphering Groundwater Potential Zones Using MIF Technique and GIS: A Study from Tirupati Area, Chittoor District, Andhra Pradesh, India. *HydroResearch* **2019,** *1*, 1–7.

Ghayoumian, J.; Saravi, M. M.; Feiznia, S.; Nouri, B.; Malekian, A. Application of GIS Techniques to Determine Areas Most Suitable for Artificial Groundwater Recharge in a Coastal Aquifer in Southern Iran. *J. Asian Earth Sci.* **2007,** *30* (2), 364–374.

Golla, V. Delineation of Groundwater Potential Zones in Sathyavedu Area, Chittoor District (Andhra Pradesh), South India, Using Geospatial Technologies. *Model. Earth Syst. Environ.* **2020,** *6*, 895–905. DOI:10.1007/s40808-020-00726-9.

Horton, R. E. Drainage Basin Characteristics. *Trans. Am. Geophys. Union* **1932,** *13*, 350–361. DOI:10.1029/TR013i001p00350.

Horton, R. E. Erosional Development of Streams and Their Drainage Basins: Hydrophysical Approach to Quantitative Morphology. *Geol. Soc. Am. Bull.* **1945,** *56*, 275–370.

Joy, J.; Kanga, S.; Singh, S. K. Kerala Flood 2018: Flood Mapping by Participatory GIS Approach, Meloor Panchayat. *Int. J. Emerg. Technol.* **2019,** *10* (1), 197–205.

Kadam, A. K.; Kale, S. S.; Pande, N. N.; Pawar, N. J.; Sankhua, R. N. Identifying Potential Rainwater Harvesting Sites of a Semi-arid, Basaltic Region of Western India, Using SCS-CN Method. *Water Resour. Manage.* **2012,** *26*, 2537–2554. DOI:10.1007/s11269-012-0031-3.

Kadam, A.; Kale, S.; Umrikar, B.; Sankhua, R.; Pawar, N. Identifying Possible Locations Construct Soil-Water Conservation Structures by Using Hydrogeological and Geospatial Analysis. *Hydrospat. Anal.* **2017,** 18–27. DOI:10.21523/gcj3.17010103.

Kaliraj, S.; Chandrasekar, N.; Magesh, N. S. Identification of Potential Groundwater Recharge Zones in Vaigai Upper Basin, Tamil Nadu, Using GIS-Based Analytical Hierarchical Process (AHP) Technique. *Arab. J. Geosci.* **2014,** *7* (4), 1385–1401.

Kaliraj, S.; Chandrasekar, N.; Ramachandran, K. K. Mapping of Coastal Landforms and Volumetric Change Analysis in the South West Coast of Kanyakumari, South India Using Remote Sensing and GIS Techniques. *Egypt. J. Remote Sens. Space Sci.* **2016**. DOI:10.1016/j.ejrs.2016.12.006.

Kanga, S.; Sheikh, A. J.; Godara, U. GIscience for Groundwater Prospect Zonation. *J. Crit. Rev.* **2020,** *7* (18), 697–709.

Kudnar, N. S.; Rajasekhar, M. A Study of the Morphometric Analysis and Cycle of Erosion in Waingangā Basin, India. *Model. Earth Syst. Environ.* **2020,** *6* (1), 311–327.

Kumar, M. G.; Agarwal, A. K.; Bali, R. Delineation of Potential Sites for Water Harvesting Structures Using Remote Sensing and GIS. *J. Indian Soc. Remote Sens.* **2008,** *36* (4), 323–334.

Kumar, T.; Gautam, A. K.; Kumar, T. Appraising the Accuracy of GIS-based Multi-criteria Decision Making Technique for Delineation of Groundwater Potential Zones. *Water Resour Manage.* **2014,** *28* (13), 4449–4466. DOI:10.1007/s11269-014-0663-6.

Kumar, T.; Kant, A.; Jhariya, G. D. C. Multi-criteria Decision Analysis for Planning and Management of Groundwater Resources in Balod District, India. *Environ. Earth Sci.* **2016**. DOI:10.1007/s12665-016-5462-3.

Kumar, S.; et al. *Delineation of Shoreline Change along Chilika Lagoon (Odisha), East Coast of India Using Geospatial Technique*, 2018.

Kumbhar, V.; Swami, V.; Kulkarni, S. Effectiveness of Watershed Management: A Means of Economic Development: A Case Study. *J. Mech. Civil Eng.* **2013,** *6* (6), 7–14.

Meraj, G.; et al. Geoinformatics Based Approach for Estimating the Sediment Yield of the Mountainous Watersheds in Kashmir Himalaya, India. *Geocarto Int.* **2018,** *33* (10), 1114–1138.

Meraj, G.; Singh, S. K.; Kanga, S.; Islam, M. N. Modeling on Comparison of Ecosystem Services Concepts, Tools, Methods and Their Ecological-economic Implications: A Review. *Model. Earth Syst. Environ.* **2021,** 1–20. DOI:10.1007/s40808-021-01131-6.

Mustafa, A. A.; Singh, M.; Sahoo, R. N.; Ahmed, N.; Khanna, M.; Sarangi, A.; Mishra, A. K. Land Suitability Analysis for Different Crops: A Multi-criteria Decision Making Approach Using Remote Sensing and GIS. *Researcher* **2011,** *3* (12), 61–84.

Nathawat, M. S.; et al. Monitoring & Analysis of Wastelands and Its Dynamics Using Multiresolution and Temporal Satellite Data in Part of Indian State of Bihar. *Int. J. Geomat. Geosci.* **2010,** 1 (3), 297–307.

Nsiah, E.; Appiah-Adjei, E. K.; Adjei, K. A. Hydrogeological Delineation of Groundwater Potential Zones in the Nabogo Basin, Ghana. *J. Afr. Earth Sci.* **2018**. DOI:10.1016/j.jafrearsci.2018.03.016.

Pandey, A. C.; Singh, S. K.; Nathawat, M. S. Waterlogging and Flood Hazards Vulnerability and Risk Assessment in Indo-Gangetic Plain. *Nat. Hazards* **2010,** *55* (2), 273–289.

Pandey, A. C.; Singh, S. K.; Nathawat, M. S.; Saha, D. Assessment of Surface and Subsurface Waterlogging, Water Level Fluctuations, and Lithological Variations for Evaluating Groundwater Resources in Ganga Plains. *Int. J. Digit. Earth* **2013,** *6* (3), 276–296.

Pyne, R. D. G. *Aquifer Storage Recovery—A Guide to Groundwater Recharge through Wells*, 2nd ed.; ASR Systems: Gainesville, FL, 2005; 608 p.

Rahman, M. A.; Rusteberg, B.; Gogu, R. C.; Ferreira, J. L.; Sauter, M. A New Spatial Multi-criteria Decision Support Tool for Site Selection for Implementation of Managed Aquifer Recharge. *J. Environ. Manage.* **2012,** *99*, 61–75.

Rahmati, O.; Nazari Samani, A.; Mahdavi, M.; Pourghasemi, H. R.; Zeinivand, H. Groundwater Potential Mapping at Kurdistan Region of Iran Using Analytic Hierarchy Process and GIS. *Arab. J. Geosci.* **2015,** *8* (9), 7059–7071. DOI:10.1007/s12517-014-1668-4.

Rais, S.; Javed, A. Identification of Artificial Recharge Sites in Manchi Basin, Eastern Rajasthan (India) Using Remote Sensing and GIS Techniques. *J. Geogr. Inf. Syst.* **2014,** *2014*.

Rajasekhar, M.; Sudarsana Raju, G.; Bramaiah, C.; Deepthi, P.; Amaravathi, Y.; Siddi Raju R. Delineation of Groundwater Potential Zones of Semi-arid Region of YSR Kadapa District, Andhra Pradesh, India Using RS, GIS and Analytic Hierarchy Process. *Remote Sens. Land.* **2018,** *2* (2), 76–86. DOI:10.21523/gcj1.18020201.

Rajasekhar, M.; Sudarsana Raju, G.; Imran Basha, U.; Siddi Raju, R.; Pradeep Kumar, B.; Ramachandra, M. Identification of Suitable Sites for Artificial Groundwater Recharge Structures in Semi-arid Region of Anantapur District: AHP Approach. *Hydrospat. Anal.* **2019a,** *3* (1), 1–11. DOI:10.21523/gcj3.19030.101.

Rajasekhar, M.; Raju, G. S.; Sreenivasulu, Y.; Raju, R. S. Delineation of Groundwater Potential Zones in Semi-arid Region of Jilledubanderu River Basin, Anantapur District, Andhra Pradesh, India Using Fuzzy Logic, AHP and Integrated Fuzzy-AHP Approaches. *HydroResearch* **2019b,** *2*, 97–108.

Rajasekhar, G.; Raju, S.; Siddi Raju, R. Assessment of Groundwater Potential Zones in Parts of the Semi-arid Region of Anantapur District, Andhra Pradesh, India Using GIS and AHP Approach. Model. *Earth Syst. Environ.* **2019c,** *5*, 1303–1317. DOI:10.1007/s40808-019-00657-0.

Raju, R. S.; Sudarsana Raju, G; Rajasekhar, M. Estimation of Rainfall-runoff Using SCS Method with RS and GIS Techniques for Mandavi Basin in YSR Kadapa District of Andhra Pradesh, India. *Hydrospat. Anal.* **2018,** *2* (1), 1–15. DOI:10.21523/gcj3.18020101.

Raju, R. S.; Raju, G. S.; Rajasekhar, M. Identification of Groundwater Potential Zones in Mandavi River Basin, Andhra Pradesh, India Using Remote Sensing, GIS and MIF Techniques. *HydroResearch* **2019,** *2*, 1–11.

Ramachandra, M.; Raghu Babu, K.; Rajasekhar, M.; Badapalli, P. K. Identification of Groundwater Potential Zones Using AHP and Geospatial Techniques in Western Part of Cuddapah Basin, Andhra Pradesh, India. *Hydrospat. Anal.* **2019,** *3* (2), 60–71. DOI:10.21523/gcj3.19030202.

Saaty, T. L. *The Analytic Hierarchy Process: Planning, Priority Setting, Resource Allocation*; McGraw-Hill: New York, 1980.

Saaty, T. L. Fundamentals of the Analytic Network Process. In *International Symposium of the Analytic Hierarchy Process (ISAHP)*, Kobe, Japan, 1999.

Saaty, T. L. Fundamentals of the Analytic Network Process—Multiple Networks with Benefits, Costs, Opportunities and Risks. *J. Syst. Sci. Syst. Eng.* **2004,** *13* (3), 348–379. DOI:10.1007/s11518-006-0171-1.

Saaty, T. L. Decision Making with the Analytic Hierarchy Process. *Int. J. Serv. Sci.* **2008,** *1* (1), 83–98.

Senthilkumar, M.; Gnanasundar, D.; Arumugam, R. Identifying Groundwater Recharge Zones Using Remote Sensing & GIS Techniques in Amaravathi Aquifer System, Tamil Nadu, South India. *Sustain. Environ. Res.* **2019,** *29* (1), 15.

Shailaja, G.; Kadam, A. K. *Integrated Geophysical, Geospatial and Multiple-criteria Decision Techniques for Delineation of Groundwater Potential Zones in a Semi-arid Hard-rock Aquifer in Maharashtra, India*, 2018.

Shankar, M. N. R.; Mohan, G. A GIS Based Hydrogeomorphic Approach for Identification of Site-specific Artificial-recharge Techniques in the Deccan Volcanic Province. *J. Earth Syst. Sci.* **2005,** *114* (5), 505–514.

Singh, S. K.; Pandey, A. C. Geomorphology and the Controls of Geohydrology on Waterlogging in Gangetic Plains, North Bihar, India. *Environ. Earth Sci.* **2014,** *71* (4), 1561–1579.

Singh, L. K.; Jha, M. K.; Chowdary, V. M. Multi-criteria Analysis and GIS Modeling for Identifying Prospective Water Harvesting and Artificial Recharge Sites for Sustainable Water Supply. *J. Clean. Prod.* **2017a**. DOI:10.1016/j.jclepro.2016.11.163.

Singh, S. K.; Mishra, S. K.; Kanga, S. Delineation of Groundwater Potential Zone Using Geospatial Techniques for Shimla City, Himachal Pradesh (India). *Int. J. Sci. Res. Dev.* **2017b,** *5* (4), 225–234.

TWB. *Deep Wells and Prudence: Towards Pragmatic Action for Addressing Groundwater Overexploitation in India*. The World Bank: Washington, DC, 2010.

Zolekar, R. B.; Bhagat, V. S. Multi-criteria Land Suitability Analysis for Agriculture in Hilly Zone: Remote Sensing and GIS Approach. *Comput. Electron. Agric.* **2015,** *118*, 300–321.

CHAPTER 4

Prioritization of Watersheds for Sustainable Water Resource Management at Basin Level Using Remote Sensing and Geographic Information System

SUBHANIL GUHA* and HIMANSHU GOVIL

Department of Applied Geology, National Institute of Technology, Raipur, India

**Corresponding author. E-mail: himgeo@gmail.com*

ABSTRACT

Remote sensing techniques have brought a new dimension in the watershed study. Solani watershed of Upper Ganga drainage basin was morphometrically analyzed to delineate the prioritization of subwatersheds (SWS) for soil conservation. Dendritic to parallel drainage pattern was noticed with a stream order ranged up to fourth order. The watershed has a coarser drainage pattern with a drainage density of 0.93 to 1.07 km/km^2. This watershed falls under the normal watershed category as the bifurcation ratio ranged from 1.66 to 2.56. The overall analysis identified SWS-3 as the most vulnerable SWS in Solani watershed and should get the most importance for soil management.

4.1 INTRODUCTION

Remote sensing technology helps us to get a lot of information from the inaccessible parts of the earth surface and it is used in almost every domain

GIScience for the Sustainable Management of Water Resources. Gowhar Meraj, Shruti Kanga, Majid Farooq, Suraj Kumar Singh, & Sudhanshu (Eds.)
© 2023 Apple Academic Press, Inc. Co-published with CRC Press (Taylor & Francis)

of field sciences (Guha et al., 2020). Many geological, hydrological, environmental, and geomorphological studies applied multispectral and hyperspectral remote sensing technique to obtain authentic results in critical analysis (Mondal et al., 2011; Govil, 2015; Khan et al., 2018; Govil et al., 2020; Guha and Govil, 2019). River geomorphology is a special field which uses satellite imageries at a large extent for any morphometric analysis. Different satellite sensors, such as Landsat, Indian Remote Sensing (IRS), Sentinel, etc. are frequently used in analyzing the geomorphological parameters (Guha, 2015a, 2015b; Khan et al., 2020).

Various types of application fields, such as prioritization of watershed, soil erosion estimation, flood frequency analysis, hydrological and geomorphological modeling, etc. are benefitted a lot through the analysis of morphometric features of a watershed (Guha, 2015a, 2015b; Guha and Dey, 2016a, 2016b). The hydrological behavior of a watershed is closely represented by geomorphological parameters. Recently, some researchers worked on morphometric analysis and watershed prioritization using remotely sensed data. Morphotectonic analysis of various watersheds of Kashmir valley was carefully examined applying remote sensing techniques (Altaf et al., 2013, 2014; Meraj et al., 2015, 2016; Bali et al., 2016). Using drainage morphometry, Gopinath et al. (2016) used a multicriteria decision-making method to determine watershed prioritization on the Kuttiyadi river basin in Kerala. A revolutionary water and land management program was used to classify the most susceptible areas in the Rmel river basin (Bouguerra and Jebari, 2017). A modern method of sub-watershed (SWS) prioritization was performed in the lower part of Bhavani basin in Tamil Nadu (Balasubramanian et al., 2017). Farhan et al. (2017) prioritized the arid agronomic catchment in Zerqa river basin in north Jordan. Arefin et al. (2020) ranked catchments for soil and water conservation at northern Bangladesh. Rao et al. (2017) recently used the remote sensing technique for morphometric analysis of Yamuna river basin around Agra, Firozabad, and Etawah district of Uttar Pradesh, India. The main objective of the present study was to interpret the morphometric parameters for prioritization of SWS of Solani watershed.

4.2 MATERIALS AND METHODS

The Solani watershed extends from 29°50′24″ N to 30°16′12″ N latitude and from 77°44′ 24″ E to 78°04′12″ E longitude and covers 914.7411 km^2 land area. This hilly watershed is a part of the foothills region in lower Himalaya.

The mean annual rainfall is around 2000 mm and most of the rainfall occurred in monsoon season. The mean annual temperature ranges from 5°C to 30°C. Landsat 8 OLI and TIRS satellite data of February 2, 2017, were used in this study.

Figure 4.1 presented the four sub-watersheds of Solani watershed (SWS-1–4). The morphometric elements were evaluated using equations provided by Altaf et al. (2013). A priority ranking was proposed based on the interpretation of eight specific morphometric parameters (bifurcation ratio, elongation ratio, texture ratio, form factor, compactness constant, stream frequency, drainage density, and circulatory ratio) for four SWS. Prioritization was determined by using the calculation of the compound parameter. Highest priority was given to the SWS having the lowest value of compound parameter.

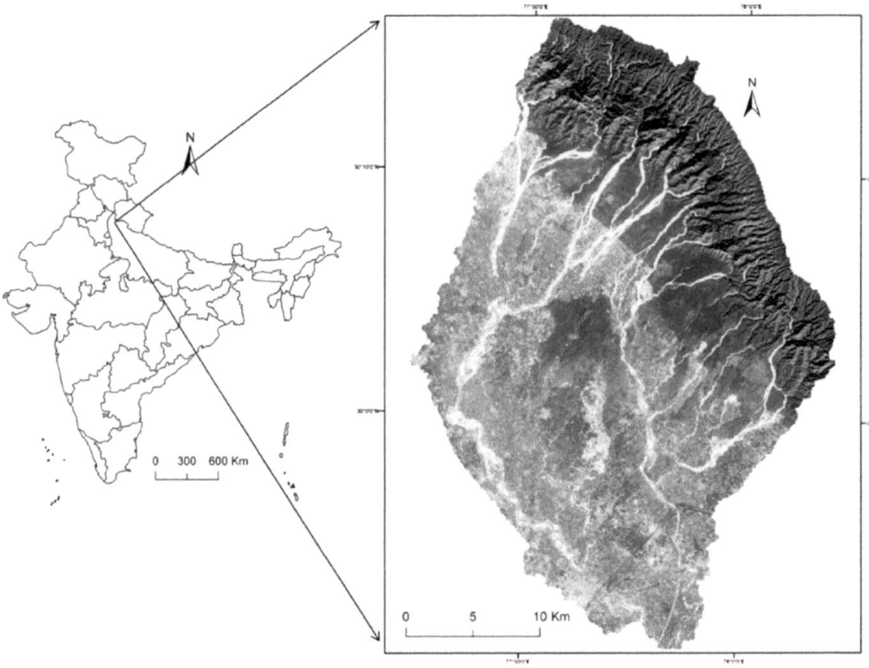

FIGURE 4.1 Location of the study area.

4.3 RESULTS AND DISCUSSION

Strahler (1964) proposed the stream ordering technique which was followed in the study. It was noticed from Table 4.1 that first-order stream had the maximum frequency. Figure 4.2 showed a semilog plot indicating the stream order and stream number relationship.

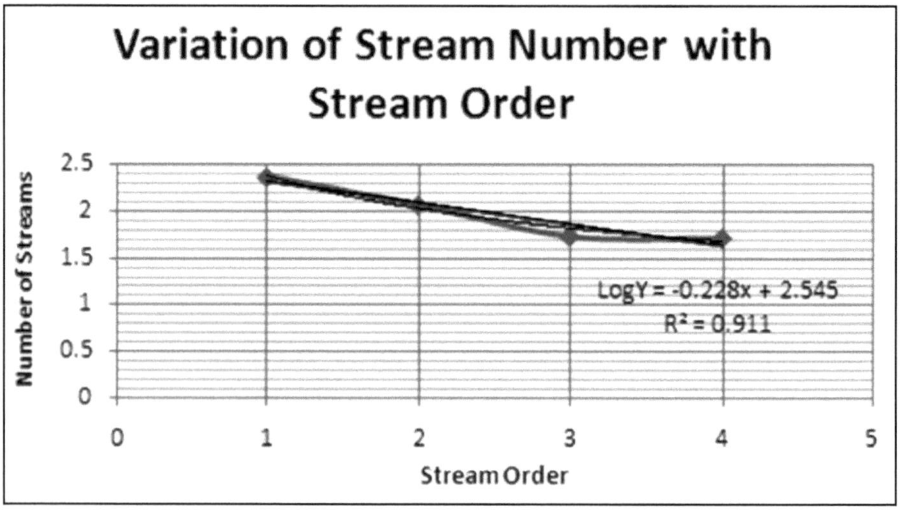

FIGURE 4.2 Relation of stream numbers with stream order.

TABLE 4.1 Stream Ordering with Number of Stream Segments.

Sl no.	SWS no.	Watershed area (km²)	Stream order				Total segments of stream
			I	II	III	IV	
1	SWS-1	129.51	36	19	13	3	71
2	SWS-2	164.47	34	18	11	3	66
3	SWS-3	270.89	71	37	7	25	140
4	SWS-4	349.85	91	44	24	22	181
Total		914.74 –	232 (50.66%)	118 (25.76%)	55 (12.01%)	53 (11.57%)	458 (100.00)

Stream length is a significant index in describing stream development. Stream length is controlled by steepness of slope. GIS technology helps us to count the number and measure the length of the streams under different orders (Figure 4.3). A Semilog plot indicated a negative slope of stream lengths with stream orders (Figure 4.4). A total of 75.43% of the total watershed stream length was occupied by first and second-order streams (Table 4.2).

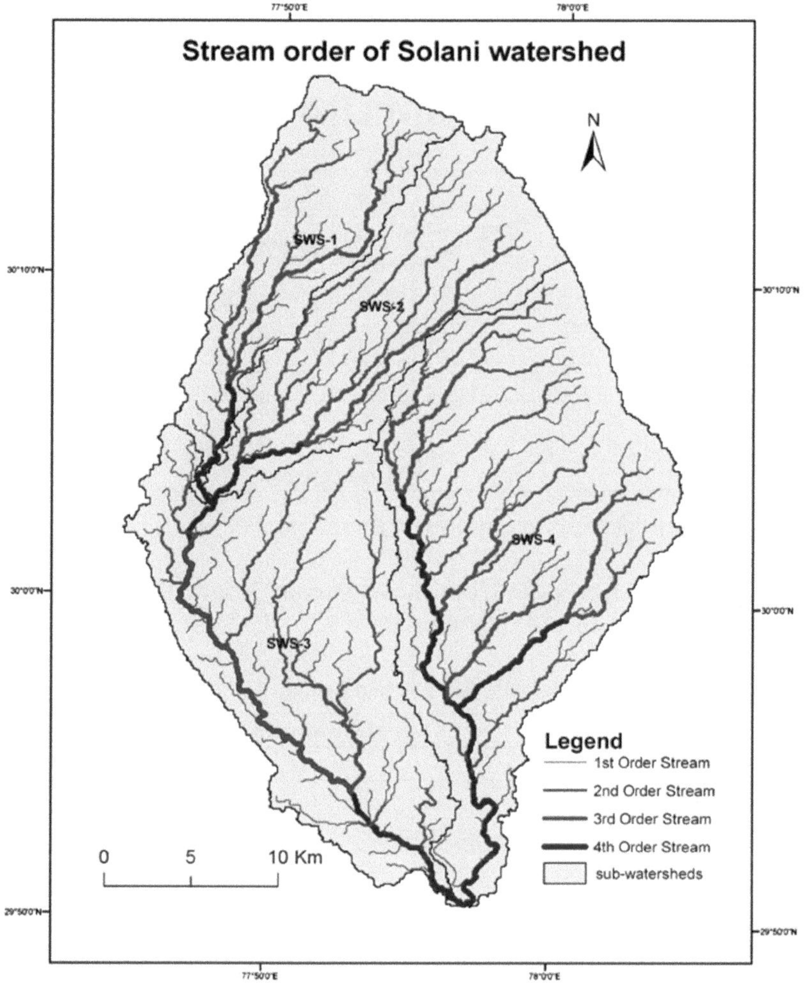

FIGURE 4.3 Drainage map with stream ordering.

TABLE 4.2 Stream Ordering with Stream Length.

Sl no.	SWS no.	Watershed area (km²)	Stream length (km)				Total stream length (km)
			I	II	III	IV	
1	SWS-1	129.51	58.81	32.56	32.02	9.53	132.92
2	SWS-2	164.47	79.50	60.52	28.50	7.79	176.33
3	SWS-3	270.89	137.31	66.08	8.96	40.45	252.81
4	SWS-4	349.85	148.74	97.40	51.83	42.75	340.73
Total		914.74	424.38 (47.01%)	256.58 (28.42%)	121.33 (13.44%)	100.52 (11.13%)	902.82 (100.00

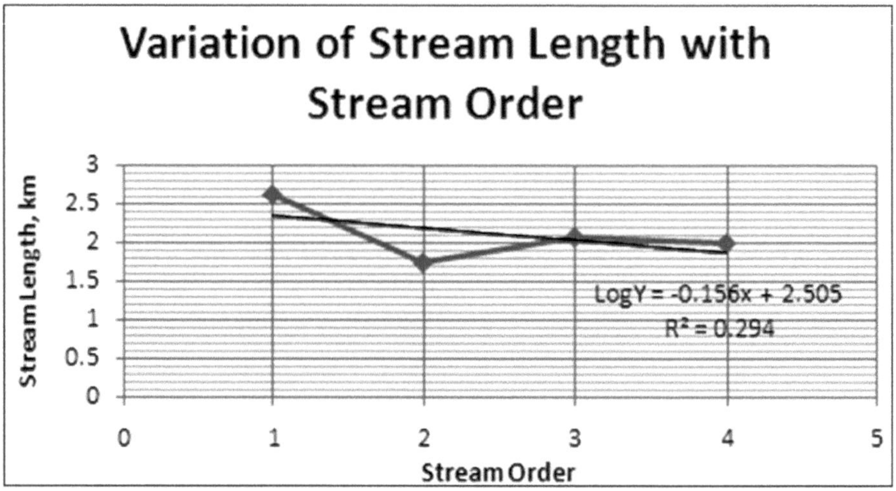

FIGURE 4.4 Relationship of stream order to stream lengths.

Strahler (1964) proposed the mean stream length that is the average length of all the streams (Table 4.3).

TABLE 4.3 Stream Ordering with Mean Stream Length.

Sl. no.	SWS no.	Mean stream length (L_{sm}) (km)			
		I	II	III	IV
1	SWS-1	1.63	1.71	2.46	3.17
2	SWS-2	2.33	3.36	2.59	2.59
3	SWS-3	1.93	1.78	1.28	1.61
4	SWS-4	1.63	2.21	2.15	1.94
Mean stream length (km)		1.88	2.26	2.12	2.33

Stages of cycle of erosion of any watershed may be stream length ratio (Table 4.4).

TABLE 4.4 SWS Wise Stream Length Ratio.

Sl no.	SWS no.	Stream length ratio (Rl)		
		II/I	III /II	IV/III
1	SWS-1	0.55	0.98	0.29
2	SWS-2	0.76	0.47	0.27
3	SWS-3	0.48	0.13	4.51
4	SWS-4	0.65	0.53	0.82
Mean stream length ratio		0.60	0.47	0.82

Schumm (1956) suggested the bifurcation ratio as the ratio among the numbers of stream sections of a stream order and the succeeding higher order. The R_b in the drainage watershed varied from 1.94 to 2.55 (Table 4.5) that indicates that all the SWS fall under a normal watershed.

TABLE 4.5 SWS Wise Bifurcation Ratio.

Sl. no.	SWS no.	Bifurcation ratio (R_b)			Regression coefficient	Mean bifurcation ratio
		I/II	II/III	III/IV		
1	SWS-1	1.89	1.46	4.33	(−)0.34	2.56
2	SWS-2	1.88	1.63	3.66	(−)0.30	2.39
3	SWS-3	1.91	5.28	0.28	(−)0.28	2.49
4	SWS-4	2.06	1.83	1.09	(−)0.18	1.66
		1.94	2.55	2.34	(−)0.25	2.279

Table 4.6 (column III) showed the drainage density ranged from 0.93 to 1.07 km/km^2 which suggests coarser drainage pattern (Figure 4.5). Table 4.6 (column IV) showed stream frequency (0.495 for the study area) as the total number of stream segments of the watershed. Smith (1950) proposed the texture ratio as the relative distance of drainage channels. The range of texture ratio for the SWS was 0.67 to 1.37 (Table 4.6, column V). Table 4.6 (column VI) showed the value of form factor (Schumm, 1956) varied from 0.20 to 0.25 indicated mostly elongated SWS. Circulatory ratio (Miller, 1953) of the SWS are ranged from 0.17 to 0.25 (Table 4.6, column VII), showed mainly elongated shape. Schumm (1956) described the elongation ratio as the ratio between the diameters of the circle equals to the watershed and the highest length of the watershed which was shown in Table 4.6 (column VIII). The ratio between the perimeter and area of any watershed is identified as compactness constant (Horton, 1945). The C_c values was ranged between 1.99 and 2.37 in the study area reflected the compact nature of the study area (Table 4.6, column IX). Horton (1945) described L_o as the initial surface water length prior to the formation of definite stream channels. Table 4.6 (column X) reveals the values varied from 0.46 to 0.53 and indicate high relief.

Schumm (1956) designated relief ratio as the peak relief to the horizontal distance along the watershed corresponding to the primary drainage. Relief and channel gradient were always related positively which was shown by the relief ratio. Table 4.7 showed the value of relief ratio (0.002–0.025) which indicated a low relief.

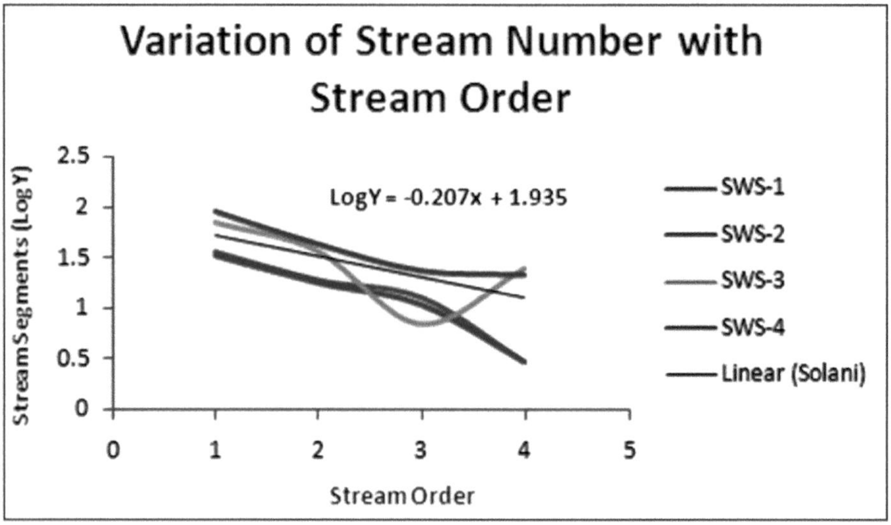

FIGURE 4.5 Relation of stream number with stream order.

TABLE 4.6 Areal Aspects of the Morphological Parameters.

Sl. no.	SWS no.	D_d	F_s	R_t	F_f	R_c	R_e	C_c	L_o
(I)	(II)	(III)	(IV)	(V)	(VI)	(VII)	(VIII)	(IX)	(X)
1	SWS-1	1.026	0.548	0.741	0.202	0.177	0.508	2.374	0.487
2	SWS-2	1.073	0.401	0.670	0.233	0.213	0.545	2.166	0.466
3	SWS-3	0.933	0.517	1.091	0.259	0.207	0.574	2.199	0.536
4	SWS-4	0.974	0.517	1.374	0.248	0.253	0.562	1.987	0.513

TABLE 4.7 Relief Aspects of the Morphological Parameters.

Sl. no.	SWS no.	Perimeter (km)	Watershed length (km)	Total relief (km)	Relief ratio
1	SWS-1	95.76	25.29	0.632	0.025
2	SWS-2	98.46	26.56	0.608	0.023
3	SWS-3	128.34	32.35	0.061	0.002
4	SWS-4	131.76	37.58	0.563	0.015

Prioritization of SWS is required to delineate the intensive erosional zones. The final map for four SWS with their priority ranking (Table 4.8) was shown in Figure 4.6. SWS-3 ranks first on prioritization (compound parameter 2.25).

TABLE 4.8 Priority Ranking of the SWS.

Sl. no.	SWS no.	R_b	D_d	F_s	R_t	F_f	R_c	R_e	C_c	Compound parameter	Priority ranking
1	SWS-1	1	2	1	3	4	4	4	1	2.50	3rd
2	SWS-2	2	1	4	4	3	2	3	3	2.75	4th
3	SWS-3	3	4	2	2	1	3	1	2	2.25	1st
4	SWS-4	4	3	2	1	2	1	2	4	2.375	2nd

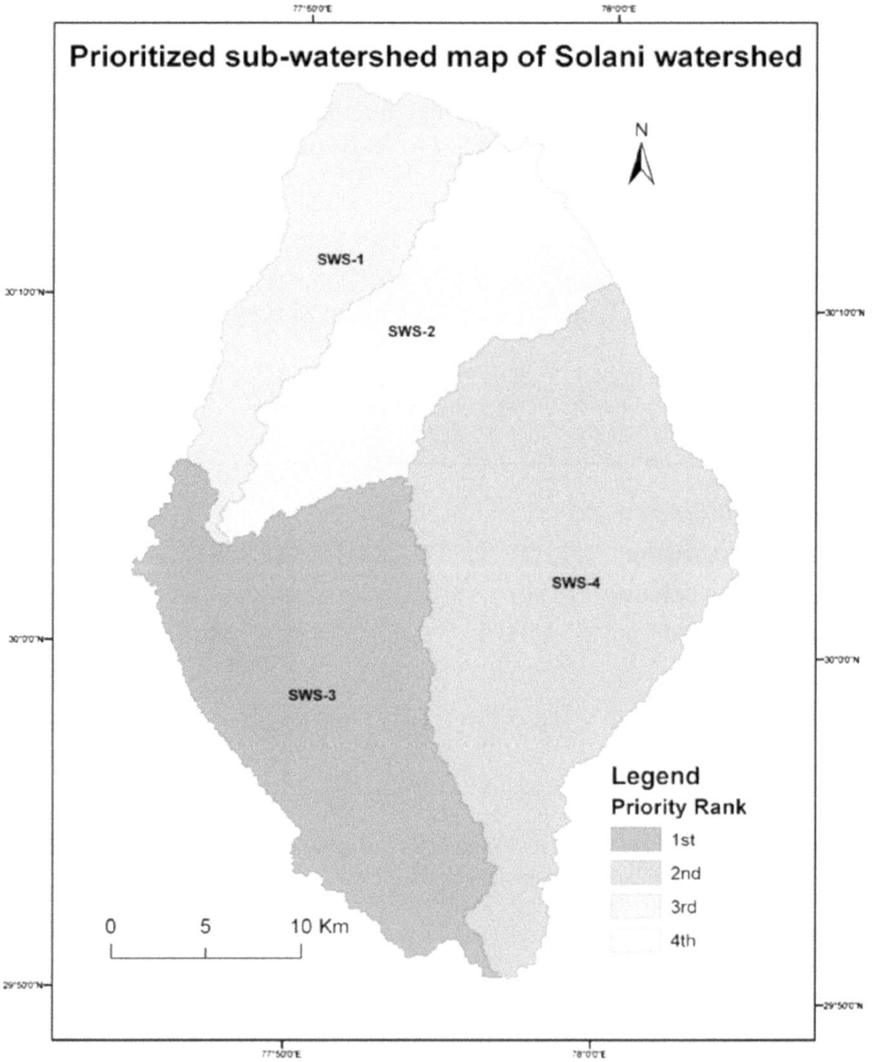

FIGURE 4.6 Priority ranking of the SWS.

4.4 CONCLUSION

The present study focused on the extraction of morphological characters of the watershed with the help of earth observation data. The watershed morphometric analysis was worked out by earth observation technique. In the watershed, lower order streams were more in number. Stream frequency is inversely related with stream order. Horton's stream ordering method was followed in the present research work. The stream length ratios were calculated for each sub-watershed. The bifurcation ratio categorized the watershed as a normal watershed. The Solani watershed is characterized by coarse drainage network. The watershed is of extended shape and of modest relief. Finally, it can be concluded that suitable soil conservation measures must be taken for SWS-3 and SWS-4 to resist rapid soil erosion.

ACKNOWLEDGMENTS

Author is thankful to United States Geological Survey (USGS).

KEYWORDS

- **morphometry**
- **prioritization**
- **remote sensing**
- **watershed**

REFERENCES

Altaf, F.; Gowhar, M.; Shakil, A. R. Morphometric Analysis to Infer Hydrological Behaviour of Lidder Watershed, Western Himalaya, India. *Geogr. J.* **2013**. DOI:10.1155/2013/178021.

Altaf, S.; Gowhar, M.; Shakil A. R. Morphometry and Land Cover Based Multi-criteria Analysis for Assessing the Soil Erosion Susceptibility of the Western Himalayan Watershed. *Environ. Monit. Assess.* **2014,** *186* (12), 8391–8412.

Arefin, R.; Mohir, M. M. I.; Alam, J. Watershed Prioritization for Soil and Water Conservation Aspect Using GIS and Remote Sensing: PCA-based Approach at Northern Elevated Tract Bangladesh. *Appl. Water Sci.* **2020,** *10*, 91.

Balasubramanian, A.; Duraisamy, K.; Thirumalaisamy, S.; Krishnaraj, S.; Yatheendradasan, R. K. Prioritization of Subwatersheds Based on Quantitative Morphometric Analysis in Lower Bhavani Basin, Tamil Nadu, India Using DEM and GIS Techniques. *Arab. J. Geosci.* **2017,** *10*, 552.

Bali, B. S.; Wani, A. A.; Khan, R. A.; Ahmad, S. Morphotectonic Analysis of the Madhumati Watershed, Northeast Kashmir Valley. *Arab. J. Geosci.* **2016,** *9*, 390.

Bouguerra, S.; Jebari, S. Identification and Prioritization of Sub-watersheds for Land and Water Management Using InVEST SDR Model: Rmel River Basin, Tunisia. *Int. J. Remote Sens.* **2017,** *10*, 348.

Farhan, Y.; Anbar, A.; Al-Shaikh, N.; Mousa, R. Prioritization of Semi-arid Agricultural Watershed Using Morphometric and Principal Component Analysis, Remote Sensing, and GIS Techniques, the Zerqa River Watershed, Northern Jordan. *Agric. Sci.* **2017,** *8* (1), 113–148.

Gopinath, G.; Nair, A. G.; Ambili, G. K.; Sweta, T. V. Watershed Prioritization Based on Morphometric Analysis Coupled with Multi Criteria Decision Making. *Arab. J. Geosci.* **2016,** *9*, 129.

Govil, H. Identification and Mapping of Hydrothermally Altered Minerals in and Around Askot Base-metal Mineralization of Kumoan Himalaya, India Using EO-1 Hyperion Data. *Int. J. Adv. Remote Sens. GIS Geogr.* **2015,** *3* (1), 1–8.

Govil, H.; Guha, S.; Diwan, P.; Gill, N.; Dey, A. Evaluation of ASTER TIR Data-Based Lithological Indices in Parts of Madhya Pradesh and Chhattisgarh State, India. In *Data Management, Analytics and Innovation: Advances in Intelligent Systems and Computing*; Sharma, N., Chakrabarti, A., Balas, V., Eds.; Springer: Singapore, 2020; pp 161–169, 1042.

Guha, S. Coastal Dynamics and Sediment Transport in Parts of Hooghly Estuary. *Int. J. Trend Res.* **2015a,** *2* (6), 2394–9333.

Guha, S. Mathematical Analysis of Solani Watershed, North India. *Int. J. Geomat. Geosci.* **2015b,** *6* (2), 1512–1529.

Guha, S.; Dey, A. Study of Morphological Changes in Parts of Hugli Estuary during 1999 to 2008 Applying Remote Sensing and GIS Technique. *Environ. Ecol.* **2016a,** *34* (4), 1813–1817.

Guha, S.; Dey, A.; Study on Morphological Changes in and Around Nayachara Tail Using Remote Sensing Techniques. *Int. J. Dis. Recov. Bus. Cont.* **2016b,** *7*, 13–26.

Guha, S.; Govil, H.; Evaluation of ASTER TIR Data-based Lithological Indices in Malanjkhand Copper Mines of Madhya Pradesh, India. *Appl. Earth Sci.* **2019,** *129* (1), 3–8.

Guha, S.; Govil, H.; Dey, A.; Gill, N. Analysing the Capability of NCI Technique in Change Detection Using High- and Medium-Resolution Multispectral Data. In *Geoecology of Landscape Dynamics: Advances in Geographical and Environmental Sciences*; Sahdev, S., Singh, R., Kumar, M., Eds.; Springer: Singapore, 2020; pp 133–147.

Horton, R. E. Erosional Development of Streams and Their Drainage Basins: Hydro-physical Approach to Quantitative Morphology. *Bull. Geol. Soc. Am.* **1945,** *56*, 275–370.

Khan, A.; Rao, L. A. K.; Yunus, A. P.; Govil, H. Characterization of Channel Planform Features and Sinuosity Indices in Parts of Yamuna River Flood Plain Using Remote Sensing and GIS Techniques. *Arab. J. Geosci.* **2018,** *11* (17), 525.

Khan, A.; Govil, H.; Kumar, G.; Dave, R. Synergistic Use of Sentinel-1 and Sentinel-2 for Improved LULC Mapping with Special Reference to Bad Land Class: A Case Study for Yamuna River Floodplain, India. *Spat. Inf. Res.* **2020,** *28*, 669–681.

Maxwell, J. C. The Bifurcation Ratio in Horton's Law of Stream Numbers. *Trans. Am. Geophys. Union* **1955,** *36*, 520.

Meraj, G.; et al. Assessing the Influence of Watershed Characteristics on the Flood Vulnerability of Jhelum Basin in Kashmir Himalaya. *Nat. Hazards* **2015,** *77* (1), 153–175.

Meraj, G.; Shakil, A. R. Sadaf, A. Inferring Land Surface Processes from Watershed Characterization. *Geostatistical and Geospatial Approaches for the Characterization of Natural Resources in the Environment*; Springer: Cham, 2016; pp 741–744.

Miller, V. C. *A Quantitative Geomorphic Study of Drainage Basin Characteristics on the Clinch Mountain Area Virginia and Tennessee*; Columbia University, Department of Geology, ONR: New York, 1953; pp 389–402.

Mondal, A.; Guha, S.; Mishra, P. K.; Kundu, S. Land Use/Land Cover Changes in Hugli Estuary Using Fuzzy C-Mean Algorithm. *Int. J. Geomat. Geosci.* **2011,** *2* (2), 613–626.

Rao, L.; Khan, A.; Govil, H. Morphometric Characterization of Yamuna River Basin Around Agra, Firozabad and Etawah Districts, Uttar Pradesh, Using Remote Sensing and GIS Techniques. *Int. Adv. Res. J. Sci. Eng. Technol.* **2017,** *129* (1), 3–8.

Schumm, S. A. Evaluation of Drainage System and Slopes in Badlands at Perth Amboy, New Jersey. *Bull. Geol. Soc. Am.* **1956,** *17*, 595–646.

Smith, K. G.. Standards for Grading Texture of Erosional Topography. *Am. J. Sci.* **1950,** *248*, 655–668.

Strahler, A. N. Quantitative Analysis of Watershed Geomorphology. *Am. Geophys. Union Trans.* **1957,** *38*, 913–920.

Strahler, A. N. Quantitative Geomorphology of Drainage Basin and Channel Networks. In *Handbook of Applied Hydrology*; Chow, V. T., Ed.; McGraw Hill Book Company: New York, 1964; Section 4.

CHAPTER 5

Use of GI Science for Understanding the Flood and Gully Erosion at Landscape Level

UGWUEGBU IKECHUKWU EMMANUEL[1] and IFEKA ADOLPHUS[2*]

[1]*Nigerian Meteorological Agency, Abuja, Nigeria*

[2]*Federal University of Technology, Akure, Ondo State, Nigeria*

[*]*Corresponding author. E-mail: adolphusifeka@yahoo.com*

ABSTRACT

Taking on the problem of flooding and gully erosion necessitates a thorough understanding of the causes that cause them. This study aims to evaluate flood and gully erosion in Imo State in terms of magnitude and extent. Primary and secondary data were obtained for model flutter and gully erosion vulnerability. For the period 1985–2015, precipitation data were obtained from NIMET. For 1986, 2000, and 2015, respectively, Landsat TM, ETM+, and OLI were procured from the USGS website. The Harmonized World Soil Database has also acquired soil data. The Nigeria Geological Survey Agency acquired a scanned geology map for Imo State. Data from the USGS Website have been collected from SRTM DEM. The renegotiation trend between 1985 and 2015 showed their total and average annual pattern. In order to determine the years of flood between 1985 and 2015, the Standard Precipitation Index (SPI) was used. The flood risk assessment was carried out using the multicriteria analysis and fuzzy logic. To determine the soil erosion, the revised Universal Soil Loss Equation (RUSLE) has been used. Criteria, such as precipitation, runoff, pitch, drainage distance, drainage

GIScience for the Sustainable Management of Water Resources. Gowhar Meraj, Shruti Kanga, Majid Farooq, Suraj Kumar Singh, & Sudhanshu (Eds.)
© 2023 Apple Academic Press, Inc. Co-published with CRC Press (Taylor & Francis)

density, soil type, types and land use, geological formation. In accordance with its impact on the Imo State, the criteria were reclassified. Due to the foggy overlay, a flood risk map was created showing that 62.9% of Imo State is situated in a low-risk region, 17.2% in the medium-risk area and 19.9% in the low-risk zone. The low-risk areas include 214 towns and villages, a medium-risk area comprises 64 towns and villages and a high-risk area comprising 52 towns or towns. The amount of soil lost is 0–123,922 ha per tube per year. There are very low, low, moderate, high and very high-risk areas on the erosion risk map. 87.3% of the Imo State lies within the very low-risk zone, 8.1% in the low-risk area of the State of Imo, 4.3% in the middle risk area, 0.3% in the high-risk area, and 0.2% in the high-risk area. In northern Imo State, soil loss was more prevalent. In the southern parts, the effects were smaller because they were low. An effective soil erosion and flood management database has been created.

5.1 INTRODUCTION

Although natural fluctuations in temperature, rainfall, and the frequency of events can all vary over time, they tend to be within the normal range in any given region. While climate variability exists on a global, regional, and local scale, there are no catastrophes hidden anywhere. In Nigeria, tropical storms generate wind gusts that are capable of taking lives and wreaking property damage totaling millions of naira during each rainy season. Heavy rains wash away thousands of hectares of agricultural land because of flash floods. People died in August of 1988, over 1800 houses were destroyed, and over 14,000 farms were inundated when Baguda Dam in Nigeria collapsed after a thunderstorm unleashed a tremendous flash flood (IPCC, 2007; Adefolalu, 2007). Flooding and gully erosion are global problems that require immediate attention (Bekele et al., 2018; Vijith and Dodge-Wan, 2019). Even though their approaches have been far from adequate, the federal and state governments have tried to devote attention to both problems in Nigeria. Erosion and flooding are significant concerns for both urban and rural areas, as the twin problems have serious impact on all aspects of productivity, including food supply, the population, and the infrastructure (Pandey et al., 2010, 2013). Because there is not much available land in urban areas, this impact assumes greater importance in areas like the Niger Delta, where due to oil and gas exploration and extraction, a large percentage of the land is polluted and unsuitable for farming.

The aim of this study is to evaluate the likelihood of flooding and gully erosion in the Imo State and to propose suitable management strategies for preventing potential occurrences. The specific goals of the study are to analyze the spatiotemporal trends of basic climatic and soil parameters that are largely responsible for disaster risks, evaluate flood, and gully erosion scenarios in the Imo State, to assess the areas vulnerable to flood and gully erosion based on multicriteria evaluation for continuous monitoring, and generate a flood and soil erosion vulnerability database for Imo State.

5.2 METHODOLOGY

5.2.1 STUDY AREA

It is located in southeast Nigeria, which is just south of latitude 4.45 degrees North and longitude 7.15° East, with an area of approximately 5100 sq. km. Also known as the tropical belt, the humid tropics feature year-round high surface air temperatures. The average minimum temperature is 23.5°C and the average maximum temperature is 32.1°C. The year is made up of two seasons, wet, and dry. Rain falls from April to October.

Okigwe, Orlu, and Owerri are all major political regions in the state, and the administrative capital is located in Owerri. There are 3927,563 people living in the state, and over 1.9 million are male, and 1.9 million are female (Okorie, 2010).

In addition to the Imo State, which forms the Ogwashi-Asaba Formation, the Bende Ameki Formation, the Benin Formation, the Imo State, and the Nsukka Formation form the other five geology formations in the Imo State. Calcaric Fluvisols, Dystric Nitosols, Gleysols, and Xanthic Ferralsols are all soil types found in this area.

5.2.2 MATERIALS AND METHODS

Precipitation data for Imo State was acquired from the NIMET for 1985–2015. The precipitation data was further analyzed to retrieve the rainfall regime over the state. The precipitation data was used to calculate the Standard Precipitation Index (SPI) and to show the trend of rainfall in Imo State. Gridded precipitation data was acquired from Era-Interim for 1985–2015. The precipitation data was interpolated to show the spatial precipitation across Imo State.

Landsat 8 (OLI) imagery that covers Imo State was acquired from USGS. The Landsat imagery acquisition date was January 15, 2015 used to generate the LULC map (Farooq and Muslim, 2014; Nathawat et al., 2010; Kumar et al., 2018).

This sensor is useful for aerosol and coastal studies because the ultra-blue band 1 is blue. Bands 9 and 8 are useful for detecting cirrus clouds. The approximate scene size for Band 8 is approximately 170 km north–south by 183 km east–west (106 mi by 114 mi).

5.2.2.1 STANDARD PRECIPITATION INDEX

SPI is a gauge that is developed for identifying and tracking drought. The SPI is calculated as the total probability of rainfall at a given location and estimation of the parameters for the gamma distribution, α and β, by going through maximum likelihood. In other words, it is possible to define a mathematical cumulative probability function to represent the rainfall distribution at the station. From this, an analyst can estimate the probability of the rainfall falling between a certain amount and zero. If this rainfall event had a low cumulative probability, this would show that we are likely to have a drought event. An anomalously wet event could be defined as a rainfall event which increases the probability on the CPF. To calculate the SPI, use the formula:

$$SPI = \frac{Xi - \overline{X}}{S} \tag{5.1}$$

5.2.2.2 NORMALIZED DIFFERENCE VEGETATION INDEX (NDVI)

NDVI is a ratio of red and near infrared reflectance, with the reflectivity measured in those wavelengths factored in. We have chosen these two spectral bands as being most affected by the green vegetation on the surface. Also, the contrast between the vegetation and the soil is at its greatest in the red and near infrared bands. NDVI transformation is computed as the relation of the measured R and NIR intensity intensities using the following formula:

$$NDVI = \frac{NIR - RED}{NIR + RED} \tag{5.2}$$

5.2.2.3 SOIL EROSION MODELING BY UNIVERSAL SOIL LOSS EQUATION (USLE)

The Universal Soil Loss Equation (USLE) is an empirically based soil erosion model developed by Wischmeier and Smith (1978). USLE was used to calculate the soil loss due to water erosion per unit area per year by the following equation:

$$A = R \times K \times LS \times C \times P \tag{5.3}$$

where A is the mean soil loss due to water erosion, R the runoff erosivity factor (MJ mm/ha/h/year), K the soil erodibility factor (Mg h/MJ/mm), L the slope length factor (dimensionless), S the slope steepness factor (dimensionless), C the land cover and management factor (dimensionless), and P is the support practice (dimensionless).

5.2.2.4 RAINFALL AND RUNOFF FACTOR (R-FACTOR)

The R-factor reflects the influence of rainfall energy and strength (Kunta and Carosio, 2007). It is the long-term annual estimate of the product of the maximum rainfall intensity in 30 minutes in mm per hour and the incident rainfall kinetic energy in MJ/ha (Wischmeier and Smith, 1978; Renard and Freimund, 1994). The rainfall's kinetic energy is the easiest way to measure rain erosivity.

Analyzing the R-factor can be done in different ways. For instance, as an example:

i. $R = 9.28 \times P—8838$ Mean annual erosivity ($KE > 25$) where P is mean annual precipitation (Morgan, 1986). (5.4)

ii. $R = 0.276 \times P \times 130$. Mean annual E130, where P is mean annual precipitation (Morgan, 1986). (5.5)

iii. $R = 0.5 \times P$ (in US unit) and $R = 0.5 \times P \times 1.73$ (in Metric unit) (Morgan, 1986). (5.6)

iv. $R = 38.5 + 0.35$ X Pr. Where Pr = the annual average rainfall (mm/year) (Morgan, 1986) (5.7)

In this investigation, the third condition was chosen for the assurance of R-value as it has been found to function admirably in tropical locales. Subsequently, every framework cell of mean yearly precipitation was determined to be dependent on the third condition to get the R-value utilizing

GIS programming (ArcGIS). The mean yearly precipitation surface was utilized to figure the R-factor utilizing the spatial expert module of the ArcGIS programming.

5.2.2.5 SOIL ERODIBILITY FACTOR (K)

The K-factor is a soil erodibility factor. The soil erodibility factor K is a proportion of the natural susceptibility of an offered soil-to-soil disintegration. The limit of soil particles to withdraw and move by means of precipitation and spillover is a marker of how delicate the soil is. In particular, soil erodibility is an element of molecule size conveyance, natural matter substance, construction, and porousness (Renard et al., 1997). The soil erodibility factor (K) was determined utilizing five attributes for each soil type. The attributes for each soil type. The attributes decided for each soil type included soil molecule size dissemination, soil structure, natural matter substance and porousness (for example term and level of waterlogging). The erodibility worth of each soil class can be determined utilizing a condition suggested by Wischmeier and Smith (1978).

$$100\,k = 2.1\,(\%silt \times (100\text{–}\%clay)^{1.14}\,(10^{-4})\,(12\text{–}\%organic\ matter) + 3.23\,(b-2) + 2.5\,(c-3)\,(2) \qquad (5.8)$$

where b is the soil structure code utilized in soil classification (very fine granular, 1; fine granular, 2; medium or coarse granular, 3; slow to moderate, 4; slow, 5; very low, 6). Inspected soils were air-dried and examined adhering to the standard systems portrayed by Wischmeier and Smith (1975). Utilizing the soil structure code portrayed by Wischmeier and Smith (1975), soil structure of the visited destinations was dictated via cautious assessment of section of undisturbed soil through profile channel of up to 40 cm profundity. The porousness class test was done to decide the penetrability of soils of the different networks. Oil tests from the different areas were placed in isolated estimating chambers and 100 ml of water was added to every one of the chambers containing soil. Perception was then made on the time taken for the deliberate amount of water to arrive at a specific level in the chamber as it invades down through the soil sample. The time was recorded and this was utilized for soil penetrability characterization as per the accompanying codes: Fast—1, Moderate to Fast—2, Moderate—3, Slow to Moderate—4, Slow—5, and Slow—6 as portrayed by Wischmeier and Smith (1978), to give a K-factor for each soil type. The raster guide of the K-factor was delivered by a computerized soil map and the erodibility worth of each soil type.

5.2.2.6 SLOPE LENGTH AND SLOPE STEEPNESS FACTORS (LS)

The geographical factor is comprised of two subfactors: a slope gradient factor and a slope length factor. To process the factor and S factor, we previously got the computerized height model and tracked down the geographical disintegration helplessness (DEM). To digitize geographical guide shapes, a DEM was built. The filled DEM, just as the slope length and slope gradient factors, were gone into the condition appeared beneath to deliver the geographical factor lattice.

$$LS = (L/22)^{0.5}\,(0.065 + 0.045S + 0.0065S^2) \quad (5.9)$$

where L = (x/22.13) m, in which x is length of slope (in m), m is 0.5 if the slope is >5%, 0.4 if between 3% and 5%, 0.3 if between 1 and 3% and 0.2 if below 1 and L is the slope length factor; $S = (0.43 + 0.30\,s + 0.043\,s^2)/6.613$, where s is the gradient (%), and S is the slope gradient factor.

5.3 RESULTS AND DISCUSSION

Results of the study are discussed in the following categories: rainfall trend, SPI, normalized difference vegetation index (NDVI), precipitation, runoff, slope, distance to drainage, drainage density, soil type, land use, geology formations, flood risk map, flooding risk database, rainfall-runoff erosivity factor (R), soil erodibility factor (K), topography factor (LS), cover management factor (C), conservation practice factor (P), soil loss and soil erosion database (Singh and Pandey, 2014; Bhatt et al., 2017; Sharma and Kanga, 2020).

5.3.1 RAINFALL TREND OVER IMO STATE

Figure 5.1(a) shows the annual rainfall amount from year 1985 to year 2015. Years with increasing rainfall pattern within this period includes 1988, 1990, 1994, 1996, 1999, 2006, 2007 and 2009. Rainfall years with decreasing pattern within this period includes 1986, 1987, 1989, 1993, 1998, 2004, 2010, and 2014 (Table 5.1). The years with peak rainfall are indicator of possible flood years within the period of 1985–2015. Figure 5.1(b) shows the monthly mean of precipitation for the period of 1985–2015. The rainfall starts in March up to the month of July. The drop of rainfall amount in the month of August is identified as the “August break” where rainfall cease for a short period. The rainfall peaks again at the month of September. The months which the rainfall peaks were the months of July and September which are indicators of possible flood months within the year.

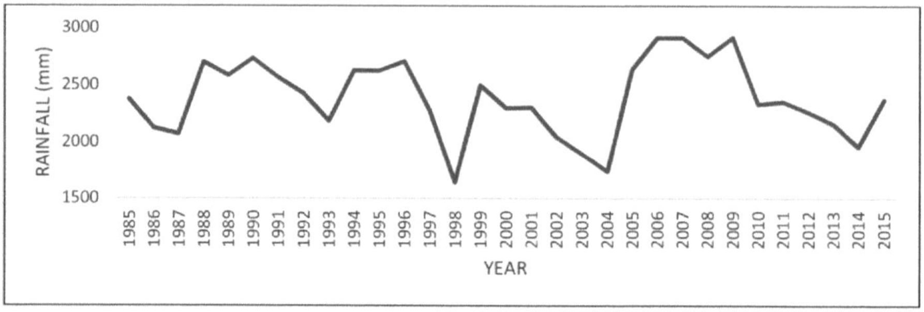

FIGURE 5.1(a) Annual rainfall amount in Imo state.

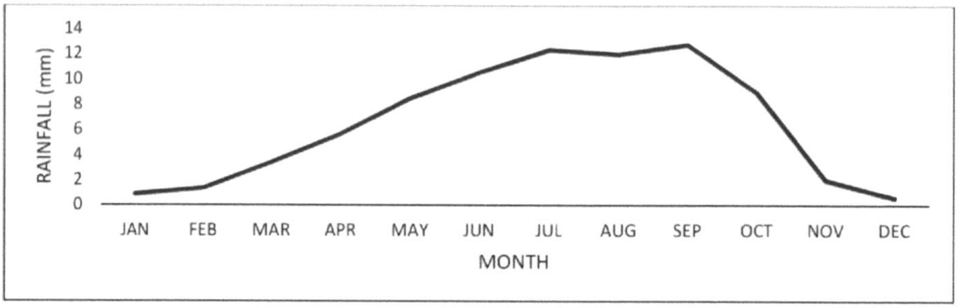

FIGURE 5.1(b) Monthly mean of precipitation in Imo state.

5.3.2 STANDARD PRECIPITATION INDEX

SPI is used to identify the flood years. Monthly SPI was calculated for the period of the 30 years. The six-month SPi below the mean indicates no flooding during these months. The six-month SPI above the mean indicates possible occurrence during these months with July as the peak period. The month with no flood includes January, February, March, April, November, and December. Months with flood include July, August, and September. July is the month with the highest flooding. Figure 5.1(c) shows the monthly SPI and Table 5.2 depicts the monthly SPI values with their corresponding classification. Figure 5.1(d) shows the Standard Precipitation Index graph for 1985–2015. Table 5.2 provides the monthly SPI and their corresponding classification. Table 5.3 gives the yearly SPI values with their corresponding classification. The SPI was carried out for precipitation data of 30 years. Precipitation was above the mean for 14 years out of the 30 years studied with 3 years showing flood signals. The flood years were identified as 2006, 2007 and 2009.

TABLE 5.1 Rainfall Annual Total and Mean.

Year	Total rainfall (mm)	Mean rainfall (mm)
1985	2376.3	6.510410959
1986	2121.1	5.795355191
1987	2070	5.686813187
1988	2698.4	7.413186813
1989	2731.9	7.092032967
1990	2581.5	7.484657534
1991	2565.1	7.027671233
1992	2424.1	6.623224044
1993	2182.8	5.980273973
1994	2626.1	7.194794521
1995	2622.3	7.184383562
1996	2705.5	7.392076503
1997	2272.5	6.803892216
1998	1641.5	4.497260274
1999	2496.5	6.839726027
2000	2296.4	7.504575163
2001	2304.2	6.312876712
2002	2042.2	6.096119403
2003	1893.7	6.229276316
2004	1735.6	4.742076503
2005	2635.3	6.973972603
2006	2914.8	7.985753425
2007	2914.8	7.985753425
2008	2749	7.531506849
2009	2916.7	7.990958904
2010	2331.6	6.387945205
2011	2351.7	6.443013699
2012	2260.7	6.19369863
2013	2156.5	5.908219178
2014	1958.4	5.365479452
2015	2369.9	6.492876712

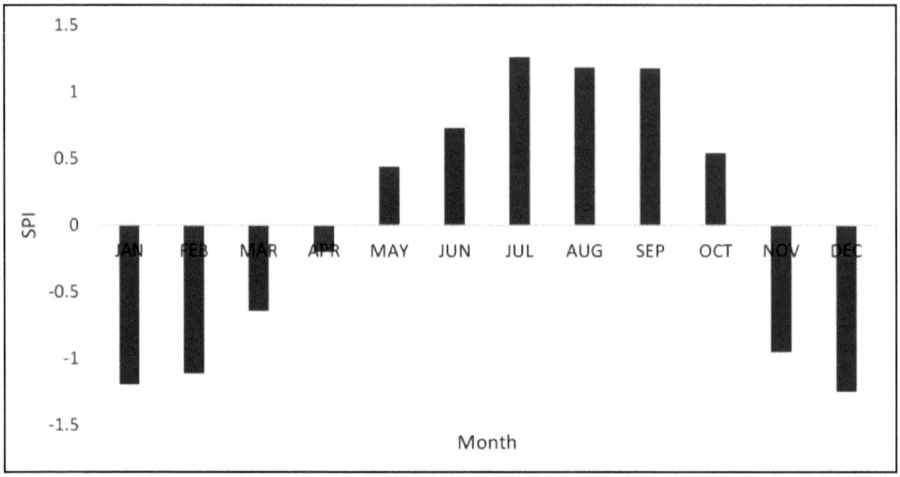

FIGURE 5.1(c) Monthly standardized precipitation index (SPI) in Imo state.

TABLE 5.2 Monthly SPI and Classification.

Month	SPI value	Classification
January	−1.19543	Moderately dry
February	−1.1118	Moderately dry
March	−0.64191	Near normal
April	−0.19842	Near normal
May	0.438897	Near normal
June	0.733528	Near normal
July	1.262757	Moderately wet
August	1.186683	Moderately wet
September	1.179929	Moderately wet
October	0.544654	Near normal
November	−0.9519	Near normal
December	−1.24699	Moderately dry

5.3.3 NORMALIZED DIFFERENCE VEGETATION INDEX (NDVI)

NDVI was generated for the years 1986, 2000, and 2015 to assess the changing pattern for vegetation in response to flooding. Figure 5.2(a) shows the NDVI for year 1986. The NDVI for year 1986 shows that the vegetation was very healthy as SPI shows that it was a near-normal year with a value of −0.78691 (Table 5.3). Figure 5.2(b) shows the NDVI for year 2000. The NDVI for year 2000 decreased compared with the value in 1986. Highest NDVI recorded was 0.296552. The vegetation was less healthy and the SPI shows that it was a near-normal year with a value of −0.26496 (Table 5.3). Figure 5.2(c) shows NDVI for the year 2015.The NDVI for the year 2015 shows that the vegetation health increased compared with the NDVI for year 1986. The highest recorded value for NDVI was 0.385839. The SPI value recorded for the year 2015 is −0.04612 (Table 5.3) which is a near-normal year.

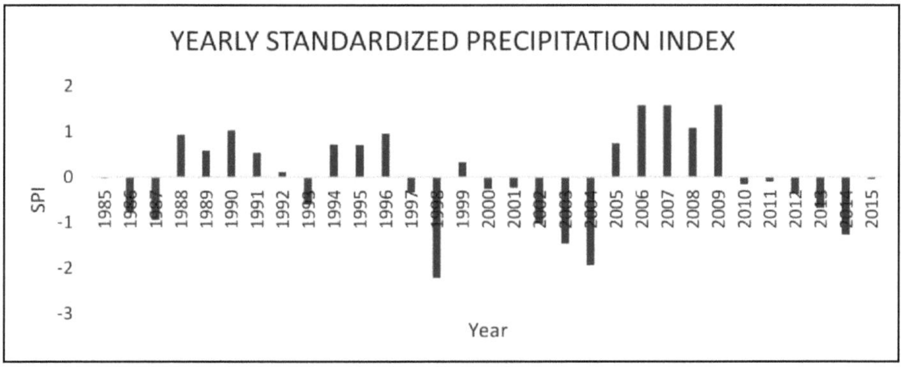

FIGURE 5.1(d) Yearly standardized precipitation index (SPI) in Imo state.

5.3.4 PRECIPITATION

Prolonged rainfall is the most common cause of flooding (Singh et al., 2017). The earth surface becomes saturated after a rainfall with long duration, at this point, the soil would no longer be able to allow water infiltrate thereby causing increased surface runoff which leads to flooding (Kanga et al., 2020a, 2020b). Figure 5.3(a) shows the spatial distribution of precipitation in Imo State. The average precipitation for Imo State for the 30-year period is 2385.39 mm. The precipitation was reclassified into two types: 3.09–3.5 mm were ranked 2 and 3.51–4.19 mm were ranked 4. Figure 5.3(b) shows the reclassification of precipitation. The ranking indicates that areas ranked as high would experience more precipitations than areas ranked as low. Hence, the possibility of more flooding in the areas ranked as high as areas ranked as low.

TABLE 5.3 Yearly SPI and Classification.

Year	SPI value	Classification
1985	−0.02707	Near normal
1986	−0.78691	Near normal
1987	−0.93905	Near normal
1988	0.931965	Near normal
1989	0.583903	Near normal
1990	1.031709	Moderately wet
1991	0.535074	Near normal
1992	0.115255	Near normal
1993	−0.6032	Near normal
1994	0.716697	Near normal
1995	0.705383	Near normal
1996	0.953105	Near normal
1997	−0.33612	Near normal
1998	−2.21488	Extremely dry
1999	0.330822	Near normal
2000	−0.26496	Near normal
2001	−0.24174	Near normal
2002	−1.02183	Moderately dry
2003	−1.46397	Moderately dry
2004	−1.93471	Severely dry
2005	0.744089	Near normal
2006	1.576282	Severely wet
2007	1.576282	Severely wet
2008	1.082623	Moderately wet
2009	1.581939	Severely wet
2010	−0.16016	Near normal
2011	−0.10031	Near normal
2012	−0.37126	Near normal
2013	−0.68151	Near normal
2014	−1.27134	Moderately dry
2015	−0.04612	Near normal

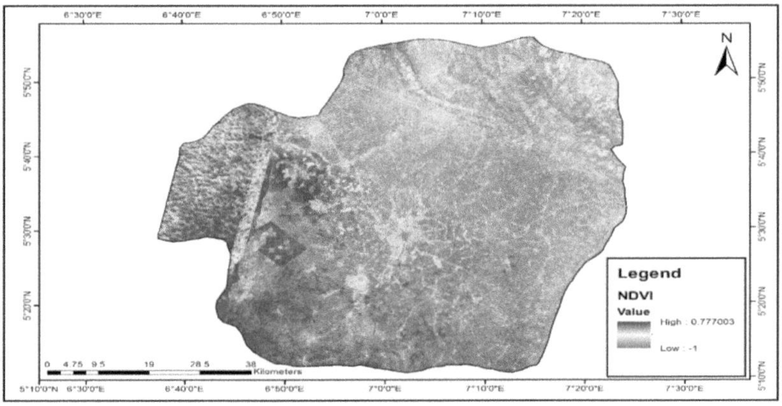

FIGURE 5.2 (a) NDVI for year 1986 in Imo state.

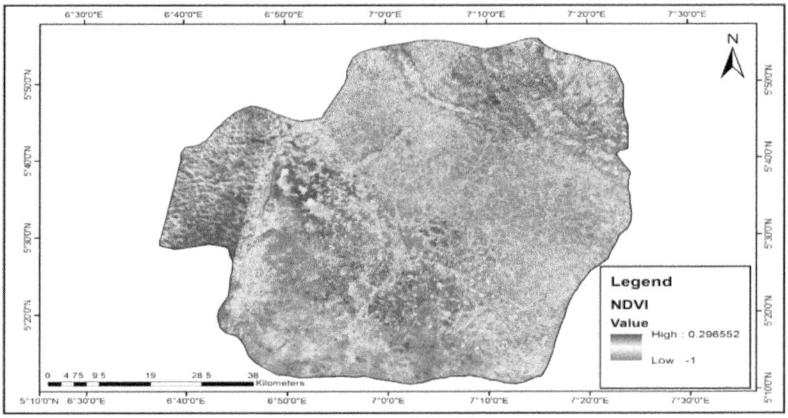

FIGURE 5.2 (b) NDVI for year 2000 in Imo state.

FIGURE 5.2 (c) NDVI for year 2015 in Imo state.

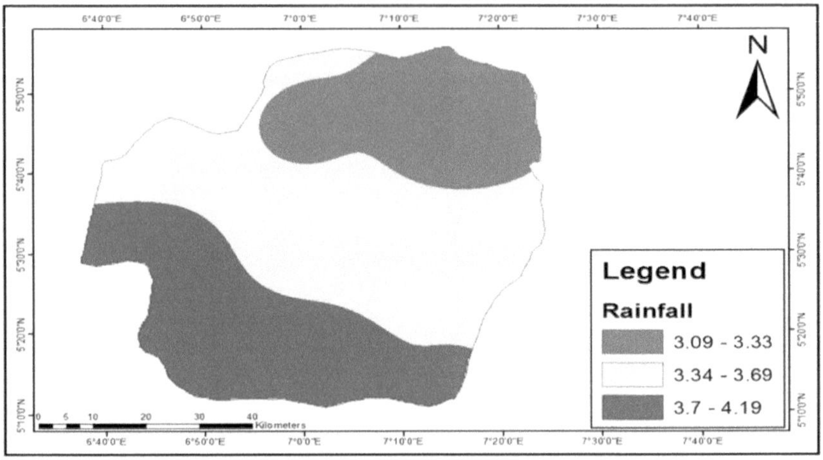

FIGURE 5.3(a) Map showing spatial distribution of precipitation in Imo State.

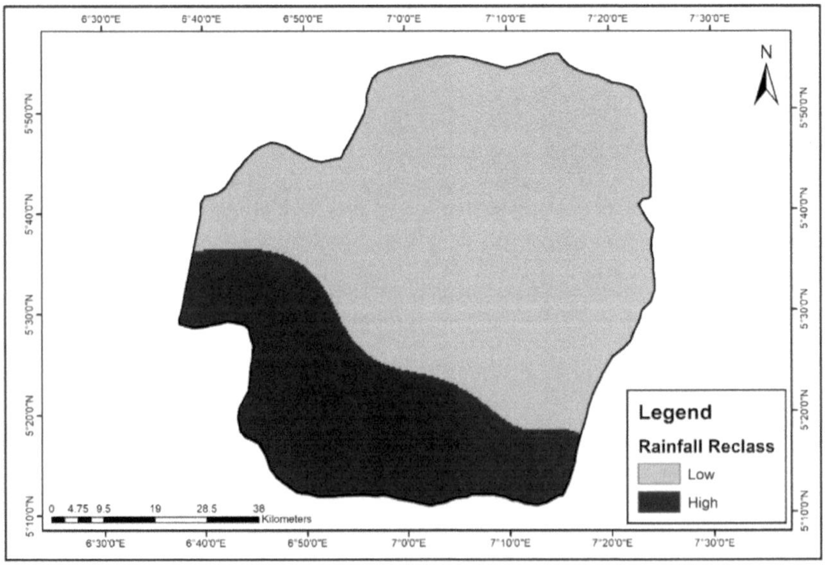

FIGURE 5.3(b) Map showing precipitation reclassification in Imo state.

5.3.5 RUNOFF

Runoff is the flow of water over the earth surface as a result of excess storm. It is also known as overland flow. ERA-Interim gridded data for surface runoff was acquired for 1979–2016. The runoff in the Imo State was generally high, it ranges from 0.94 to 1.29 mm (Figure 5.4(a)). The runoff was reclassified into two types: runoff ranging from 0.94 to 1 mm was ranked low and runoff ranging from 1.1 to 1.29 was ranked high as shown in Figure 5.4(b).

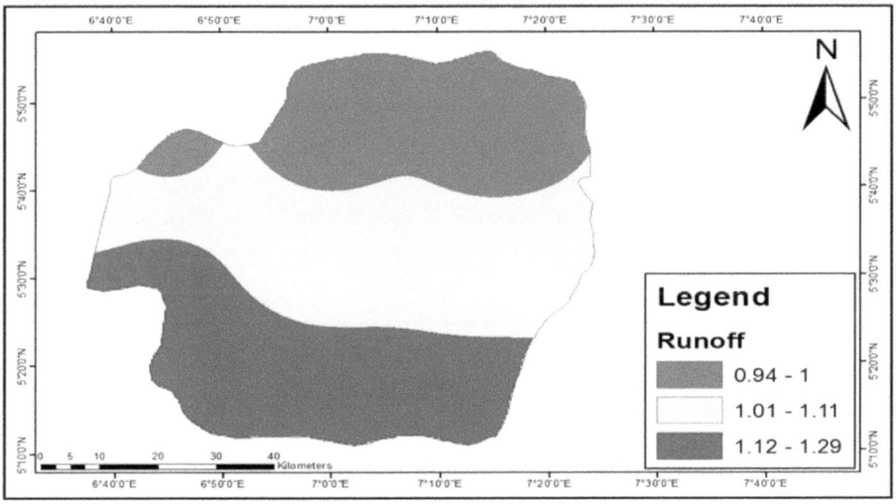

FIGURE 5.4(a) Map showing spatial distribution of runoff in Imo state.

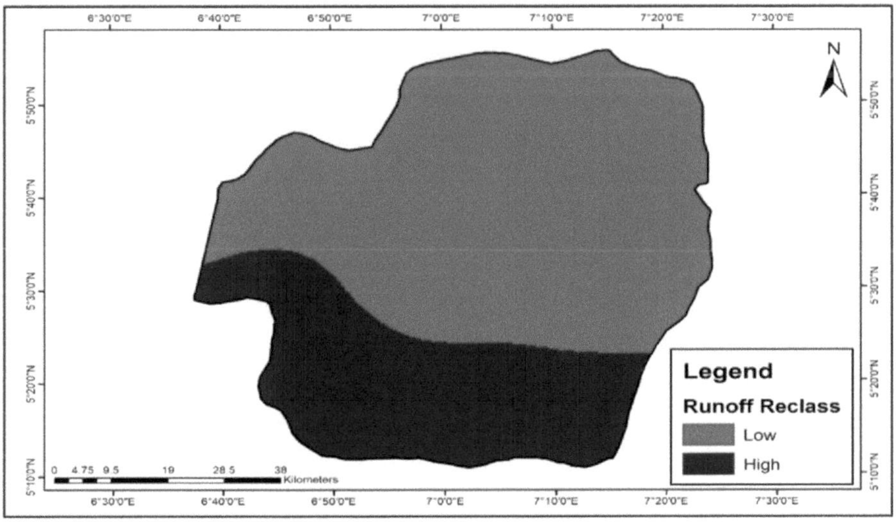

FIGURE 5.4(b) Map showing runoff reclassification in Imo state.

5.3.6 SLOPE

Figure 5.5(a) shows the slope map of Imo State. The slope distribution within the study area ranges from 0° to 24.2°. The range 0–5 represent areas that are flat, the range 5.1–10 represents areas that have gentle slopes and the range 10.1–24.2 represents areas with steep slopes. Water from rainfall tends to flow from a region of steep slope as runoff to flat regions where the water gathers to become flood. The slope factor was ranked based on its impact in causing flood. Figure 5.5(b) shows the slope reclassification. Areas with lower slope values, flat or almost flat areas were ranked highest as they support flood by allowing water to gather in them. Areas with gentle slope are ranked medium as they have little impact in causing flood. Steep slope is ranked lowest as they do not allow water gather thereby, they do not support flooding.

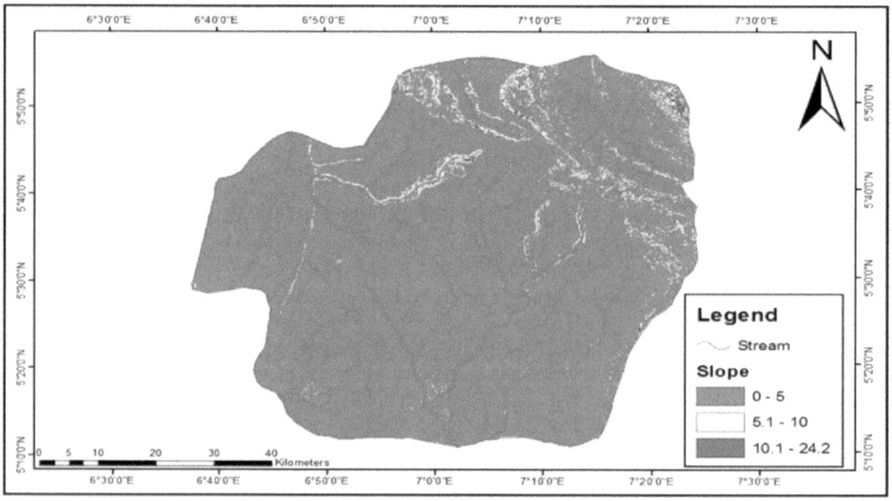

FIGURE 5.5(a) Slope map of Imo state.

5.3.7 DISTANCE TO DRAINAGE

The drainage pattern within the Imo State was generated from shuttle Radar Topography Mission (SRTM) Imagery using hydrology analysis tool in ArcGIS. Figure 5.6(a) shows the distance to stream map. The distance to drainage ranges from 0 to 2648 m. Areas that are closer to drainage systems (0–30 m) are prone to flooding. This is especially because extreme rainfall causes river water to overflow their banks entering into land with the areas close to the river has been most vulnerable. Distance to drainage was classified based on the level of vulnerability to flood. Distance to drainage was classified into three classes. The first class was set at 30 m and ranked as high based on the limit set by the Nigerian government through the Town and Regional Planning (Building Plan) regulation, 1986. The second class was set at 350 m and was ranked as medium in the event where flood goes beyond the 30 meters limit. The third class was set at distances greater than 350 meters and was ranked as low. Figure 5.6(b) shows the distance to drainage reclassification map.

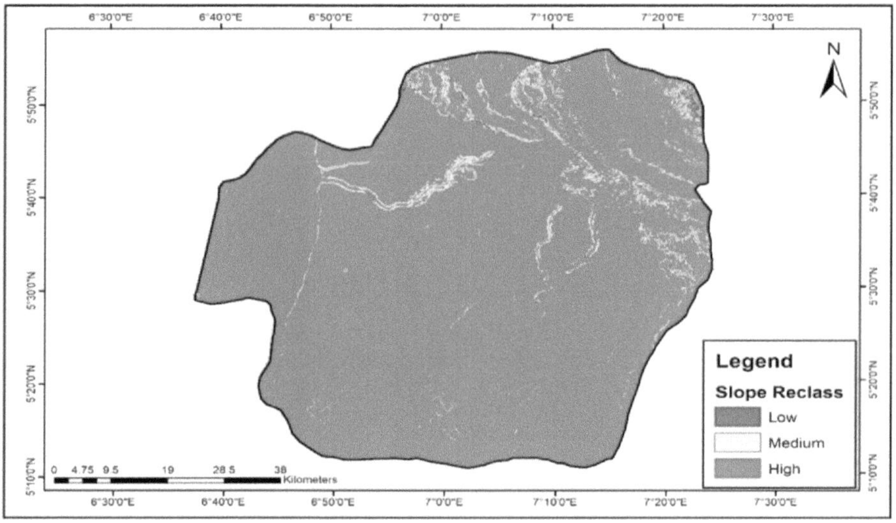

FIGURE 5.5(b) Map showing slope reclassification in Imo state.

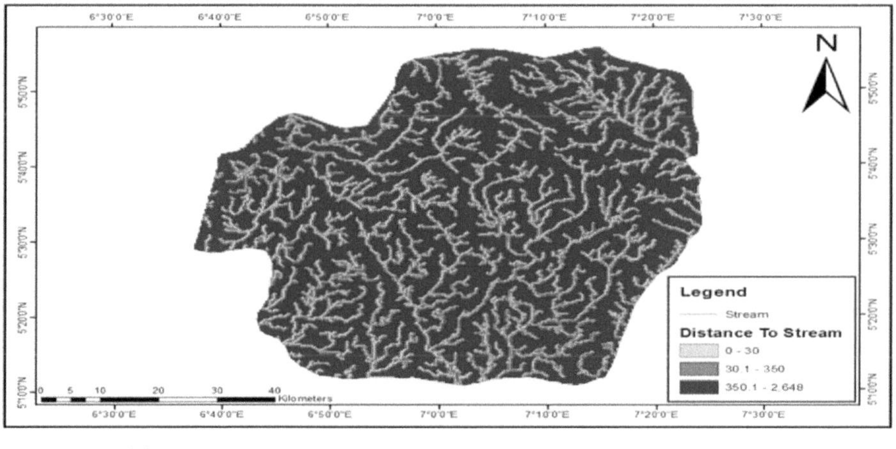

FIGURE 5.6(a) Map of distance to stream in Imo State.

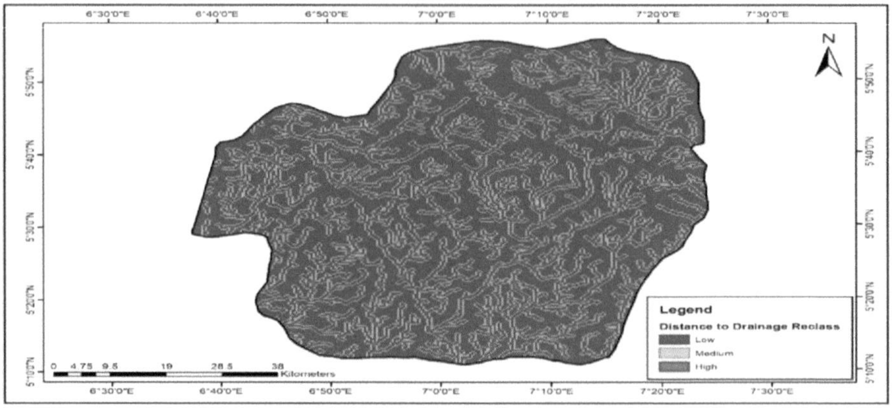

FIGURE 5.6(b) Map of distance to drainage reclassification in Imo state.

5.3.8 DRAINAGE DENSITY

Drainage density for the study area was generated from the drainage network using the density function in ArcGIS software. Water tends to gather in areas of high drainage density which makes it to overflow its bank into dry land thereby causing flood. Figure 5.7(a) shows the map of drainage density in Imo State. The drainage density was classified into three types: low, medium, and high density. Places with high drainage density indicate high flood vulnerability, medium drainage density indicates a moderate level of flood vulnerability and low drainage density indicates low flood vulnerability. Figure 5.7(b) shows the map of drainage density reclassification. The high drainage density areas were ranked highest as they create high flood vulnerability if there is excess precipitation. The medium drainage density areas were ranked medium because they cause little flood. The low drainage density areas were ranked lowest as they cause no flooding.

5.3.9 SOIL TYPE

Figure 5.8(a) shows the distribution of soil types in the Imo State. There are four types of soil in the study area. These include Calcaric Fluvisols, Dystric Nitosols, Gleysols, and Xanthic Ferralsols. The soil types were ranked based on their infiltration properties which is identified in their drainage class (FAO, 2012). Figure 5.8(b) shows the map of soil type reclassification. Calcaric Fluvisols and Calcaric Gleysols were ranked highest in causing flood because they have poor drainage which would enhance runoff and allow water gather at the surface. Dystric Nitosols were ranked medium as they drain moderately well. They contain high amount of sand content which would allow water infiltration thereby reducing flood. Xanthic Ferralsols are ranked lowest as they are well drained. They allow easy infiltration of water since their sand content are very high.

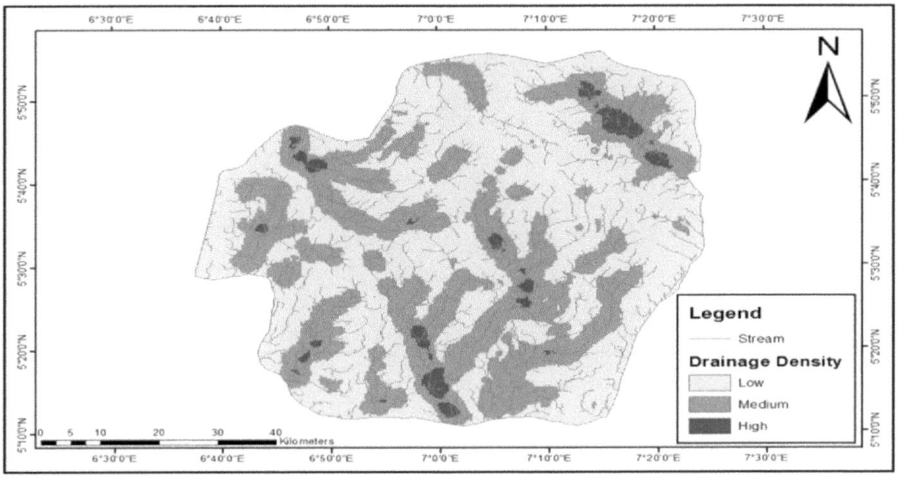

FIGURE 5.7(a) Map of drainage density in Imo state.

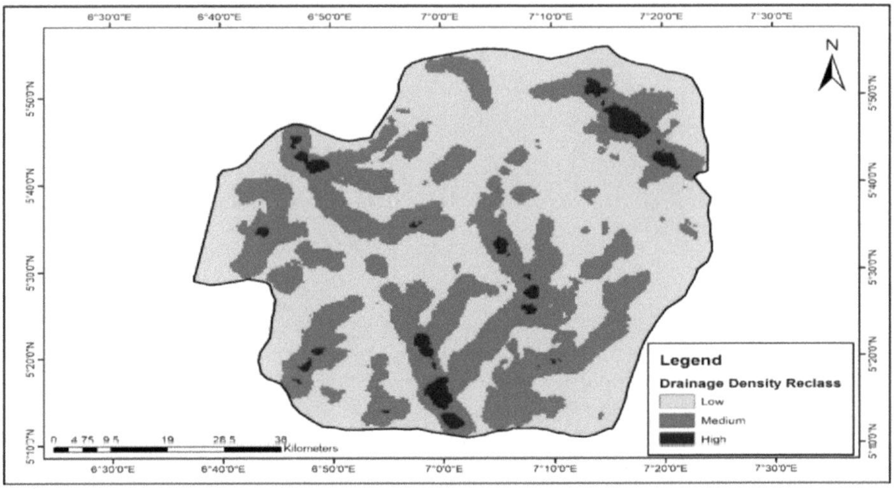

FIGURE 5.7(b) Map of drainage density reclassification in Imo state.

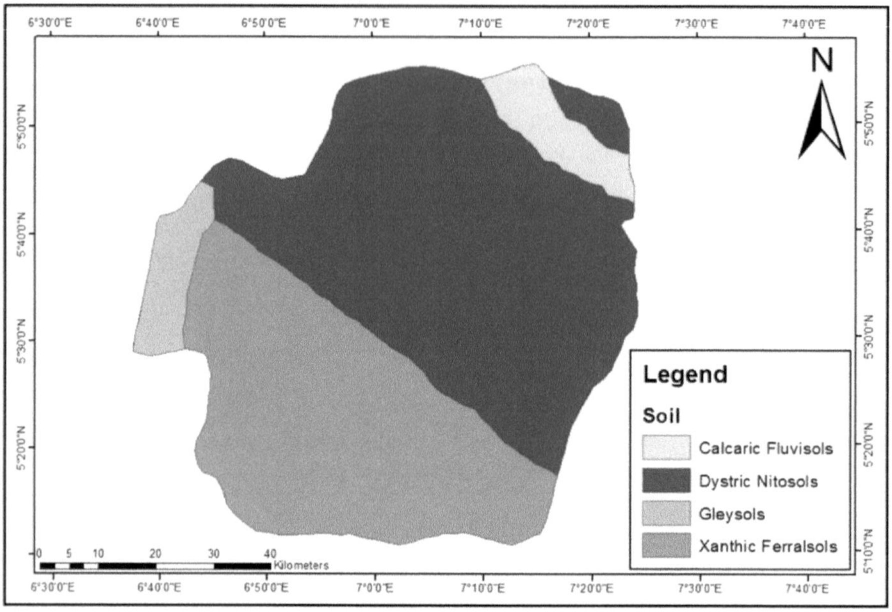

FIGURE 5.8(a) Map of Imo state soil types.

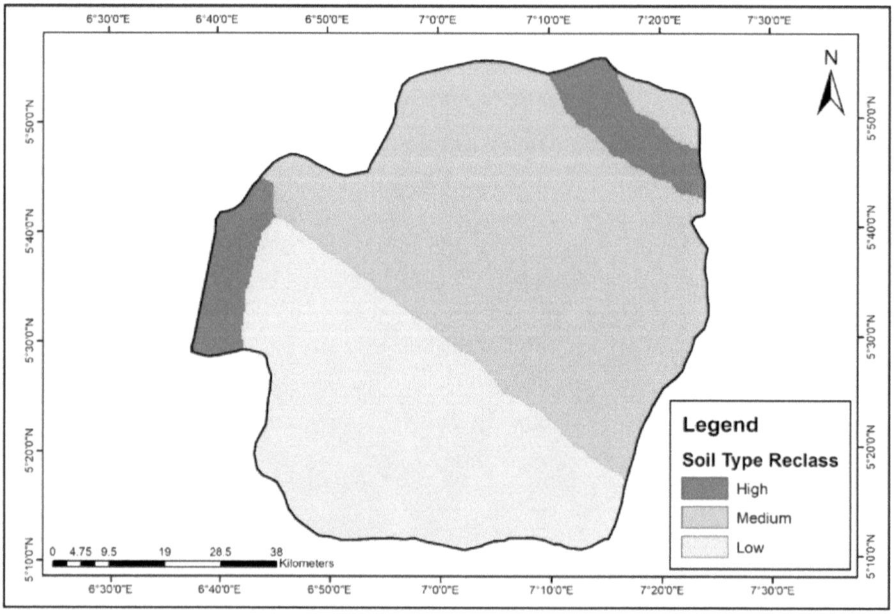

FIGURE 5.8(b) Map of soil type reclassification in Imo state.

5.4 LAND USE

Figure 5.9(a) shows the land use map of Imo State. Five land use types are present within the study area: built up, vegetation, marsh, rock outcrop, and water body. Figure 5.9(b) shows the land use reclassification map. Built up, marsh, and water body areas were ranked the highest because the presence of water body is what causes flood, marshy areas are where water gathers to make it muddy and built up is where humans live with their properties. Vegetation was ranked medium because farm lands and grass lands could also be affected by floods. Rock outcrop was ranked lowest because there is no collateral damage to them in the event of flood.

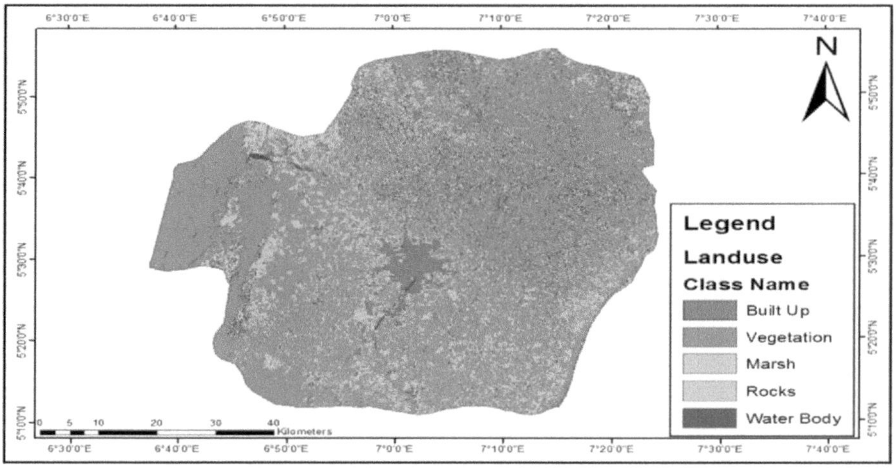

FIGURE 5.9(a) Map of Imo State land use.

5.4.1 GEOLOGY FORMATIONS

Figure 5.10(a) shows the geological formations of the Imo State. Six geology types are present in the study area which include: Alluvium, Bende Ameki Formation, Benin formation, Imo Shale, Nsukka formation, and Ogwashi-Asaba formation. Figure 5.10(b) shows the geology formation reclassification. The formations were reclassified based on their permeability properties. Permeable formation types would allow water infiltrate through them thereby reducing surface runoff. Impermeable formations enhance surface runoff which causes flooding. Therefore, Bende Ameki Formation, Benin

Formation, Imo Shale, Nsukka Formation, and Ogwashi-Asaba Formation were ranked as high and Alluvium was ranked as low.

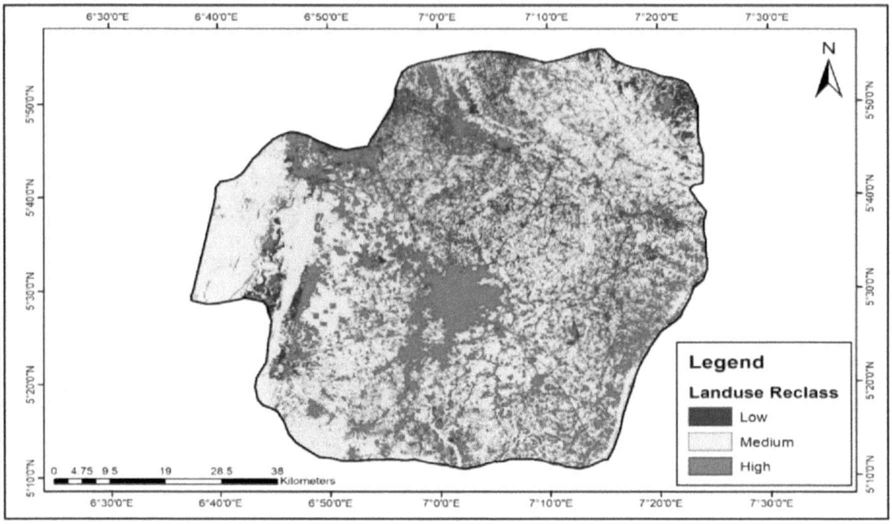

FIGURE 5.9(b) Map of land use reclassification in Imo state.

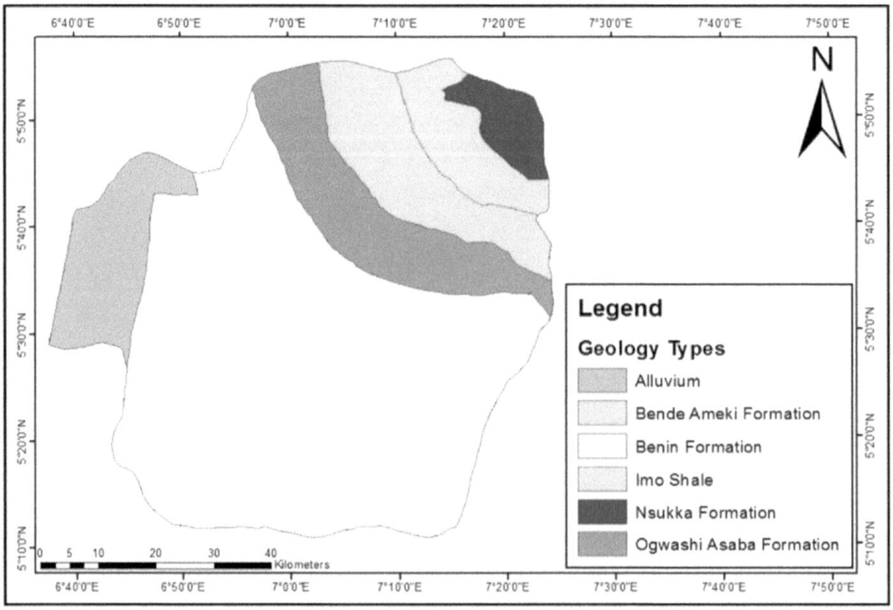

FIGURE 5.10(a) Map of Imo state geology in Imo state.

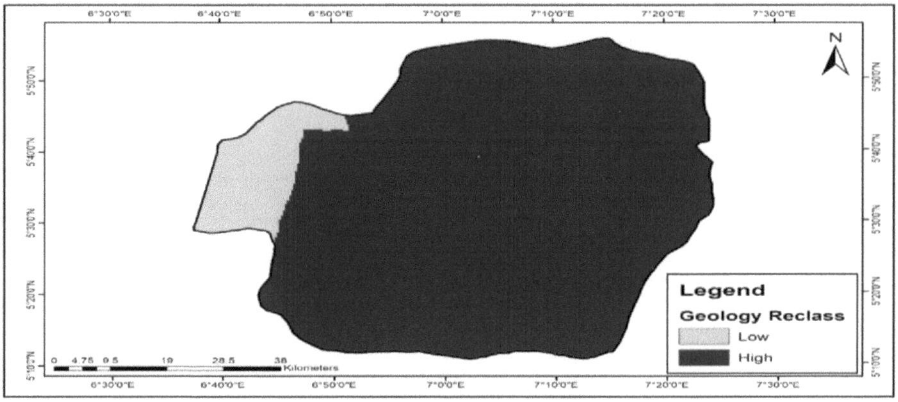

FIGURE 5.10(b) Map of geology formations reclassification in Imo state.

5.4.2 FLOOD RISK MAP

The various criteria were overlaid using their fuzzy membership layers and the flood risk map was therefore generated. Table 5.4 shows the fuzzy membership class for each criterion. The flood risk map was classified into low, medium, and high-risk zones. The low-risk area covers an area of 3357.59 km^2, medium-risk area covers an area of 1052.92 km^2 and high-risk area covers 781.36 km^2. Imo State covers a total area of 5332.49 km^2. 62.9% of the Imo State lies in the low-risk area, 17.2% lies in the medium-risk area and 19.9% lies in the high-risk area. Figure 5.11(a) shows the flood risk spatial distribution map. There are 354 towns and villages in the Imo State. Two hundred and fourteen towns and villages lie in the low vulnerability area, 64 towns and villages lie in the medium vulnerable area and 52 towns and villages lie in the highly vulnerable areas (see Table 5.5). Figure 5.11(b) shows the flood risk map overlaid on the Local Government Area.

5.4.3 FLOODING RISK DATABASE

A database was generated for flood vulnerability in the Imo State. Table 5.5 shows the levels of flood risk and the towns within each flood risk levels. Fifty two towns and villages are within the high flood risk zones indicating that there would be higher damages to lives, properties, and infrastructures in these areas, 64 towns and villages are within the medium flood risk zone indicating an average level of damage that would be caused by flood in these areas to live properties and infrastructures, and 214 towns and villages are within the low flood risk zones indicating least damages to lives, properties, and infrastructures in these areas.

TABLE 5.4 Fuzzy Membership for Each Criteria.

Criteria	Classes	Fuzzy membership
Precipitation	3.09–3.33	Fuzzy large
	3.34–3.69	
	3.7–4.19	
Runoff	0.94–1	Fuzzy large
	1.01–1.11	
	1.12–1.29	
Slope	0–5	Fuzzy MS small
	5.1–10	
	10.1–24.2	
Distance to drainage	0–30	Fuzzy small
	30.1–350	
	350.1–2648	
Drainage density	Low (0)	Fuzzy large
	Medium (2)	
	High (4)	
Soil type	Calcaric fluvisols (high)	Fuzzy large
	Dystric nitosols (medium)	
	Gleysols (high)	
	Xanthic ferralsols (low)	
Land use	Built up (high)	Fuzzy large
	Vegetation (medium)	
	Marsh (high)	
	Rocks (low)	
	Water body (high)	
Geology formation	Alluvium (low)	Fuzzy large
	Bende Ameki formation (high)	
	Benin formation (high)	
	Imo shale (high)	
	Nsukka formation (high)	
	Ogwashi-Asaba formation (high)	

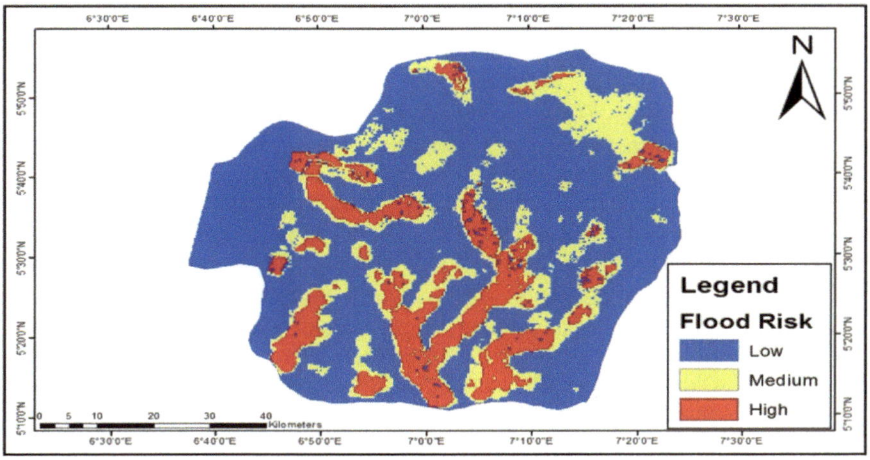

FIGURE 5.11(a) Flood risk map in the Imo State.

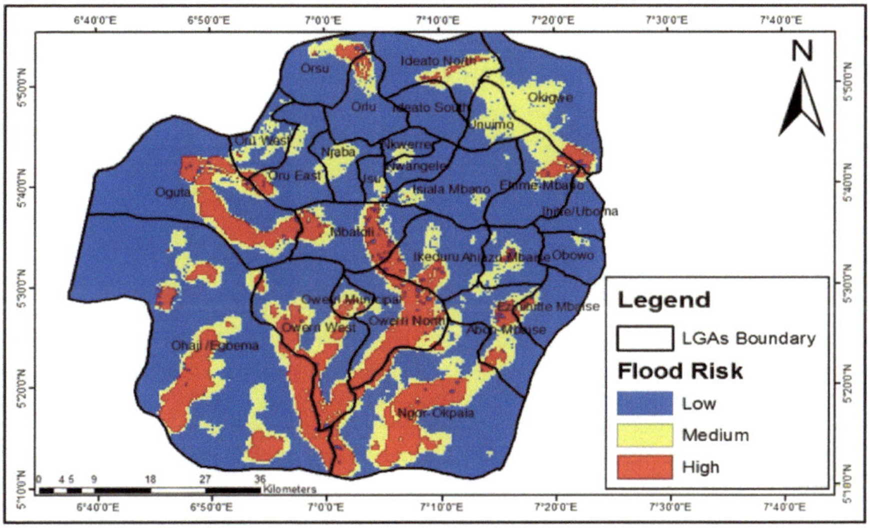

FIGURE 5.11(b) Flood risk map showing LGAs in Imo state.

5.4.4 *RAINFALL-RUNOFF EROSIVITY FACTOR (R)*

The average precipitation was calculated for the 30 years period. The average precipitation in Imo State for the 30 years period is 2385.39 mm. When fitted into Eq. (5.4), the R-factor was calculated as 2063.36. This indicates that the amount of rainfall and runoff within the period 1985–2015 in the Imo

TABLE 5.5 Flood Risk Database of Imo State.

Flood risk	Affected towns and villages
High	Upe, Umu Oye, Umuonhie, Umu Ogii, Umununu, Omunoha, Umu Neke, Umu Nama, Umunahu, Umu Ikoro, Umuhu, Umuehere, Umu Dim, Ugwuaku, Owelu, Ovoro, Orisaeze, Oramirukwa, Onicha, Omu Oye, Okitankwo, Ogwuaku, Oguta Lake, Oguta, Ochicha, Oboama, Obilokwa Mbieri, Obeke, Nzerem, Nwaorie, River Njaba, Ngor, Mbutu, Mberichi, Mba river, Imeabiam, ihuo, Ihiagwa, Ezeze River, Egwe, Azaea Egbelu, Awbana, Amorie, Amatta, Abor, Umu Akpa, Uborum, Okohia, and Owu
Medium	Umu oma, Umu Okrika, Omuofa, umu Obom, Umu Numu, Umuma, Umulogho, Umu Ekwune, Umuebe Ogwa, Umu Akan, Umuaghadi Inyishia, Umu Abiala, Uchu, Ubaha, Owerri, Oru, Olakwo, Okwudo, Okpofe, Okohia, Ogwa, Ochicha Ngugo, Obolo, Obofia, Obite, Obiohoro osu, Norie, Nnarambe, Nkume, Ngbele, Nempe, Nekede, Nduche, Ndimoko, Ndiawa, Ndianiche, Iyeochara river, Isuobiangwu, Inyiogugu, Ikembara, Ihite, Idem Ogwa, Ibeku, Eziama, Emekeobibi, Ekwuoke, Ekwerazu Ogwa, Efuru river, Awkwa river, Awka Agbobu, Amucha, Amiri, Amibo, Amafor, Alum river, Akoko, Abuba river, Abala, and Apane
Low	Yinku, Uzuaba, Uzombe, Urualla, Umuzi, omu Ozu Isu, Umu Owa, Umu Osu, Umuosi Amumara, Umu Opara, Umu Omeke, Umuokwara, Umu Okpukpara, Umuoke, Umuohiri, Umu Ohie Uku, Umu Ogu, Umuodoma, Umu Nelu, Umunakanu, Umu Nachi, Umu Lawlaw, Umukani, Umukabia, Umu lago, Umuhu, Umu-Ezoalo-Ofor, Umu Ezegu, Umu Ezeala, Umu Eze, Umu Ewere, Umu Eshi, Umu Elemai, Umuelagwa, Umueke Umuduru, Umu Echem, Umuduru Okwe, Umu Duru, Umu Dike, Umudi, Umuawuchi, Umuaweke, Umu Awa, Umuariam, Umuakpu, Umu Agwo, Umuagu Okwelle, Umuagu, Umuouwa, Ugiri, Udo, Ubomiri, Owakiki, Oteru, Osunachu, Ossu, Osonachu, Osina, Osiam Lake, Orogwe, Orlu, Orji, Oratta, Opete, Oparanadim, Onwakiko, Onicha Uma Awada, Omo Akpo, Omicha Umu Awada, Okwuzu, Okwuohia, Okweli, Okwe, Okuku, Okpuala, Okporo, Okporku, Okigwi, Ohoba, ohia river, Ohekelem, Ogeh, Ogbor Ugiri, Ogbe Lulu, Obwe, Obufie, Obudi, Obor Ovoro, Obolo, Obokwe, Obohia, Obizi, Obiohuru, Obiogo, Obinze, Obigwo, Obibi, Obiakpo, Obeki, Obaku, Nkwesi, Nkwerre, Nguru, Ngugu, Ndizuogu, Ndiuche, Nde Ohinhin, Ndarshi, Naze, Mpam, Mbwelli, Mbieri, Mbidi, Mbeke, Lorji, Lagwa, Izombe, Itu, Isu Okpu, Isu njaba, Isiekenesi, irete, Loumama, Lile, Ikpem, Ihube, Ihitte Ogwa, Ihitte, Ihitte owerri, Ihite Nansa, Ihioma, Igitemasa, Ife, Ibinta, Ibeme, Eziudo, Eziobo, Ezinachi, Ezike, Eziama, Ezenu, Okwe, Ezeborgu, Eyinja, Emii, Emekukwu, Elugu Nkumato, Elelem, Ekwurazu, Ekwe, Ehume, Egbu, Egbelu, Ebu, Dimnaeze, Dikenafai, Awomama, Awlu, Awka, Awban, Awarra, Awaka, Avutu, Atta, Anghara, Anafis, Amuzu, Amopara, Amawzari, Amaumara, Amaruru Ikeduru, Amaraka, Amanduba, Amala, Amakohia, Ama Inyi, Amaimo, Amaibo, Amafor, Amaeze, Amaebu Ogwa, Alulu, Akukwa, Akpuru, Akokwa, Akabo, Aguneze, Agbobu, Agbaja isu, Achi, Abor Isu, Umuochi, and Umu Ayanmele.

State was high enough to cause flood. High rainfall causes localized rivers to overflow their banks and drainage system becomes filled up thereby causing increased surface runoff which leads to flooding.

5.4.5 SOIL ERODIBILITY FACTOR (K)

Figure 5.12(a) shows the spatial distribution map of K-factor for each soil type. The K-factor is calculated using Eq. (5.6). Table 5.6 shows the K-factor for each soil type in the Imo State. The erodibility of the various soil types determines the ease with which rainfall and runoff detach soil particles. Higher erodibility factors indicate that the soil particles are easily eroded. Gleysols, Xanthic Ferralsols, and Calcaric Fluvisols have low erodibility factors indicating that the driving force required to detach soil particles would be higher since the soil particles are compacted together. Dystric Nitosols have the highest soil erodibility in Imo State causing quick detaching of soil particles in these areas.

TABLE 5.6 Shows the K Factor for Each Soil Type in the Imo State.

Soil type	K-factor
Dystric nitosols	0.8
Calcaric fluvisols	0.05
Xanthic ferralsols	0.07
Gleysols	0.06

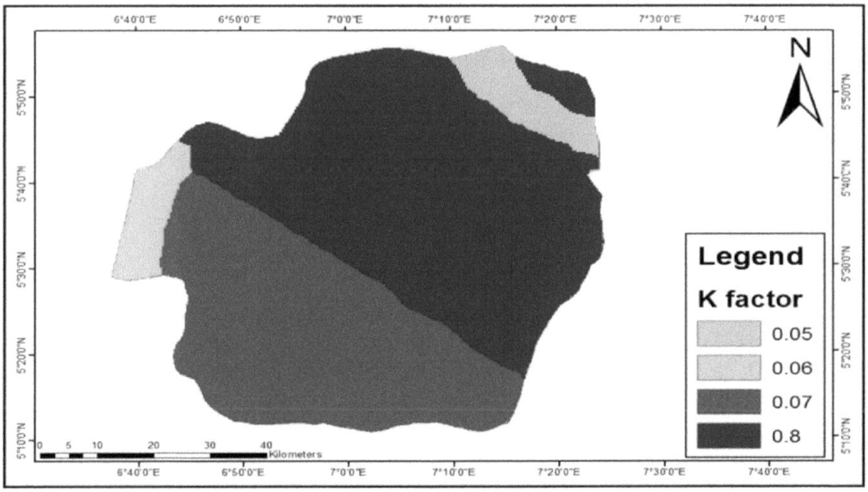

FIGURE 5.12(a) Map of K-factor in Imo state.

5.4.6 TOPOGRAPHY FACTOR (LS)

Figure 5.12(b) shows the map of LS factor. The LS factor ranges from 0 to 179.894. The LS factor indicates the slope length and slope steepness. High LS factor shows that such areas have steep slopes and long slope lengths. Steep slopes increase the speed of surface runoff. Increased surface runoff speed over a long distance would cause more soil particles to be detached over a wider area. This implies that areas with lower LS factor in Imo State have lesser soil erosion effect while areas with high LS factor in Imo State experiences more adverse effects of soil erosion. The LS factors were higher around hilly areas which is dominant in Northern parts of Imo State.

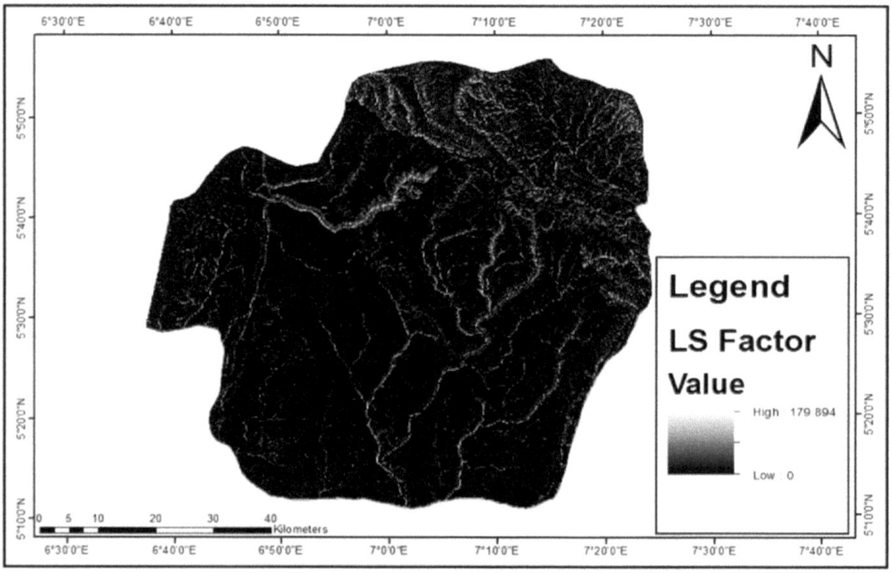

FIGURE 5.12(b) Map showing LS factor in Imo state.

5.4.7 COVER MANAGEMENT FACTOR (C)

Figure 5.12(c) shows the map of C-factor. The generated C-factor ranges between 0.25 and 1.15. The C-factor is used to indicate the effect of vegetation cover on soil erodibility. Vegetation protects the soil from direct contact with rainfall. Thick vegetation reduces the speed of rain drops on the soil causing less detachment of soil particles. The C-factors were higher in areas with bare and exposed surfaces. Vegetation areas have moderate C-factor values. Built up and water body has the lowest C-factor values. This shows that soil erosion is lesser in built up and vegetation areas and higher in bare and exposed surfaces.

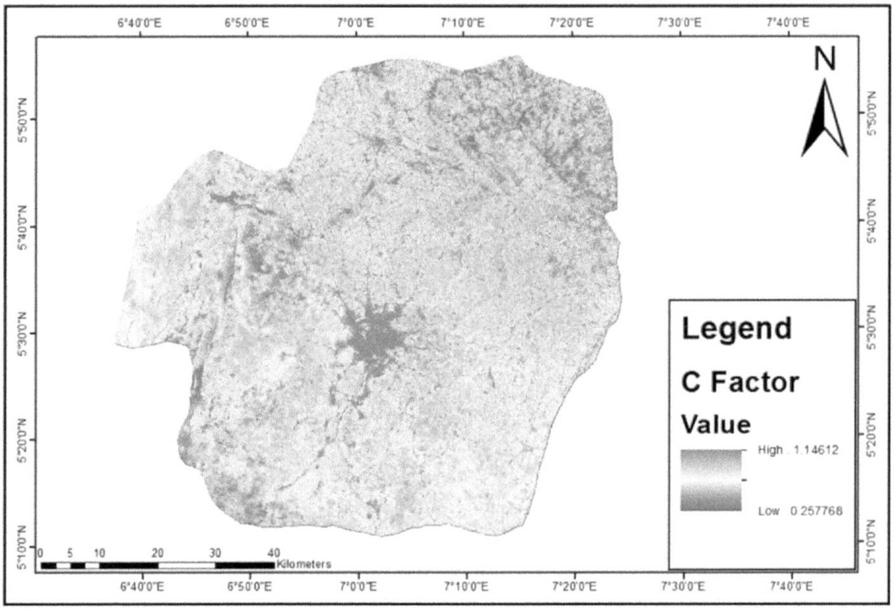

FIGURE 5.12(c) Map of C-factor in Imo state.

5.4.8 CONSERVATION PRACTICE FACTOR (P)

Soil conservation methods reduces the effect of soil erosion in areas where they have been practiced. Where there are no soil conservation practices, the soil is exposed to soil erosion. The P-factor value for Imo State was set at 1 since there is no specified management practices used in the region. This indicates that the soils in Imo State are left bare and exposed to all factors causing soil erosion.

5.4.9 SOIL LOSS

Figure 5.12(d) shows the amount of soil loss in tonnes per hectares per year in the Imo State. The soil loss ranges from 0 to as high as 123,922 ha/t/year. Soil erosion activities are more prevalent in the northern part of Imo state and toward the south. The southern Imo State experiences little erosion activities as these areas are low-lying region. The southern Imo State has a flat or almost flat terrain and this is responsible for the reduced effect of erosion in those parts. Figure 5.12(e) shows the soil erosion risk map for Imo State indicating existing gully erosion sites extracted from the previous literatures (Amanbagara et al., 2015). Sixty (60) gully erosion sites were identified. The gully sites were overlaid over the Local Government Areas to identify the Local Governments affected. The erosion risk map was classified into very low, low, moderate, high, and very high-risk zones. The very low-risk area covers an area of 4653.18 km^2, low-risk area covers an area of 429.53 km^2, moderate risk area covers an area of 227.05 km^2, high-risk area covers 14.26 km^2, and very high-risk area covers 8.47 km^2. Imo State covers a total area of 5332.49 km^2. Approximately, 87.3% of the Imo State lies in the very low-risk area, 8.1% of the Imo State lies in the low-risk area, 4.3% lies in the moderate risk area, 0.3% lies in the high-risk area and 0.2% lies on the very high-risk area as shown in Figure 5.12(e–f).

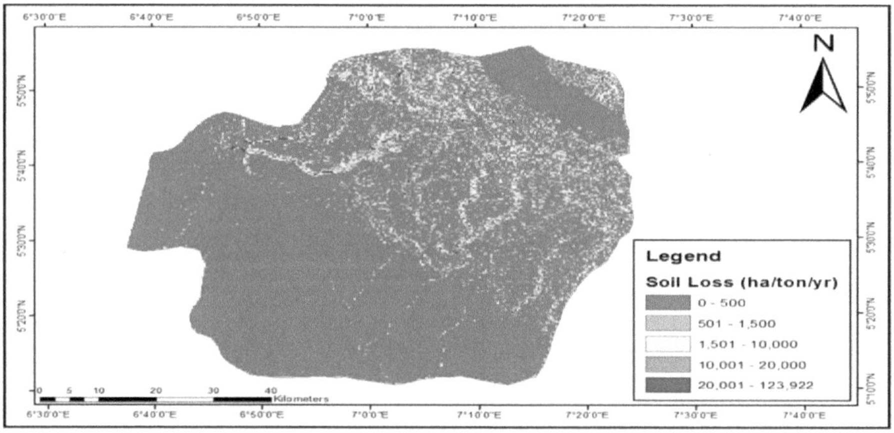

FIGURE 5.12(d) Map of soil loss in Imo state.

TABLE 5.7 Area and Percentage of Soil Erosion Risk.

Risk	Area (km^2)	Percentage (%)
Very low	4653.18	87.3
Low	429.53	8.1
Moderate	227.05	4.3
High	14.26	0.3
Very high	8.47	0.2

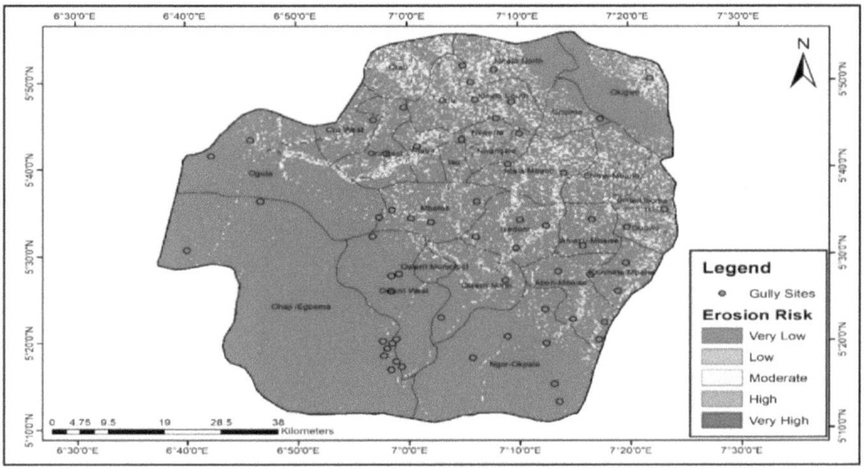

FIGURE 5.12(e) Map showing gully erosion sites in Imo state.

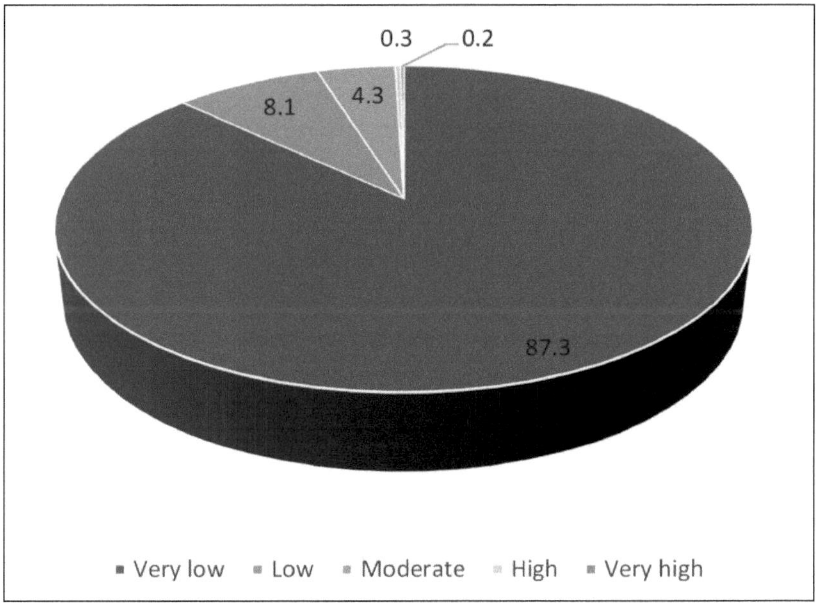

FIGURE 5.12(f) Map showing percentage gully erosion risk in Imo state.

5.5 SOIL EROSION DATABASE

A soil erosion database was created for Imo State. Table 5.8 shows the soil loss (ha/t/year) classes with the corresponding affected towns and villages.

One hundred and six towns and villages are lies within the 0–500 ha/t/year soil loss area. These are the least affected areas. Thirty-seven towns and villages lie within the 50.1–1500 ha/t/year soil loss area. The effect of soil erosion here are more than the previous class. Twenty-four towns and villages lie within the 1500–10,000 ha/t/year soil loss area. The effect of soil erosion in these towns and villages are more than the previous two classes. One town lies within the 10,001–20,000 ha/t/year soil loss area. One town also lies within the 20,001–123,922 ha/t/year soil loss area. These areas are the most affected with soil erosion.

5.6 CONCLUSION

Rainfall trend and SPI indicated that 2006, 2007, and 2009 are flood years. Spatial distribution of rainfall and runoff in the Imo State increases from North to South. The biodiversity of Imo State has evidently been degraded given the huge loss in vegetation as shown by the Normalized Difference Vegetation Index (NDVI) values. Rainfall water tends to move from steep slopes and gathers in flat areas causing flood while steep slope increases the effect of soil erosion as more soil is eroded in steep slopes. The topography of northern part of Imo State shows that they have steep slopes and the southern parts are almost flat. These features indicate that soil erosion is being prevalent in the northern parts and floods are prevalent in the southern parts of Imo State. Loose soil, such as dystric nitosols has high erodibility which leads to soil erosion, and poorly drained soil types are the causes of flood, such as calcaric fluvisols, and Gleysols within the study area. Impermeable surfaces in Imo State promote flooding as indicated by the geological structure of Imo State.

Meteorological and geophysical parameters were combined to delineate the flood and gully erosion risk assessment. Geographic Information System (GIS), fuzzy logic, and Revised Universal Soil Loss Equations (RUSLE) were integrated to carry out the risk assessment. GIS and fuzzy logic using multiple criteria were used to determine flood risk zones. The flood risk was classified into three types: low, medium and high. 62.9% of the Imo State lies in the low-risk zones, 17.2% lies in the medium-risk zones and 19.9 % lies in the high-risk zones. A database of towns and villages with each risk zone was generated. The database would provide useful information to stakeholders, policy and decision makers for effective flood preparedness, response, mitigation, and management.

TABLE 5.8 Soil Erosion Database of Imo State.

Soil loss (ha/t/year)	Towns and villages affected
0–50.0	Abekenta, Abor, Agbobu, Akalovo, Amaeze, Amafor, Amanduba, Amaruru Ikeduru, Amiri, Anghara, Asa, Awban, Awo, Azara Egbelu, Ebu, Egbu, Ekwuoke, Ekwurazu, Elelem, Emekukwu, Eyinja, Ezenu Okwe, Ezike, Ezinachi, Ibi, Ibinta, Ihitte, Ikembara, Ikperejere, Imeabiam, Isu Okpu, Isuobiangwu, Lorji, Mbeke, Iyeochara River, Mba River, Mbeke, Mberichi, Mbidi, Mbwelli, Nde Ohinhin, Ndiawa, Nekede, Ngugu, Nguru, Nsu, Nzerem, Obeki, Obinze, Obofia, Obokwe, Obozotto, Ohekelem, Ohoba, Ohozotto, Okohia, Okporo, Okuku, Okwe, Okwele, Okweli, Omo Akpo, Opete, Oramirukwa, Orji, Orodo, Orogwe, Osiam, Ossu, Owelu, Ubomini, Uborum, Udo, Ugiri, Umouwa, Umu Akam, Umu Akpa, Umu Ayanmele, Umu Dike, Umu Duru, Umu Ekwune, Umu Elemai, Umu Ewere, Umu Ezeala, Umu Nachi, Umu Nam, Umu Nelu, Umu Ogli, Umu Okrika, Umu Oma, Umu Opara, Umu Opee, Umu Owa, Umu Ozu Isu, Umuakpu, Umuhu, Umukabia, Umuna, Umunahu, Umuofa, Umuosi Amumara, Upe, and Uzuaba
50.1–1500	Abala, Akpuru, Alike, Amakohia, Amopara, Ata, Atta, Avutu, Awaka, Awkwa River, Awo Idemiri, Awo Ndemini, Dimnaeze, Egwe, Elugu Nkumato, Ikpem, Inyiogugu, Ndiuche, Obibi, Obiohoro Osu, Obiohuru, Obohia, Obolo, Okitankwo, Okwuohia, Osonachu, Umu Agwo, Umu Awa, Umu Eze, Umu Kabia, Umu Ogu, Umu Okpukpara, Umu Osu, Umuaga Okwelle, Umuariam, and Umuoke
1501–10,000	Umuodoma, umunakanu, Umudi, Omu Omeke, Umu Numu, Umu Ezegu, Umu Ezeala Ofor, Ubaha, Onicha, Okwudo, Ohoba, Ochicha ngugo, Obilokwu Mbieri, Nduche, Itu, Ihube, Ihite Nansa, Igitemasa, Eziudo, Dikenafai, Awka, Amucha, Achi, and Abor Isu
10,001–20,000	Idem Ogwa
20,001–123,922	Awomama

KEYWORDS

- **flood**
- **gully erosion**
- **rainfall trend**
- **risk area**

REFERENCES

Adefolalu, D. O. Climate Change and Economic Sustainability in Nigeria. In *Paper 142 Presented at International Conference on Climate Change and Economic Sustainability,* 143 Nnamdi Azikiwe University; Enugu: Nigeria, 2007; pp 12–14.

Amanbagara, G. T.; Njoku, J. D.; Obenade, M. Applying Satellite Remote Sensing and GIS Tools in the Study of Gully Erosion. *J. Sci. Res. Rep.* **2015,** *4* (3), 253–264. ISSN: 2320–0227.

Bekele, B.; Wu, W.; Legesse, A.; Temesgen, H.; Yirsaw, E. Random and Systematic Land Use/Land Cover Transitions in Semi-arid Landscapes of Ethiopian Central Rift Valley Lakes Region (East Africa). *Appl. Ecol. Environ. Res.* **2018,** *16,* 3993–4014.

Bhatt, C. M.; et al. Satellite-Based Assessment of the Catastrophic Jhelum Floods of September 2014, Jammu & Kashmir, India. *Geomat. Nat. Hazards Risk* **2017,** *8* (2), 309–327.

FAO. *Harmonized World Soil Database (version 1.2)*; Food Agriculture OrganizationL Rome, Italy; IIASA, Laxenburg, Austria, 2012. http://webarchive.iiasa.ac.at/Research/LUC/External-World-soil-database/html/.

Farooq, M.; Mohammad, M. Dynamics and Forecasting of Population Growth and Urban Expansion in Srinagar City—A Geospatial Approach. *Int. Arch. Photogramm. Remote Sens. Spat. Inf. Sci.* **2014,** *40* (8), 709.

IPCC. *Climate Change 2007: Impacts, Adaptation, and Vulnerability. The Fourth Assessment Report,* Vol. 1, 2007.

Kunta, K.; Carosio, A. *Effects of Geographic Information Quality on Soil Erosion Prediction*; Institute of Geodesy and Photogrammetry Swiss Federal Institute of Technology: Zurich, 2007; p 60. ISBN 978-3-906467-84-9.

Kanga, S.; et al. Modeling the Spatial Pattern of Sediment Flow in Lower Hugli Estuary, West Bengal, India by Quantifying Suspended Sediment Concentration (SSC) and Depth Conditions Using Geoinformatics. *Appl. Comput. Geosci.* **2020a,** 100043.

Kanga, S.; Sheikh, A. J.; Godara, U. GIscience for Groundwater Prospect Zonation. *J. Crit. Rev.* **2020b,** *7* (18), 697–709.

Kumar, S.; et al. Delineation of Shoreline Change along Chilika Lagoon (Odisha); East Coast of India Using Geospatial Technique, 2018.

Morgan, R. *Soil Erosion and Conservation*; Copublished in the United States with John Wiley and Sons: New York, 1986; p 298. ISBN 0-582-30158-0.

NGSA. *Nigerian Geological Survey Agency Report*; 2015.

Nathawat, M. S.; et al. Monitoring & Analysis of Wastelands and Its Dynamics Using Multiresolution and Temporal Satellite Data in Part of Indian State of Bihar. *Int. J. Geomat. Geosci.* **2010,** *1* (3), 297–307.

Okorie, F. C. *Great Ogberuru in Its Contemporary Geography and Population Density*; Cape Publishers: Owerri, Nigeria, 2010; p 64.

Pandey, A. C.; Singh, S. K.; Nathawat, M. S. Waterlogging and Flood Hazards Vulnerability and Risk Assessment in Indo-Gangetic Plain. *Nat. Hazards* **2010,** *55* (2), 273–289.

Pandey, A. C.; Singh, S. K.; Nathawat, M. S.; Saha, D. Assessment of Surface and Subsurface Waterlogging, Water Level Fluctuations, and Lithological Variations for Evaluating Groundwater Resources in Ganga Plains. *Int. J. Digital Earth* **2013,** *6* (3), 276–296.

Renard, K.; Freimund, J. Using Monthly Precipitation Data to Estimate the R-Factor in USLE. *J. Hydrol.* **1994,** *157* (1), 287–306.

Renard, K.; Foster, G.; Weesies, G.; McDool, D.; Yoder, D. Predicting Soil Erosion by Water: A Guide to Conservation Planning with the Revised Universal Soil Loss Equation (RUSLE); Agricultural Handbook 703, USDA-ARS, 1997.

Singh, S. K.; Pandey, A. C. Geomorphology and the Controls of Geohydrology on Waterlogging in Gangetic Plains, North Bihar, India. *Environ. Earth Sci.* **2014,** *71* (4), 1561–1579.

Singh, S. K.; Mishra, S. K.; Kanga, S. Delineation of Groundwater Potential Zone Using Geospatial Techniques for Shimla City, Himachal Pradesh (India). *Int. J. Sci. Res. Dev.* **2017,** *5* (4), 225–234.

Sharma, A.; Kanga, S. *Surface Runoff Estimation of Sind River Basin Using SCS-CN Method and GIS Technology*; 2020.

Vijith, H.; Dodge-Wan, D. Modelling Terrain Erosion Susceptibility of Logged and Regenerated Forested Region in Northern Borneo through the Analytical Hierarchy Process (AHP) and GIS Techniques. *Geoenviron. Disasters* **2019,** *6* (1), 8.

Wischmeier, W. H.; Smith, D. D. *Predicting Rainfall Erosion Losses from Cropland East of the Rocky Mountains. Handbook No. 282*; USDA, Washington, DC, 1975.

Wischmeier, W. H.; Smith, D. D.; *Predicting Rainfall Erosion Losses: Agriculture Handbook No. 537*, USDA Science and Education Administration, 1978.

CHAPTER 6

Mapping Groundwater Potential Zones Using Satellite Remote Sensing and GIS

UZAIR AHMED[1*], ABID SARWAR[2], FAZLI AMIN KHALIL[3], SHAZIA GULZAR[4], NADIA QAYUM[5], and RIMSHA QURESHI[1]

[1]*Department of Geography and Geomatics, University of Peshawar, Pakistan*

[2]*Directorate of Soil Survey, Agriculture Department, Govt. of KP, Khyber Pakhtunkhwa, Pakistan*

[3]*Department of Electronics, University of Peshawar, Peshawar, Pakistan*

[4]*The Fazal Haq College, Mardan, Pakistan*

[5]*Department of Meteorology, COMSATS University, Islamabad, Pakistan*

[*]*Corresponding author. E-mail: uzairahmedsparc770@gmail.com*

ABSTRACT

The accessible groundwater supplies are scarce fulfill the water conditions for consumption, agricultural and industrial use; therefore, the requirement for groundwater has enlarged. Groundwater is well thought out, the key segment of the world's freshwater resources. The study is an effort to identify the possible groundwater zones in Matta tehsil, Swat. ArcGIS 10.4 was used for geospatial analysis. The methodology implemented is referred as multi influencing factor (MIF) techniques. Seven parameters, that is, Land Use/Land Cover, geology, soil, rainfall, lineament density, drainage density, and slope were used for delineation. The groundwater potential zones were classified into four classes very high, high, good, low, and poor. The study suggests that very high ground water potential zones

GIScience for the Sustainable Management of Water Resources. Gowhar Meraj, Shruti Kanga, Majid Farooq, Suraj Kumar Singh, & Sudhanshu (Eds.)
© 2023 Apple Academic Press, Inc. Co-published with CRC Press (Taylor & Francis)

(GWPZ) cover an area of 8.86 km sq., high, 115.27 km sq., 335.53 km sq. having good and 210.48 km sq^2 has the poor potential of groundwater. The study showed that remote sensing and GIS provided efficient tools for mapping promising sites for groundwater exploration. MIF technique is an appropriate approach for the delineation of GWPZ, which can be applied for further research in different areas.

6.1 INTRODUCTION

Water is the basic need for life on earth and significantly affects all phases of our ecosystem (Randhir, 2012). Existing surface water resources are insufficient to fulfill the needs of drinking, irrigation, and industrial use; therefore, the demand for groundwater is increasing. Several factors like high population density, uneven distribution of water resources, space and time demand, economic development, and climate change lead to increased consumption and surface water shortages (Kaliraj et al., 2014; Krishnamurthy et al., 1996; Murthy, 2000; Obi Reddy et al., 2000; Rashid et al., 2013; Selvam et al., 2014; Singh et al., 2017a). It is envisaged that groundwater accounts for about 30.1% of worldwide freshwater; therefore, it is one of the most vital sources of freshwater and is widely used for domestic purpose.

Only 3% of the earth's water is classified as freshwater, and the remaining 97% is normal saline, which cannot be used for drinking water. In available freshwater, 68.7% are stored in the form of ice sheets and glaciers (Glieck, 1996). Due to rapid industrialization and population growth, the demand for freshwater resources in the world has increased significantly. Therefore, groundwater extraction has become an integral part of many water management methods, especially in rural areas (Senanayake et al., 2016).

The term "groundwater potential" refers to the amount of groundwater available in a region and is a function of several hydrological and hydrogeological factors. From the perspective of hydrogeological exploration, the term can also be defined as the possibility of groundwater in a region. Although there is always a lot of freshwater on the earth, there is not always water at the time and place, and it is not always suitable for all uses (Al-abadi and Al-shamma, 2014; Bhat et al., 2014; Meraj et al., 2015). Groundwater is an essential natural resource for the consistent and economic supply of consumable water resource in both manmade and natural environments. Therefore, it performs a vital role in human well-being, as well as that of some marine and land ecosystems. Currently, groundwater shares approximately 34% of the overall annual water supply and is a valuable freshwater source. So, an estimation for this resource is extremely substantial for the viable administration of groundwater systems (Dar et al., 2011; Krishna Kumar et al., 2012; Magesh et al., 2011). Geographic information systems (GIS) and remote sensing (RS) are effective tools for delineating potential groundwater areas.

In recent years, the widespread use of satellite data, as well as traditional maps and corrected ground truth data, has made it easier to establish baseline information for potential areas of groundwater (Tiwari and Rai, 1996). Remote sensing with its advantages of spatial, spectral, and temporal availability of data covering large and inaccessible areas within a short time has become a very handy tool in assessing, monitoring, and conserving groundwater resources. Satellite data provides quick and useful baseline information on the parameters like geology, geomorphology, land use/land cover, lineaments, etc. Controlling the occurrence and movement of groundwater (Selvam et al., 2014; Kanga et al., 2017a, 2017b, 2021; Rather et al., 2018; Hassanin et al., 2020). In delineation of groundwater potential zone various factors such as lineament, faults (Nag and Ghosh, 2013; Senthil and Shankar, 2014; Singh et al., 2014; Taheri et al., 2015), rainfall, soil, lithology, and soil texture (Magesh et al., 2012), slope, elevation (Magesh et al., 2011) drainage systems (Dabral et al., 2013; Rassam et al., 2008), and groundwater table distribution (Arkoprovo et al., 2012) play a crucial role.

6.2 OBJECTIVES

The study consists of one main objective to delineate potential areas of groundwater through a multiinfluencing approach (MIF).

6.3 MATERIALS AND METHODS

6.3.1 STUDY AREA

Matta tehsil is located in northern Hindu Kush range of district Swat, Khyber-Pakhtunkhwa, Pakistan, its geographical coordinates are from 34° 36′ 59″ to 35° 44′ 51″ N Longitude and from 72° 29′ 52″ to 72° 09′ 52″ E Latitude. Matta is a rural area with majority of rural population. It has a total area of 683 sq. km. Matta is a local word which means clay soil. It is located about 20 km from the central city of Mingora. Matta subdivision is the largest county of Swat tehsil and illustrated in Figure 6.1.

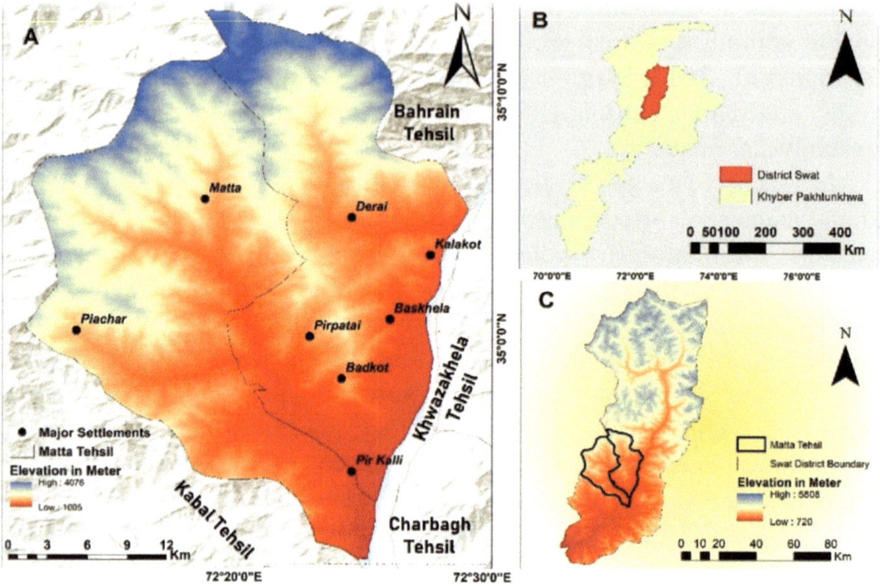

FIGURE 6.1 Location map: (A) the study area map of Tehsil Matta, district Swat, Pakistan; (B) Khyber Pakhtunkhwa map showing location of district Swat; and (C) district Swat map showing location of study area.

6.3.2 METHODOLOGY

The study consists of one main objective to delineate potential areas of groundwater through a multi influencing factor approach. For this method, the required data is collected from different departments and then analyzed in ArcGIS 10.4.1. A method called weighted overlay analysis, in MIF techniques were employed to calculate the weights of different influencing parameters (Figure 6.2).

MIF model is based on the seven parameters, which are geology, lineament, soil, slope, rainfall, land use/land cover, and drainage density (Singh et al., 2017b; Kanga et al., 2020a, 2020b). For the identification of groundwater, first, we select these parameters; second, after the brief literature scores and weights were assign to these selected parameters. Each individual parameter for each thematic layer will be ranked and weighted based on major, minor, and no influence to groundwater potential. Score 2 were given to the major, influence (A), score 1 were given to the minor influence (B) and score 0 were given to no influence. The cumulative score ($A + B$) of both major (A) and minor (B) influence are calculated according to formula which is listed in Table 6.1. This relative effect is further used to calculate the weight of each influencing parameter (Adham et al., 2010; Gumma and Pavelic, 2013; Magesh et al., 2012; Mukherjee et al., 2012; Rashid et al., 2012; Shaban et al., 2006; Yeh et al., 2009). Formula through which the weight for each influencing factor was calculated is

$$(A + B)/\sum(A + B) \times 100$$

where A represents the major influence and B represents the minor influence.

After assigning the weightage to parameters based on their influence, weightage was divided equally and assigned as rank to each subclass of all the seven influencing factors. Table 6.2 shows the weight and rank of sub classes of each influencing factors. ArcGIS 10.4.1 was employed for further geospatial analysis. All the thematic layers of influencing factors with scores integrated using the spatial analysis extension in ArcMap. After running the weighted overlay tool in the spatial analyst extension, the output result was categorized into three classes (high, moderate, and low).

TABLE 6.1 Minor and Major Influence of Parameters.

Parameter	Major effect (A)	Minor effect (B)	Proposed relative effect ($A + B$)	Proposed weight of each influencing parameter
Land use/land cover	2 + 2 + 2	1 + 1	8	20
Lineament density	2 + 2	1 + 1	6	15
Slope	2 + 2	1 + 1	6	15
Drainage density	2 + 2	1 + 1	6	15
Soil	2	1 + 1	4	10
Geology	2	1 + 1	4	10
Rainfall	2 + 2	1 + 1	6	15
			$\sum 40$	100

6.4 RESULT AND DISCUSSION

6.4.1 DRAINAGE DENSITY

SRTM digital elevation model (DEM) with spatial resolution of 30 m is used to generate slope, drainage, and elevation of study area. For slope and elevation, there are special tools available in ArcGIS 10.4.1. The drainage density describes the water runoff and groundwater infiltration in an area.

For this study, the drainage network and drainage density data have been prepared from SRTM DEM with 30 m resolution. Hydrology tool were used for the extraction of stream network from which the drainage density was calculated. The regions were classified in very low (0–0.71), low (0.71–1.42), moderate (1.42–2.14), high (2.14–2.85) drainage density zones, as shown in Figure 6.3A. Area with low drainage density indicates comparatively lower infiltration and low runoff and vice versa. The zones with low-to-moderate drainage density are considered excellent groundwater potential zones (GWPZ) and were assigned a higher weightage value.

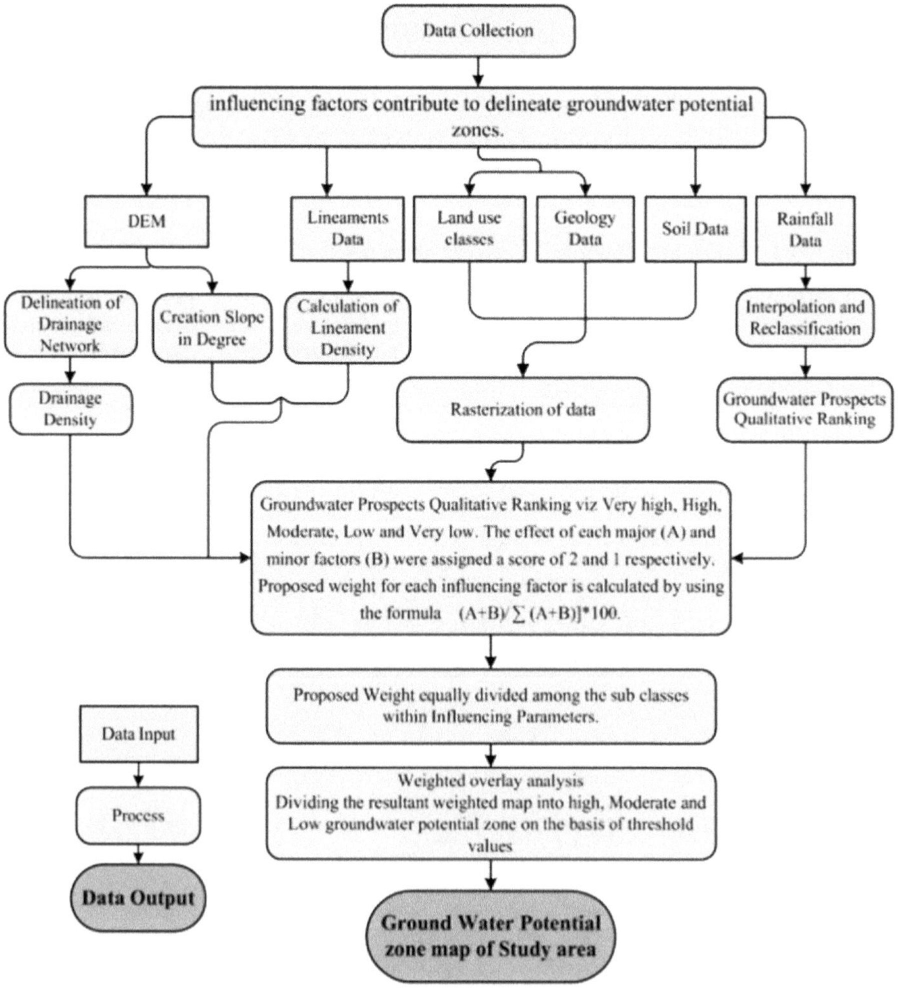

FIGURE 6.2 A flow chart for delineating the groundwater potential zone.

TABLE 6.2 Description of Different Parameters Influencing the Groundwater Recharge in the Study Area.

Influencing parameters	Sub-classes within influencing parameter	Qualitative rank	Proposed weight of each influencing parameter	Groundwater prospects (quantitative rank)
Land use/land cover	Water body	Very high	20	8
	Agriculture land	High		5
	Forest	Moderate		4
	Settlements	Low		2
	Barren land	Very low		1
Rainfall in mm	119–134	High	15	7
	135–147	Moderate		5
	148–158	Low		2
	159–175	Very low		1
Slope in degree	0–10	High	15	7
	10–20	Moderate		5
	20–30	Low		2
	30–40	Very low		1
Drainage density in km/sq. km	0–0.71	High	15	7
	0.71–1.42	Moderate		5
	1.42–2.14	Low		2
	2.14–2.85	Very low		1
Geology	Sedimentary rocks	High	10	6
	Metamorphic rocks	Low		1
	Igneous rock	Moderate		3
Lineament density in km/ sq km.	0–0.2	High	15	7
	0.2–0.7	Moderate		5
	0.7–1.3	Low		2
	1.3–3.4	Very low		1
Soil	Mainly loamy very Shallow soil	High	10	4
	Mainly loamy shallow soil	Moderate		3
	Loamy and clayey soil	Low		2
	Rock outcrop and Loamy very shallow soil	Very low		1

6.4.2 ELEVATION

Elevation is referring to height of space from mean sea level. Water tends to store at low topography rather than at higher topography. Higher the elevation lesser the groundwater potential and vice versa. Moreover, it also have role in the water table, the study area is located between 988 and 4,083 m above sea level. The elevation map is shown in Figure 6.3B.

6.4.3 GEOLOGY

Geology is among the major factors which play an important role in the distribution and occurrence of groundwater. In the present study area, three types of rocks namely sedimentary rock, igneous rock, and metamorphic rock were found. Geologically, the study area is dominated by mostly igneous and metamorphic rocks. The type of rock and geology greatly affects the availability of groundwater, some of the rocks highly resist the water to percolate underground and greatly affect the recharge, while some rocks have the capacity to allow the water and play its role in the recharge of water underground. For the present study, the geology map is derived from the existing geological map of northern Pakistan prepared by M. P. Searle and M. Asif Khan. The map was digitized from the existing map and then reclassified into major rock types. The geology map is shown in Figure 6.3C.

6.4.4 LAND USE/LAND COVER

Land use/land cover plays a crucial role in groundwater recharge for the present study land cover land use layer was prepared from Landsat7 (2017) imagery having 30 m resolution. Supervised classification method has been done through the ArcGIS 10.4.1 software. The supervised classification method has the result of the study covered by five different classes such as vegetation, Barren land, Urban, water, and forest cover. Land use/land cover map is shown in Figure 6.3D. The image classification of the study area shows that settlement was speared over the Tehsil, particularly near the waterbodies and vegetation area. Waterbodies are a continuous and outstanding source of recharge to the ground; therefore, high weightage is given to this class.

6.4.5 LINEAMENT DENSITY

Lineaments are pointers of subsurface faults and breaks that influence the existence of groundwater acting as canals and reservoirs. The lineament density of an area can ultimately expose the groundwater potential since the presence of lineament usually signifies the permeable zones. The liniment of the study area is generated from Landsat8 Pan sharpened reflected band, PCI Geomatica 2017 software was employed for the extraction of lineament while lineament density was determined from the density analysis tool in ArcGIS 10.4.1. Lineament density map is shown in Figure 6.3E. The region was classified into four classes, such as very low (0–0.271), low (0.271–0.751), moderate (0.751–1.333), and high (1.33–3.14). Furthermore, the zone with high density was categorized as high groundwater potential and assigned as higher weightage value.

6.4.6 RAINFALL

Rainfall is a major source for groundwater availability through the water cycle. The study area has been classified into five zones based on the equal intervals then the suitable weightage has been assigned for each class. The rainfall data of total six metrological stations such as Kalam, Saidu Sharif, Malam Jabba, Drosh, and Dir were collected from the regional metrological office, Peshawar. The data were available from year 2005 to 2015. The rainfall data are spatially interpolated in ARC map 10.4.1. The interpolated data were reclassified into four classes, i.e., 119–133 mm, 133–147 mm, 147–161 mm, and 161–174 mm. A higher weight value was assigned to high rainfall class because of its potential to recharge the aquifers. The rainfall map is shown in Figure 6.3F.

6.4.7 SLOPE

Slope is one of the most important factors directly affects the penetration of precipitation. Besides, sheer slopes cause lesser amount of recharge as water runs quickly from the surface during rainfall. It allows scarce time to permeate to the surface and recharge the inundated zone. For this study, the slope map has been prepared from SRTM DEM with 30 m resolution using Arc-GIS spatial analyst tool. The slope map is shown in Figure 6.3G. The slope has been classified into five classes, i.e., very low (0–13.6), low (13.6–27.2), moderate (27.2–40.9), high (40.9–54.5), and very high (54.5–68.2).

6.4.8 SOIL

The soil map of the study area was derived from the Soil Survey of Pakistan, and the properties and effectiveness of soil type in groundwater recharge were derived through literature review. The thematic layer on soil for the study area reviles for main classes, that is, mainly loamy very shallow soil, mainly loamy shallow soil, loamy and clayey soil, rock outcrop and loamy very shallow soil, loamy and clay. Weights are assigned based on the composition and soil water holding capacity. Rocky outcrops with shallow soil and soil with clay and clayey sand have poor permeability and have been given a low weightage. Mainly loamy soil has given high weight value because of its high porosity and permeability.

After assigning the suitable weightage to main parameters with respect to their availability in the field; finally, the map of groundwater potential zones is prepared showing the groundwater scenario of the study area as shown in Figure 6.3H. The result of weighted overlay analysis has been classified based on the availability of groundwater into four classes, that is, very high, high, good, low, and poor.

6.5 VERIFICATION OF RESULTS

A field survey was conducted to collect the water table depth throughout the study area. A handheld GPS was used to collect the coordinates of 40 different wells, tubewells, and hand pumps and their depth were noted in feet. According to our survey, the water depth is ranging from very low 7 ft to very high 105 ft. The well depth was divided into three classes like (0–24), (25–53), and (54–105) feet. The well data were overlaid on top of the delineated groundwater potential zones layer as shown in Figure 6.5. These well data were used as reference point to calculate the accuracy. The cross-verification method was used to check the accuracy of the weighted overlay result. According to well depth in the study area, the groundwater potentiality was high in those areas where the table is quite shallow and near. Union councils Thana, Sakhra, Behaand Shwar were fell in very high GWPZ and the water table of these union councils are also very shallow; and on other hand, the union councils like Matta, Drushkhela, and Bedara were fell in very poor GWPZ and the water table of these union councils is also very Deep.

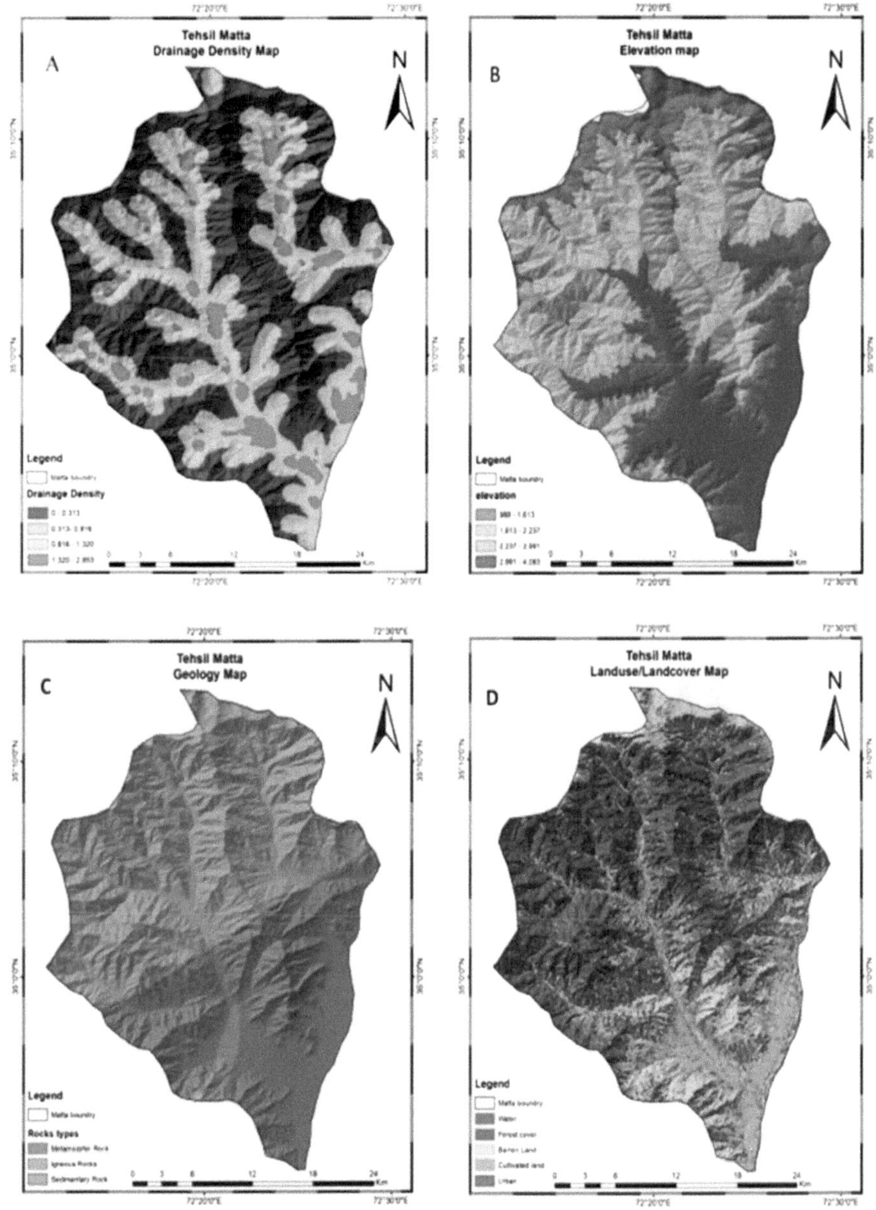

FIGURE 6.3 *(Continued)*

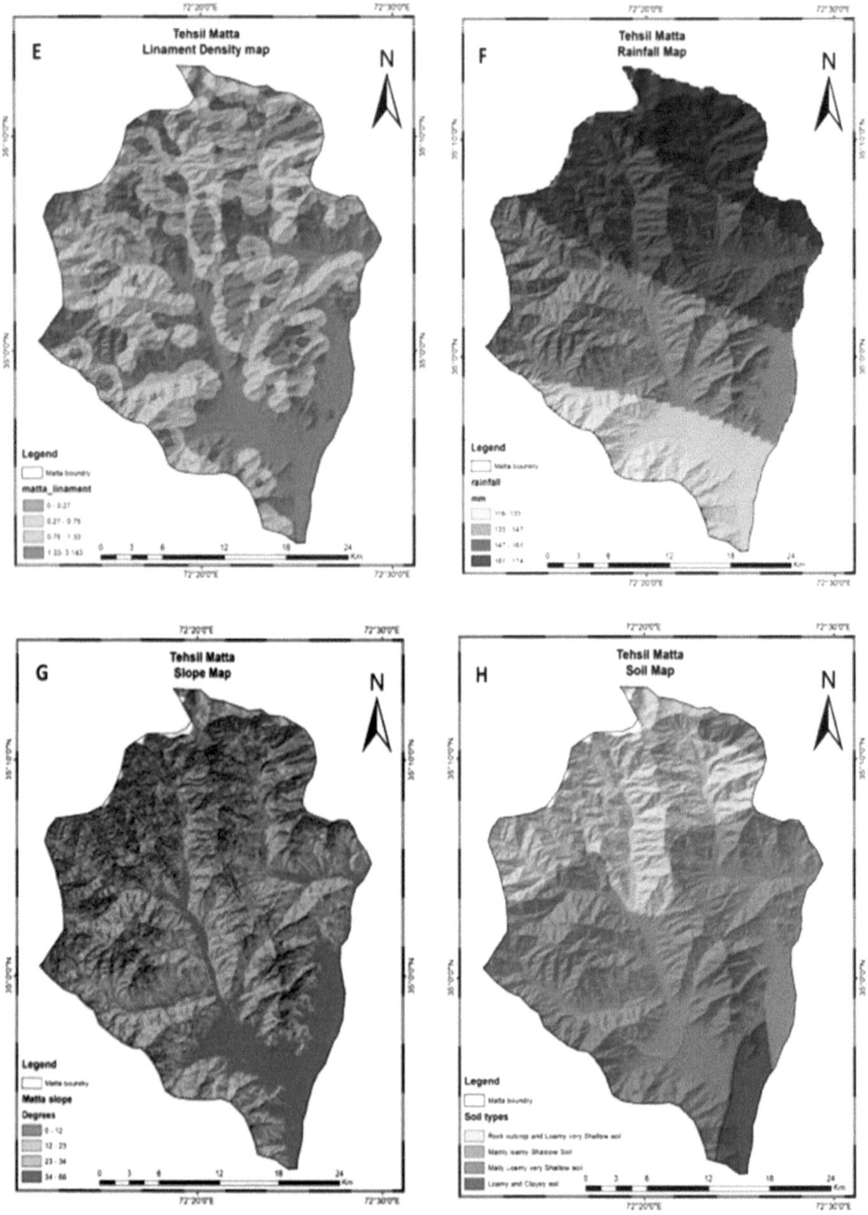

FIGURE 6.3 All parameters maps involved in the study. (A) drainage density map, (B) elevation map, (C) geology map, (D) land use/land cover map, (E) lineament density map, (F) rainfall map, (G) slope map, and (H) soil map.

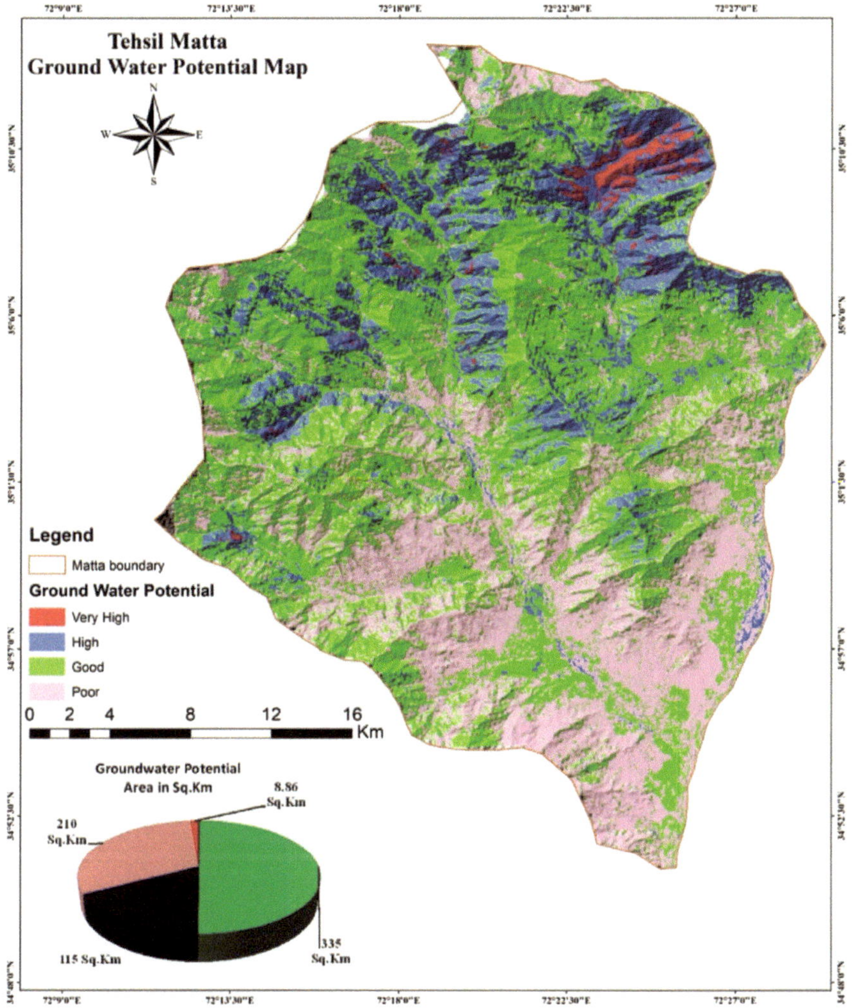

FIGURE 6.4 Groundwater potential zones map, Tehsil Matta, Swat.

The overlay analysis discloses that most of the wells with shallow and moderate groundwater depth are well within the very high and high groundwater potential zone.

6.6 CONCLUSION

GIS and RS has showed to be powerful and cost-effective technique for delineating groundwater potential in zones of Matta tehsil, Swat. The study

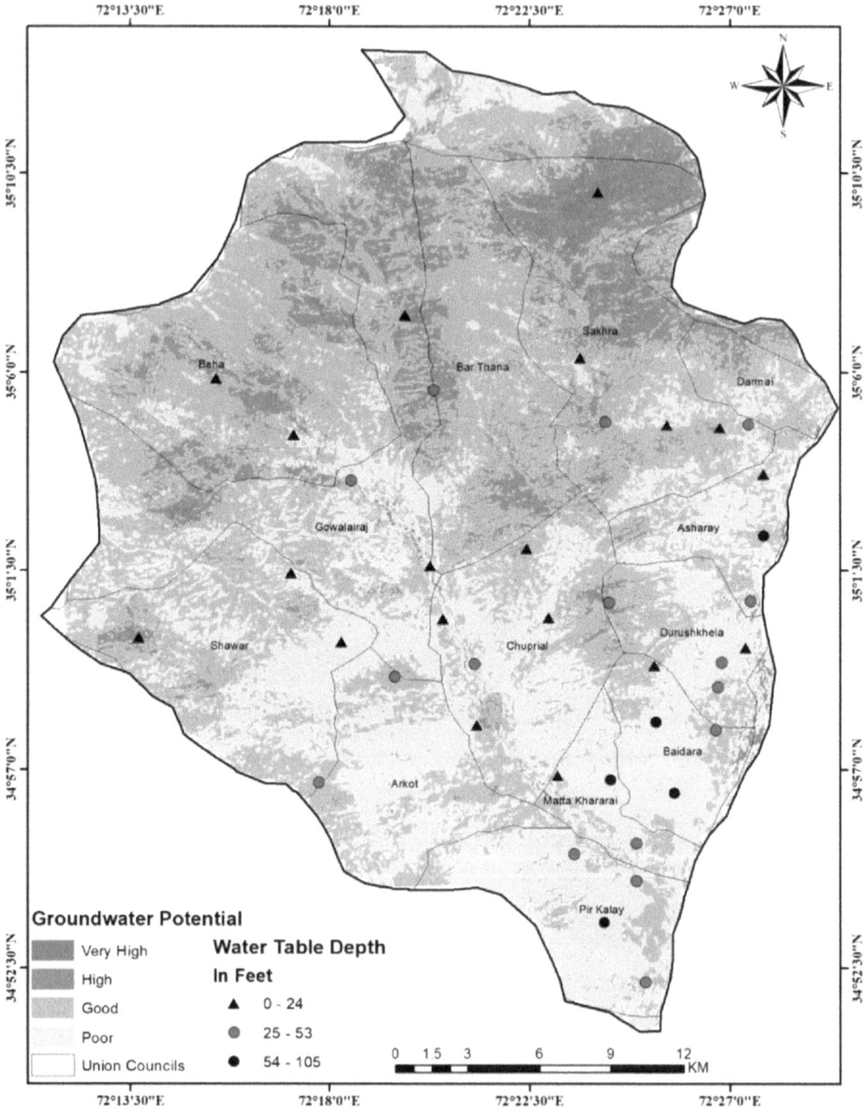

FIGURE 6.5 Map shows the groundwater potential zones in Matta Tehsil, Swat overlay by water depth.

discloses that integration of seven thematic layers such as drainage density, slope, rainfall, soil, geology, lineament density, and land use/land cover give direct information to local authorities and organizers about the areas appropriate for groundwater exploration. Delineating the groundwater potential zones in Matta tehsil using remote sensing, GIS, and MIF techniques is found efficient to minimize the time, labor, and finance and thereby enables quick decision-making for sustainable water resources management. Satellite imageries, topographic maps, and GPS data were used to prepare the thematic layers. The result of weighted overlay analysis has been classified based on the availability of groundwater into four classes, that is, very high, high, good, low, and poor. From the result of classification, it has been found that the area of 8.86 km sq. has very high, 115.27 km sq. having high, 335.53 km sq. has good, 210.48 km sq^2 having poor potential of groundwater. The GWPZ map was shown in Figure 6.4.

The use of RS and GIS in groundwater potential zone mapping is one of the main methods and technology for effective and controlled improvement of groundwater resources. These maps are utilized by specialists, organizers, policy, and decision makers to create and oversee groundwater.

KEYWORDS

- **MIF**
- **GWPZ**
- **remote sensing**
- **GIS**
- **groundwater**

REFERENCES

Adham, M. I.; Jahan, C. S.; Mazumder, Q. H.; Hossain, M. M. A.; Haque, A. M. Delineation of Groundwater Potential Zones in Arid Region of India—A Remote Sensing and GIS Approach. *J. Geol. Soc. India* **2010,** *75* (2), 432–438.

Al-abadi, A.; Al-Shamma, A. Groundwater Potential Mapping of the Major Aquifer in Northeastern Missan Governorate, South of Iraq by Using Analytical Hierarchy Process and GIS. *J. Environ. Earth Sci.* **2014,** *4* (10), 125–150.

Arkoprovo, B.; Adarsa, J.; Prakash, S. S. Delineation of Groundwater Potential Zones using Satellite Remote Sensing and Geographic Information System Techniques: A Case Study from Ganjam District, Orissa, India. *Res. J. Recent Sci.* **2012,** *1* (9), 59.

Bhat, S. A.; Meraj, G.; Yaseen, S.; Pandit, A. K. Statistical Assessment of Water Quality Parameters for Pollution Source Identification in Sukhnag Stream: An Inflow Stream of Lake Wular (Ramsar Site), Kashmir Himalaya. *J. Ecosyst.* **2014,** 1–18. https://doi.org/10.1155/2014/898054.

Dabral, S.; Sharma, N.; Bhatt, B.; Joshi, J. P. A Geospatial Technique for Demarcating Ground Water Recharge Potential Zones: A Study of Mahi-Narmada Inter Stream Region, Gujarat. *Int. J. Geomat. Geosci.* **2013,** *4* (1), 177–185.

Dar, I. A.; Sankar, K.; Dar, M. A. Deciphering Groundwater Potential Zones in Hard Rock Terrain Using Geospatial Technology. *Environ. Monit. Assess.* **2011,** *173* (1–4), 597–610. https://doi.org/10.1007/s10661-010-1407-6.

Glieck, P. H. Basic Water Requirements for Human Activities: Meeting Basic Needs. *Water Int.* **1996,** *21* (2), 83–92.

Gumma, M. K.; Pavelic, P. Mapping of Groundwater Potential Zones across Ghana Using Remote Sensing, Geographic Information Systems, and Spatial Modeling. *Environ. Monit. Assess.* **2013,** *185* (4), 3561–3579. https://doi.org/10.1007/s10661-012-2810-y.

Hassanin, M.; Kanga, S.; Farooq, M.; Singh, S. K. Mapping of Trees Outside Forest (ToF) from Sentinel-2 MSI Satellite Data using Object-based Image Analysis. *Gujarat Agric. Univ. Res. J.* **2020,** *45*, 207–213.

Kaliraj, S.; Chandrasekar, N.; Magesh, N. S. Identification of Potential Groundwater Recharge Zones in Vaigai Upper Basin, Tamil Nadu, Using GIS-based Analytical Hierarchical Process (AHP) Technique. *Arab. J. Geosci.* **2014,** *7* (4), 1385–1401. https://doi.org/10.1007/s12517-013-0849-x.

Kanga, S.; Singh S. K.; Sudhanshu. Delineation of Urban Built-up and Change Detection Analysis Using Multi-temporal Satellite Images. *Int. J. Recent Res. Asp.* **2017a,** *4* (3), 1–9.

Kanga, S.; Kumar, S.; Singh, K. S. Climate Induced Variation in Forest Fire Using Remote Sensing and GIS in Bilaspur District of Himachal Pradesh. *Int. J. Eng. Comput. Sci.* **2017b,** *6* (6), 21695–21702.

Kanga, S.; Sheikh, A. J.; Godara, U. GIscience for Groundwater Prospect Zonation. *J. Crit. Rev.* **2020b,** *7* (18), 697–709.

Kanga, S.; et al. Modeling the Spatial Pattern of Sediment Flow in Lower Hugli Estuary, West Bengal, India by Quantifying Suspended Sediment Concentration (SSC) and Depth Conditions Using Geoinformatics. *Appl. Comput. Geosci.* **2020a,** *8*, 100043.

Kanga, S.; Rather, M. A.; Farooq, M.; Singh, S. K. GIS based Forest Fire Vulnerability Assessment and Its Validation Using Field and MODIS Data: A Case Study of Bhaderwah Forest Division, Jammu and Kashmir (India). *Indian For.* **2021,** *147* (2), 120–136.

Krishna Kumar, S.; Chandrasekar, N.; Seralathan, P.; Godson, P. S.; Magesh, N. S. Hydrogeochemical Study of Shallow Carbonate Aquifers, Rameswaram Island, India. *Environ. Monit. Assess.* **2012,** *184* (7), 4127–4138. https://doi.org/10.1007/s10661-011-2249-6.

Krishnamurthy, J.; Venkatesa Kumar, N.; Jayaraman, V.; Manivel, M. An approach to Demarcate Ground Water Potential Zones through Remote Sensing and a Geographical Information System. *Int. J. Remote Sens.* **1996,** *17* (10), 1867–1884. https://doi.org/10.1080/01431169608948744.

Magesh, N. S.; Chandrasekar, N.; Soundranayagam, J. P. Morphometric Evaluation of Papanasam and Manimuthar Watersheds, Parts of Western Ghats, Tirunelveli District, Tamil Nadu, India: A GIS Approach. *Environ. Earth Sci.* **2011,** *64* (2), 373–381. https://doi.org/10.1007/s12665-010-0860-4.

Magesh, N. S.; Chandrasekar, N.; Soundranayagam, J. P. Delineation of Groundwater Potential Zones in Theni District, Tamil Nadu, Using Remote Sensing, GIS and MIF Techniques. *Geosci. Front.* **2012,** *3* (2), 189–196. https://doi.org/10.1016/j.gsf.2011.10.007.

Meraj, G.; Romshoo, S. A.; Yousuf, A. R.; Altaf, S.; Altaf, F. Assessing the Influence of Watershed Characteristics on the Flood Vulnerability of Jhelum Basin in Kashmir Himalaya. *Nat. Hazards* **2015,** *77* (1), 153–175. https://doi.org/10.1007/s11069-015-1605-1.

Mukherjee, P.; Singh, C. K. Mukherjee, S. Delineation of Groundwater Potential Zones in Arid Region of India—A Remote Sensing and GIS Approach. *Water Resour. Manage.* **2012,** *26* (9), 2643–2672. https://doi.org/10.1007/s11269-012-0038-9.

Murthy, K. S. R. Ground Water Potential in a Semi-arid Region of Andhra Pradesh—A Geographical Information System Approach. *Int. J. Remote Sens.* **2000,** *21* (9), 1867–1884. https://doi.org/10.1080/014311600209788.

Nag, S. K.; Ghosh, P. Delineation of Groundwater Potential Zone in Chhatna Block, Bankura District, West Bengal, India Using Remote Sensing and GIS Techniques. *Environ. Earth Sci.* **2013,** *70* (5), 2115–2127. https://doi.org/10.1007/s12665-012-1713-0.

Obi Reddy, G. P.; Chandra Mouli, K.; Srivastav, S. K.; Srinivas, C. V.; Maji, A. K. Evaluation of Ground Water Potential Zones Using Remote Sensing Data—A Case Study of Gaimukh Watershed, Bhandara District, Maharashtra. *J. Indian Soc. Remote Sens.* **2000,** *28* (1), 19–32. https://doi.org/10.1007/BF02991858.

Randhir, T. O. Water for Life and Ecosystem Sustainability. *J. Earth Sci. Clim. Change* **2012,** *03* (01), 1–2. https://doi.org/10.4172/2157-7617.1000e107.

Rashid, M.; Lone, M. A.; Ahmed, S. Integrating Geospatial and Ground Geophysical Information as Guidelines for Groundwater Potential Zones in Hard Rock Terrains of South India. *Environ. Monit. Assess.* **2012,** *184* (8), 4829–4839. https://doi.org/10.1007/s10661-011-2305-2.

Rashid, I.; et al. Assessing the Impact of Anthropogenic Activities on Manasbal Lake in Kashmir Himalayas. *Int. J. Environ. Sci.* **2013,** *3* (6), 2036–2047.

Rassam, D. W.; Pagendam, D. E.; Hunter, H. M. Conceptualisation and Application of Models for Groundwater-surface Water Interactions and Nitrate Attenuation Potential in Riparian Zones. *Environ. Model. Softw.* **2008,** *23* (7), 859–875. https://doi.org/10.1016/j.envsoft.2007.11.003.

Rather, M. A.; et al. Remote Sensing and GIS Based Forest Fire Vulnerability Assessment in Dachigam National Park, North Western Himalaya. *Asian J. Appl. Sci.* **2018,** *11* (2), 98–114.

Selvam, S.; Magesh, N. S.; Sivasubramanian, P.; Soundranayagam, J. P.; Manimaran, G.; Seshunarayana, T. Deciphering of Groundwater Potential Zones in Tuticorin, Tamil Nadu, Using Remote Sensing and GIS Techniques. *J. Geol. Soc. India* **2014,** *84* (5), 597–608. https://doi.org/10.1007/s12594-014-0167-2.

Senanayake, I. P.; Dissanayake, D. M. D. O. K.; Mayadunna, B. B.; Weerasekera, W. L. An Approach to Delineate Groundwater Recharge Potential Sites in Ambalantota, Sri Lanka Using GIS Techniques. *Geosci. Front.* **2016,** *7* (1), 115–124. https://doi.org/10.1016/j.gsf.2015.03.002.

Senthil, K. G. R.; Shankar, K. Assessment of Groundwater Potential Zones Using GIS. *Front. Geosci.* **2014,** *2* (1), 1–10.

Shaban, A.; Khawlie, M.; Abdallah, C. Use of Remote Sensing and GIS to Determine Recharge Potential Zones: The Case of Occidental Lebanon. *Hydrogeol. J.* **2006,** *14* (4), 433–443. https://doi.org/10.1007/s10040-005-0437-6.

Singh, P.: Gupta, A.: Singh, M. Hydrological Inferences from Watershed Analysis for Water Resource Management Using Remote Sensing and GIS Techniques. *Egypt. J. Remote Sens. Space Sci.* **2014,** *17* (2), 111–121. https://doi.org/10.1016/j.ejrs.2014.09.003.

Singh, S. K.; Kanga, S. Role of Geoinformatics in Site Suitability Analysis of Infrastructures Using PRA Approach. *Am. Int. J. Res. Sci. Technol. Eng. Math.* **2017a,** *1*(18), 81–85.

Singh, S. K.; Mishra, S. K.; Kanga, S. Delineation of Groundwater Potential Zone Using Geospatial Techniques for Shimla City, Himachal Pradesh (India). *Int. J. Sci. Res. Dev.* **2017b,** *5* (4) 225–234.

Taheri, K.; Gutierrez, F.; Mohsieni, H.; Raeisi, E.; Taheri, M. Sinkhole Susceptibility Mapping Using the Analytical Hierarchy Process (AHP) and Magnitude-Frequency Relationships: A Case Study in Hamadan Province, Iran. *Geomorphology* **2015,** *234*, 64–79. https://doi.org/10.1016/j.geomorph.2015.01.005.

Tiwari, A.; Rai, B. Hydromorphogeological Mapping for Groundwater Prospecting Using Landsat-MSS Images—A Case Study of Part of Dhanbad District, Bihar. *J. Indian Soc. Remote Sens.* **1996,** *24* (4), 281–285. https://doi.org/10.1007/BF03026236.

Yeh, H. F.; Lee, C. H.; Hsu, K. C.; Chang, P. H. GIS for the Assessment of the Groundwater Recharge Potential Zone. *Environ. Geol.* **2009,** *58* (1), 185–195. https://doi.org/10.1007/s00254-008-1504-9.

CHAPTER 7

Evaluating the Watershed Geomorphometry for Assessing Hydrological Response Using GIScience

GUHAN VELUSAMY[1*], GEETHALAKSHMI VELLINGIRI[2], MUGILAN RAMAN[3], VINOTHKANNA SHANMUGAM[4], SENTHILRAJA KANDASAMY[2], DHANYA PRAVEEN[2], and KUMARASEN PALANIAPPAN[5]

[1]*Agro Climate Research Centre, Tamil Nadu Agricultural University, Coimbatore, India*

[2]*Directorate of Crop Management, Tamil Nadu Agricultural University, Coimbatore, India*

[3]*Department of RS&GIS, Tamil Nadu Agricultural University, Coimbatore, India*

[4]*Department of Geography, Madurai Kamaraj University, Madurai, India*

[5]*Water Technology Centre, Tamil Nadu Agricultural University, Coimbatore, India*

[*]*Corresponding author. E-mail: guhanthiran@gmail.com*

ABSTRACT

This study is based on the evaluation of the hydrological response of Parambikulam Aliyar basin in Tamil Nadu. Various morphometric parameters are used to evaluate the watershed characteristics responsible for ascribing specific hydrological responses. The study area exhibits a fourth-order drainage basin with the highest stream order length of 304.7 km. The slope

GIScience for the Sustainable Management of Water Resources. Gowhar Meraj, Shruti Kanga, Majid Farooq, Suraj Kumar Singh, & Sudhanshu (Eds.)
© 2023 Apple Academic Press, Inc. Co-published with CRC Press (Taylor & Francis)

gradient varies from 0 to 75.67°, and a high degree of slope is noticed in the southern part of the study area. The aspect map of the study area reveals that the Northwest-facing slopes are predominant in the southern part of the PAP basin. Differences in slope and topographic conditions may explain the variance in stream length ratio. The results show that the watersheds of this basin imbibe it with different geoenvironmental characteristics that shall help in landscape management at any spatiotemporal. scale.

7.1 INTRODUCTION

The evaluation of a river basin's hydrological characteristics is considered a requirement for basin management plans. It involves a thorough examination of the basin's morphometric characteristics, such as stream order, volume, drainage density, and bifurcation ratio, among others. The drainage of a basin is regarded as one of the most basic elements of the fluvial landscape. Morphometric analyses were used for basin characterization in various research findings (Staley et al., 2014). Traditional approaches such as topographic sheets and field observations may be used to delineate drainage networks within a basin or subbasin. Alternative advanced approaches such as geoprocessing instruments, Geographical Information Systems (GIS), and remote sensing (Mustafa et al., 2012) are used in current situations. The conventional method's main drawback is that it's difficult to cover a large area and inspect stream networks from field level observations (Wang et al., 2016). Digital elevation models (DEMs) are considered to be the competent systematic method to find out the stream flow from higher to lower elevation (Laks et al., 2017). By using the Geoprocessing tools, it will create a computerized automatic stream module with supporting thematic layers (Bocher et al., 2018). The results obtained from the Geoprocessing tools can be used for auxiliary morphometric analysis (Vieceli et al., 2015). The current paper uses a Geographic Information System (GIS) to define the morphometric characteristics of the Parambikulam Aliyar basin in order to recognize the hydrological characteristics results.

7.2 STUDY AREA

Parambikulam Aliyar basin is an inter-state water distribution project located in the south-western part of India collaborating between Kerala and Tamil Nadu. This project diverts the water from the eight rivers. The PAP basin area lies in coordinates of 10° 10′ 00″ N to 10° 57′ 20″ N latitude, 76° 43′ 00″ E to 77° 12′ 30″ E longitudes and covering 2388.72 sq. km area (Figure 7.1).

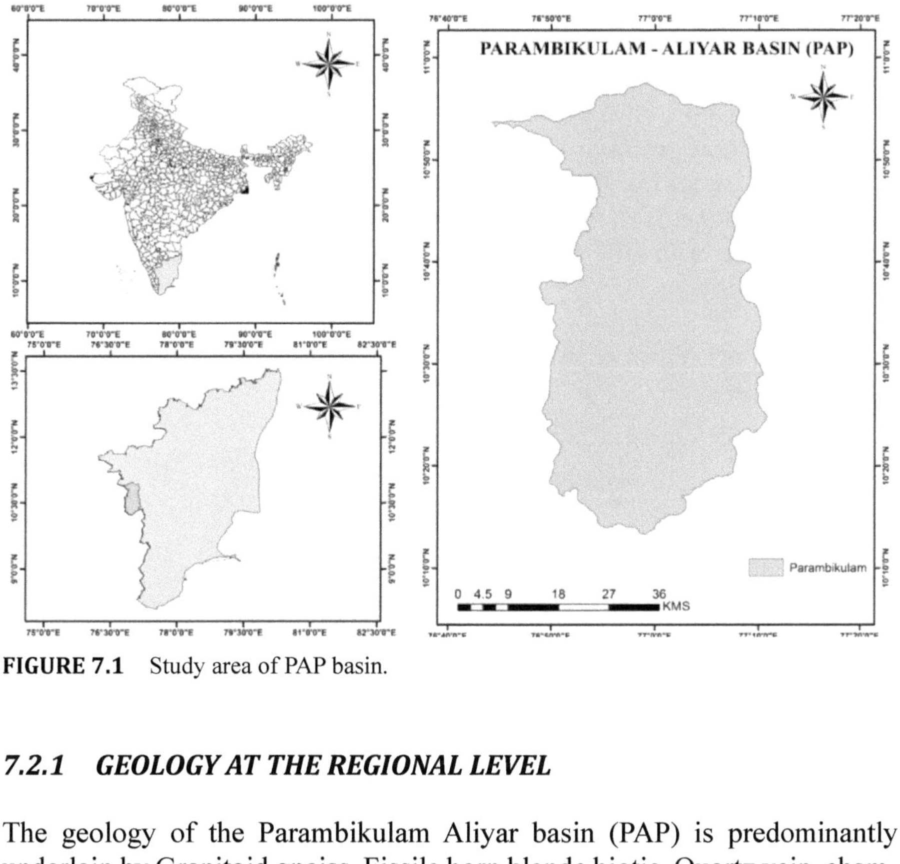

FIGURE 7.1 Study area of PAP basin.

7.2.1 GEOLOGY AT THE REGIONAL LEVEL

The geology of the Parambikulam Aliyar basin (PAP) is predominantly underlain by Granitoid gneiss, Fissile horn blendc biotic, Quartz vein, champion gneiss, granite (Gr2), granitoid, sand, and silt. Granitoid gneiss makes up the majority of the Parambikulam Aliyar basin. The geological features are digitized using ArcGIS 10.5 software. The detailed geological map of the PAP basin is shown in Figure 7.2.

7.3 EXPERIMENTAL METHODS AND MATERIALS

In this study, the DEM data for the Parambikulam Aliyar basin were obtained from the Shuttle Radar Topographic Mission (SRTM) with 30 m resolution using the hydrological tool in ArcGIS. The entire hydrological analysis was carried out by using ArcGIS version 10.5. The Geoprocessing tool requires a drainage area in hectares, DEM as the input parameters for morphometric analysis. The output is obtained in the form of the drainage basin polygon by considering its area in square kilometer from the DEM data and other required files for assessing the morphometric characteristics. The stream classification was classified based on Strahler (1952). According to this classification, sections with no tributaries were termed as first-order stream. When two first-order stream sections come together, it's referred to as a second-order stream, and so on. The PAP basin's highest stream order was calculated as a fourth-order stream. Manual corrections were made by combining streams in the same order with different nodes; otherwise, the stream order error would be unpredictable.

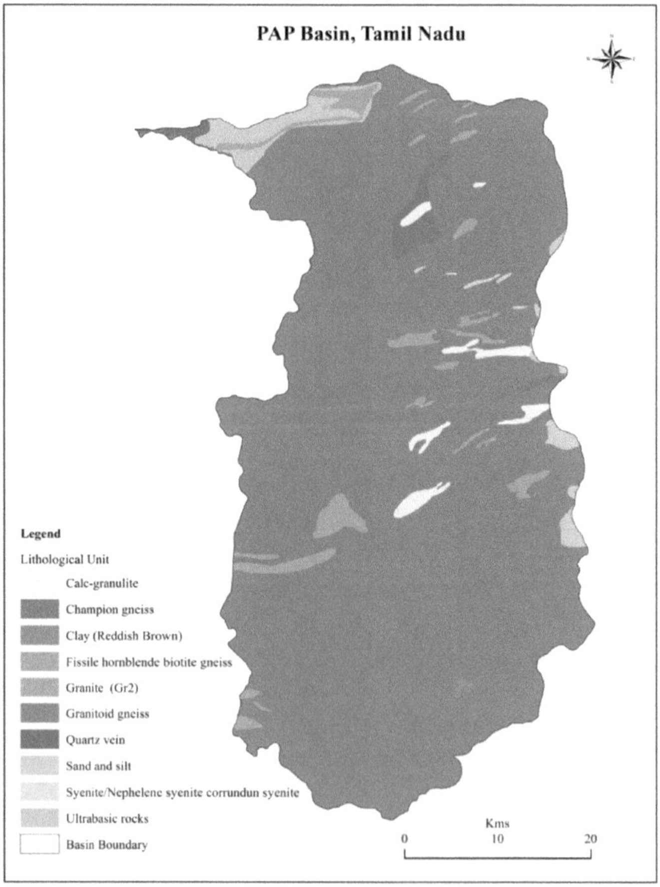

FIGURE 7.2 Geology of PAP basin.

7.4 RESULTS AND DISCUSSION

The Parambikulam Aliyar basin had a total drainage area of 2385.97 km^2. Shuttle Radar Topographic Mission (SRTM)—DEM with 30 m resolution was used in this study (Figure 7.3). The development of drainage networks mainly depends on factors like geological features, rainfall other than the endogenic and exogenic influences. Morphometric analysis of a basin is attained through measurement of linear, relief, and aerial aspects (Nag and Chakraborty, 2003). In order to evaluate the morphometric characteristics of the PAP basin, various parameters like stream number, stream length ratio,

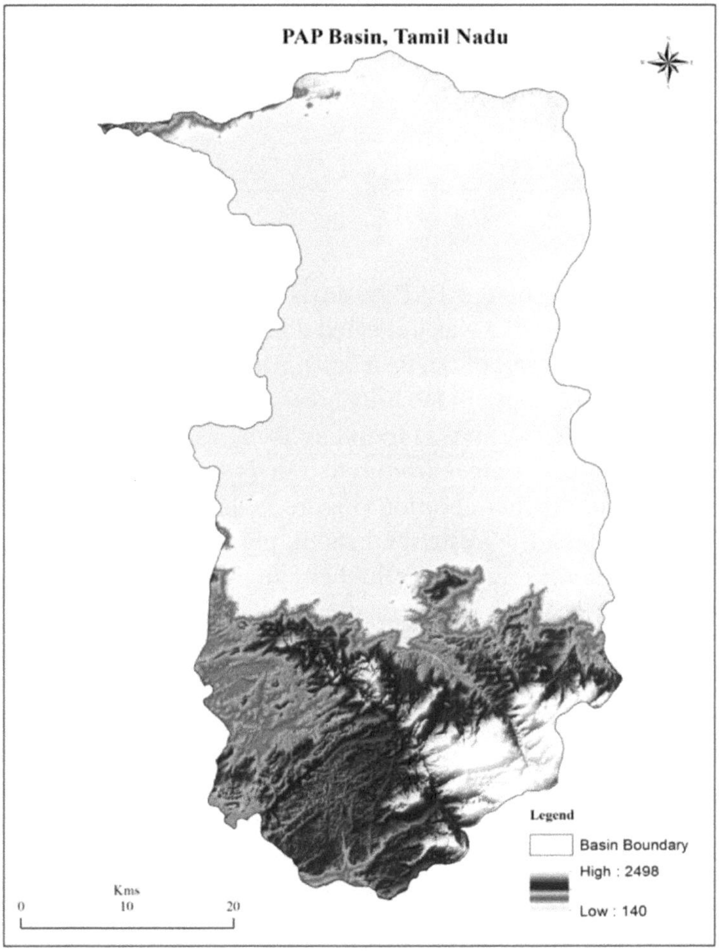

FIGURE 7.3 DEM of PAP basin.

stream order, drainage texture, drainage density, stream frequency, and bifurcation ratio were carried by following the standard mathematical formulae as given in Table 7.1.

TABLE 7.1 Description of Morphometric Factors and their Mathematical Formulations Used in PAP Basin.

Parameters	Formulae	References
Stream order (u)	Hierarchical rank	Strahler (1964)
Stream length (Lu)	Length of the stream	Horton (1945)
Mean stream length (Lsm)	Lsm = Lu/Nu	Strahler (1964)
Stream length ratio (RL)	RL = Lu/Lu − 1	Horton (1945)
Bifurcation ratio (Rb)	Rb0Nu/Nu + 1	Schumm (1956)
Mean bifurcation ratio (Rbm)	Rbm = Average of bifurcation ratios of all order	Strahler (1957)
Drainage density (Dd)	Dd = Lu/A	Horton (1945)
Drainage texture (T)	T = Dd × Fs	Smith (1950)

Source: Adapted from Mangan et al., 2019.

The aspects map of the PAP basin is shown in Figure 7.4. From the graphical interpretation, it was indicated that northwest facing slopes mainly occur in the southern part of the PAP basin whereas other direction slopes are also found to be in meagre in the hilly areas of the PAP basin. The remaining two-thirds of the PAP basin was found to be flat as shown in Figure 7.4.

The slope gradient ranges from 0 to 75.67°, with a high degree of slope in the study area's southern portion (Figure 7.5). The steep gradient is due to the presence of the hilly Western Ghats. Rapid runoff and increased erosion rates have occurred as a result of higher slope degrees, with poor recharge capacity. The PAP basin has a gross drainage area of 2385.97 km^2.

The longest stream order is 304.7 km long. The study area's mean stream length (Lsm) ranges from 5.5 to 5.8 km. In the basin, the value of stream length (RL) ranges from 1.0 to 1.1 (Table 7.2). Because of variations in slope and topographic conditions, the RL between successive stream orders varies, and it has a positive relationship with the basin's surface flow discharge and erosional level. Table 7.2 shows the bifurcation ratio (Rb) values for the PAP basin, which range from 0.7 to 1.0. In the study region, a higher Rb value indicates that the drainage pattern is under tight structural control, while a lower value indicates that the basin is less affected by structural disturbances (Strahler, 1964). It should be noted that the Rb values vary from one order to the next. These variations are determined by the drainage basin's geological and lithological growth (Strahler, 1964). The research area's mean bifurcation ratio (Rbm) is 0.9. The total length of streams of all orders per drainage area is referred to as drainage density, and it shows the closeness of channel spacing (Horton, 1945). Climate, slope, rock types, vegetation cover, soil texture, and runoff intensity all affect Dd, which is a measure of fluvial dissection (Verstappen, 1983). The consistency and characteristics of surface runoff are directly influenced by the amount of precipitation and slope gradient. The PAP basin has a drainage density of 0.2 km^2.

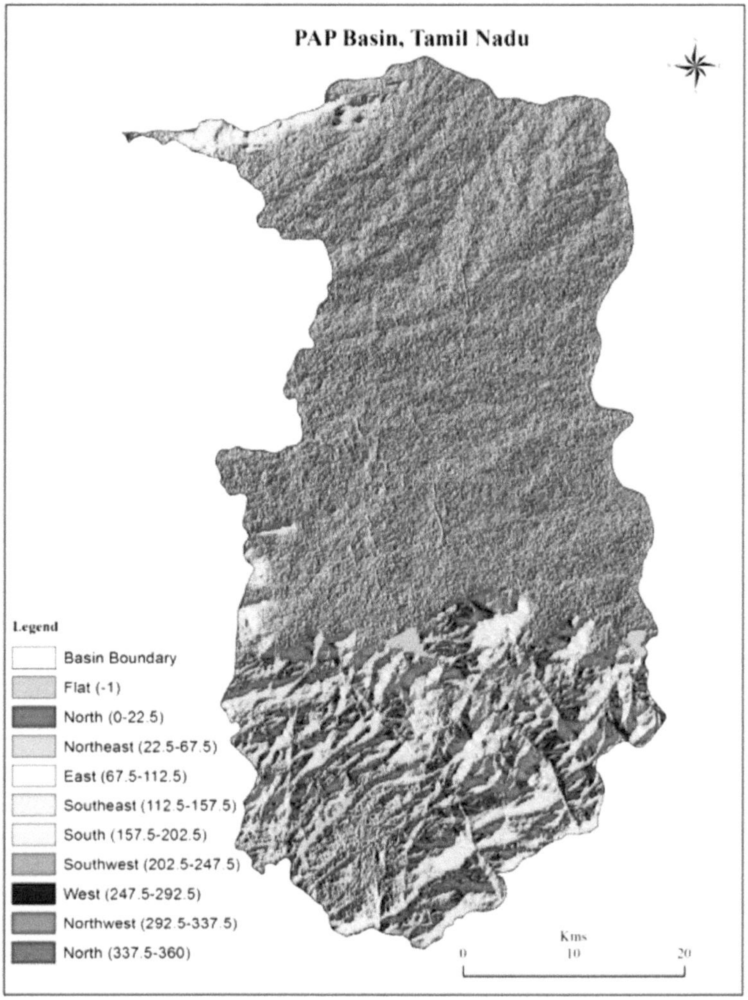

FIGURE 7.4 Aspect map of the study area.

Drainage texture, according to Horton (1945), is the total number of stream segments of all orders per perimeter of a field. This parameter is important for the basin's geomorphic configuration, which refers to the relative spacing of drainage channels. Climate, rainfall, vegetation, lithology, slope, infiltration potential, and stage of development are all factors that influence drainage texture (Smith, 1950). The drainage texture value in the research area is 5.5.

TABLE 7.2 Morphometric Parameters of PAP Basin Drainage System Derived from DEM.

Parameters	Parambikulam Aliyar project basin			
	I	II	III	IV
Stream order (u) (total)	55	35	6	2
Stream length(Lu) (km)	304.7	192.9	32.3	11.6
Mean stream length (Lsm) (km)	5.5	5.5	5.4	5.8
Stream length ratio (RL)	1.0	1.0	1.0	1.1
Bifurcation ratio (Rb)	1.0	1.0	0.9	0.7
Mean bifurcation ratio (Rbm)	0.9	–	–	–
Total basin area (A) (km^2)	2386.0	–	–	–
Drainage density(Dd) (km/km^2)	0.2	–	–	–
Drainage texture (T)	5.5	–	–	–

7.5 CONCLUSION

Application of GIS modeling merged with remote sensing data has proven to be an efficient tool for landscape morphometric analysis. The computed morphometric characteristics of the PAP basin reveal that the first-order streams dominate this region. The study area exhibits a fourth-order drainage basin with the highest stream order length of 304.7 km. The slope gradient varies from 0 to 75.67°, and a high degree of slope is noticed in the southern part of the study area. The aspect map of the study area reveals that the Northwest-facing slopes are predominant in the southern part of the PAP basin. Differences in slope and topographic conditions may explain the variance in stream length ratio. The drainage density implies that the existence of subsurface strata is permeable with a very fine drainage texture, and the bifurcation ratio in the study area indicates that geological structures have little effect on drainage networks. The results of this study are useful for landscape management in the PAP basin in Tamil Nadu, and the extraction model can be used to perform basin-level morphometric analysis at any spatiotemporal scale.

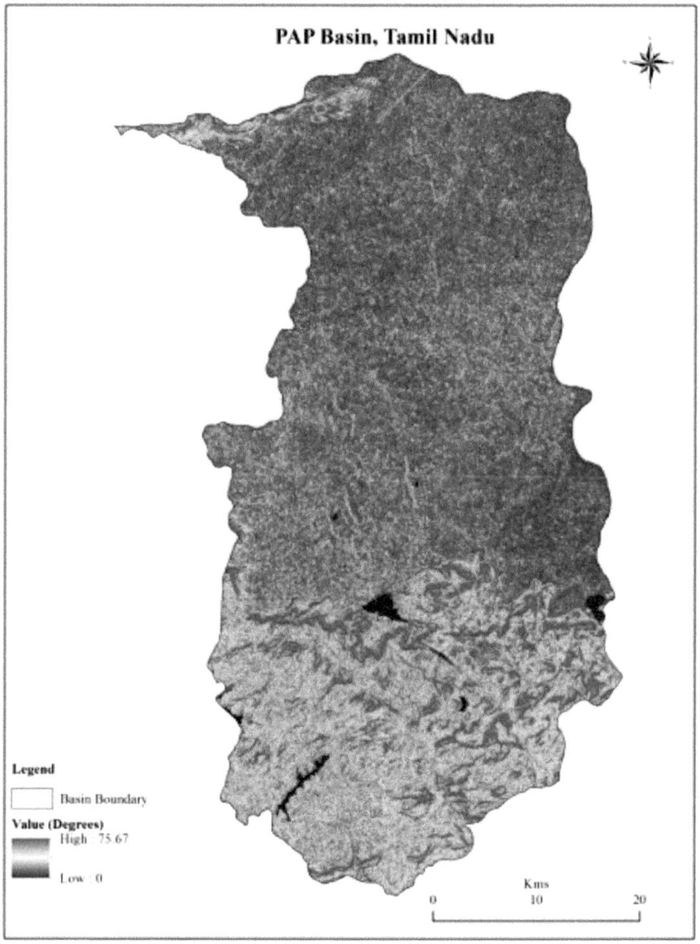

FIGURE 7.5 Slope map of the study area.

KEYWORDS

- **Parambikulam Aliyar basin**
- **digital elevation model (DEM)**
- **morphometric analysis**
- **arcGIS-10.5**

REFERENCES

Altaf, F.; Gowhar, M.; Shakil, A. R. Morphometric Analysis to Infer Hydrological Behaviour of Lidder Watershed, Western Himalaya, India. *Geogr. J.* **2013.** https://doi.org/10.1155/2013/178021.

Altaf, S.; Gowhar, M.; Shakil, A. R. Morphometry and Land Cover Based Multi-criteria Analysis for Assessing the Soil Erosion Susceptibility of the Western Himalayan Watershed. *Environ. Monit. Assess.* **2014,** *186* (12), 8391–8412.

Bocher, E.; Petit, G.; Bernard, J.; Palominos, S. A Geoprocessing Framework to Compute Urban Indicators. The MApUCE Tools Chain. *Urban Clim.* **2018,** *24*, 153–174.

Gowhar, M.; et al. Assessing the Influence of Watershed Characteristics on the Flood Vulnerability of Jhelum Basin in Kashmir Himalaya. *Nat. Hazards* **2015,** *77* (1), 153–175.

Gowhar, M.; Shakil, A. R.; Sadaf, A. Inferring Land Surface Processes from Watershed Characterization. In *Geostatistical and Geospatial Approaches for the Characterization of Natural Resources in the Environment*; Springer: Cham, 2016; pp 741–744.

Horton, R. E. Erosional Development of Streams and Their Drainage Basins: Hydrophysical Approach to Quantitative Morphology. *Bull Geol. Soc.* **1945,** *56*, 275–370.

Laks, I.; Sojka, M.; Walczak, Z.; Wróżyński, R.; Possibilities of Using Low Quality Digital Elevation Models of Floodplains in Hydraulic Numerical Models. *Water* **2017,** *9* (4), 283.

Mangan, P., Haq, M.A. & Baral, P. Morphometric analysis of watershed using remote sensing and GIS—a case study of Nanganji River Basin in Tamil Nadu, India. Arab J Geosci 12, 202 (2019). https://doi.org/10.1007/s12517-019-4382-4

Mustafa, I. S.; Din, N. M.; Ismail, A.; Omar, R. C. Use of Geographic Information System (GIS) Tools as Network Planning Aid for Antenna Mounting in Remote Areas. In *2012 International Symposium on Telecommunication Technologies* 2012; pp 11–15.

Nag, S. K.; Chakraborty, S. Influence of Rock Types and Structures in the Development of Drainage Network in Hard Rock Area. *J. Ind. Soc. Remote Sens.* **2003,** *31* (1), 25–35.

Schumm, S. A. Evolution of Drainage Systems and Slopes in Badlands at Perth Amboy, New Jersey. *Bull. Geol. Soc. Am.* **1956,** *67*, 597–646.

Smith, K. G. Standards for Grading Texture of Erosional Topography. *Am. J. Sci.* **1950,** *248*, 655–668.

Staley, D. M.; Wasklewicz, T. A.; Kean, J. W. Characterizing the Primary Material Sources and Dominant Erosional Processes for Post-fire Debris-flow Initiation in a Headwater Basin Using Multi-temporal Terrestrial Laser Scanning Data. *Geomorphology* **2014,** *214*, 324–338.

Strahler, A. N. Dynamics Basis of Geomorphology. *Bull. Geol. Soc. Am.* **1952,** *63*, 923–938.

Strahler, A. N. Quantitative Analysis of Watershed Geomorphology. *Trans. Am. Geophys. Union* **1957,** *38*, 913–920.

Strahler, A. N. Quantitative Geomorphology of Drainage Basins and Channel Networks. *Handbook of Applied Hydrology*; McGraw Hill Book Company: New York, 1964.

Verstappen, H. *The Applied Geomorphology*; International Institute for Aerial Survey and Earth Science: Enschede, 1983.

Vieceli, N.; Ortolin, T. A.; Mendes, L. A.; Bacarim, G.; Cemin, G.; Schneider, V. E. Morphometric Evaluation of Watersheds in Caxias do Sul City, Brazil, Using SRTM (DEM) Data and GIS. *Environ. Earth Sci.* **2015,** *73* (9), 5677–5685.

Wang, L.; Xiong, Y.; Wang, Z.; Qiao, Y.; Lin, D.; Tang, X.; Van Gool, L. Temporal Segment Networks: Towards Good Practices for Deep Action Recognition. In *European Conference on Computer Vision*, Springer, UK, 2016; pp 20–36.

CHAPTER 8

Web-Based Geospatial Approach for Identification of Groundwater Depletion: A Tool for Sustainable Water Resource Management

VINAY GANGARE[1], SUDARSHAN SINGH PARMAR[2], and VINEESHA SINGH[3*]

[1]*SISL Infotech, New Delhi, India*

[2]*School of Electronics Devi Ahilya University, Indore, MP, India*

[3]*Department of Earth Sciences, Barkatullah University, Bhopal, MP, India*

[*]*Corresponding author. E-mail: Vgangare@gmail.com*

ABSTRACT

This paper deals with the approach and the methods that are referred to some activities carried out in the framework of adapting the Web-based technology for sharing Geospatial data. This work aims to how the tabular-based information data can be converted into geospatial information and sharing through the Web, identifies groundwater depletion hotspots, and communicates them online. The study area covers some western parts of India, which are currently facing groundwater scarcity. According to the annual report of the CGWB of India, many blocks in Rajasthan, Madhya Pradesh, and Gujarat states have been declared as highly groundwater-depleted zones. This study gives a block-level scenario of groundwater depletion. The groundwater depletion hotspots were identified using the premonsoon and postmonsoon groundwater level data of the Central Ground Water Board by geospatial analysis.

GIScience for the Sustainable Management of Water Resources. Gowhar Meraj, Shruti Kanga, Majid Farooq, Suraj Kumar Singh, & Sudhanshu (Eds.)
© 2023 Apple Academic Press, Inc. Co-published with CRC Press (Taylor & Francis)

Distributed or Web-based GIS applies the dynamic client/server concept for performing GIS analysis tasks. Web-enabled GIS provides an ideal structure for sharing online geospatial data, and with geo-enabled processing, users can access geospatial information through the Web interface. A GIS-based approach is capable of taking information by extracting features and objects from different satellites Images and distributed GIS approach is to share this information with the help of Web technology with various geoprocessing algorithm to fulfill users and decision maker's requirement. The Web-based interface specification provides mechanisms to identify the spatially referenced data required by the process, initiate the process, and manage the output from the process so that the client can access it.

8.1 INTRODUCTION

The Web-GIS is an Internet-based GIS. Many different terms such as Internet GIS, GIS online, Distributed Geographic Information (DGI), and Web-based GIS are used to describe Web-GIS. Web-based GIS and the Internet are not necessarily synonymous. The Web-based GIS uses the Web as a client, but Internet GIS does not necessarily use the Web as the sole client; it can use another client. There are many Web Services for processing Geospatial data online (Michaelis and Daniel, 2009).

WPS has not been widely adopted yet, but it has potential in the broader context of geoinformation (GI). The Spatial database may not be in the form of Geospatial (Foerster and Stoter, 2006). It may be in other forms like Excel sheets and tabular data linked to the surface of the earth. This means the information that contains some geospatial information as standard data GIS and Remote sensing technology can change the way and represent it in the Geospatial form for user interaction. Groundwater is water mainly derived from rain and snowmelt that infiltrates the topsoil and slowly percolates to the water table and is found deep in the earth in the fissures and spaces in soil, sand, and rock. Natural recharge is a process in which water is injected into the underground reservoir. Aquifers soil, sand, and rock layers located underground that hold groundwater and slowly release it over time. Sand, sandstone, and fractured rock, like limestone, are among the aquifers that may be present. This kind of permeable material has large spaces of interconnected material that allow water to pass through. It changes according to topography and climate, depth of the water table below the surface. In moist or semiarid areas, the water table is generally between 0 and 50 ft underneath the ground surface. The water table in some desert areas can be literally thousands of feet below ground. In this study, we studied the block level spatial variations of groundwater depletion in some western parts of India (including Gujrat, Rajasthan, Haryana, and Punjab from west Madhya Pradesh). We used Kriging, a spatial interpolation technique of Inverse Distance Weighted (IDW), to estimate the spatial distribution of the groundwater depletion on the block level. The groundwater depletion maps of the entire study area will be prepared using GIS spatial interpolation techniques, which will help monitor and manage the groundwater to a safe level. After the spatial interpolation technique, the generated spatial layer deploys through geoserver and layer.

8.2 GROUNDWATER DEPLETION

Depth to water level is generally between 10 and 40 m in large parts of the North-West states. Water levels have been measured as deep as 20–40 m bgl in the east, with measurements of greater than 40 m bgl in the west. The water level in Delhi and Rajasthan can be as high as 40 m bbl. A comparison of the depth of water level in the Northwestern part of the country during the monsoon of 2013 with decadal averages discloses that the water level has decreased in the Northwestern part of the country (CGWB, 2009). The state of Rajasthan, Punjab, Haryana, and Delhi are the states most likely to experience a decline in the water level of at least 4 m. Looking at the water level from November 2013 (or Plates XX) in comparison to the average level from 2002 to 2012, it appears that on the whole, the water level has declined in the northwestern, eastern, and northeastern regions of the country (Government of India, 2012).

8.3 CONCEPT OF WEB GIS

Desktop GIS refers to either a standalone program with no information exchange between computers or share these GIS resources within Local Area Network (LANs). GIS programmes that work on the Internet or in wireless networking environments are referred to as Web GIS (Open Geospatial Consortium, n.d.). GIS access to data, spatial information exchange, and GIS modeling/processing are all impacted by the internet. It has been widely accepted in governmental, nongovernmental agencies, GIS professionals, GIS vendors, and Defense and Security and Educational institutions (Figure 8.1).

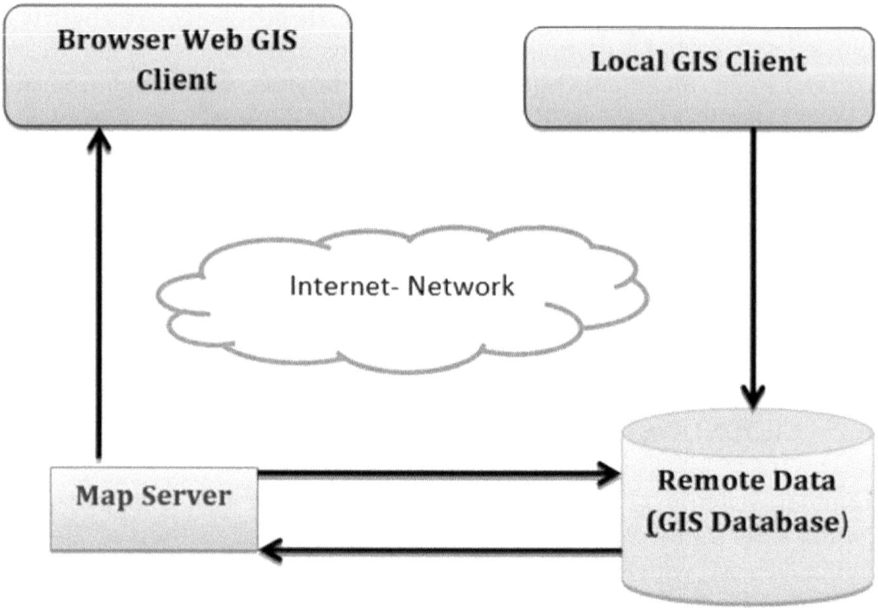

FIGURE 8.1 Concept of Web GIS.

8.4 STUDY AREA

The study area is chosen for the regional scale groundwater depletion analysis in northwest India, including Ganga, Yamuna, Chambal, Narmada, Luni, and Indus basin. The area is bounded by 75°36′43.42″E and 74°20′8.89″E longitudes and 25°13′19.15″N and 23°45′52.82″N longitudes covering an area of 191,878.40 sq. km (Figure 8.2). The state of Punjab (50,362 sq. km) and the Union Territory of Chandigarh (114 sq. km) lies in the Indo-Gangetic plains. The two broad geomorphic entities, the Siwalik foothills in the north-eastern part of Punjab and alluvial fills of the Indus drainage basin, physiographically characterize the area. The area is covered by the Siwalik Super group's rocks (Tertiary) and Quaternary Alluvium and Aeolian Deposit. The state of Haryana (44,212 sq. km), is northwest bounded by Punjab, Himachal Pradesh, and Uttarakhand in the north and northeast, Delhi and Uttar Pradesh in the east and southeast, and by Rajasthan in the west and southwest.

The area is physiographically defined by two broad geomorphic entities: the Siwalik foothills in northeastern Punjab and alluvial fills of the Indus drainage basin. The rocks of the Siwalik Supergroup (tertiary) and the Quaternary Alluvium and Aeolian Deposit cover the area. Structural and

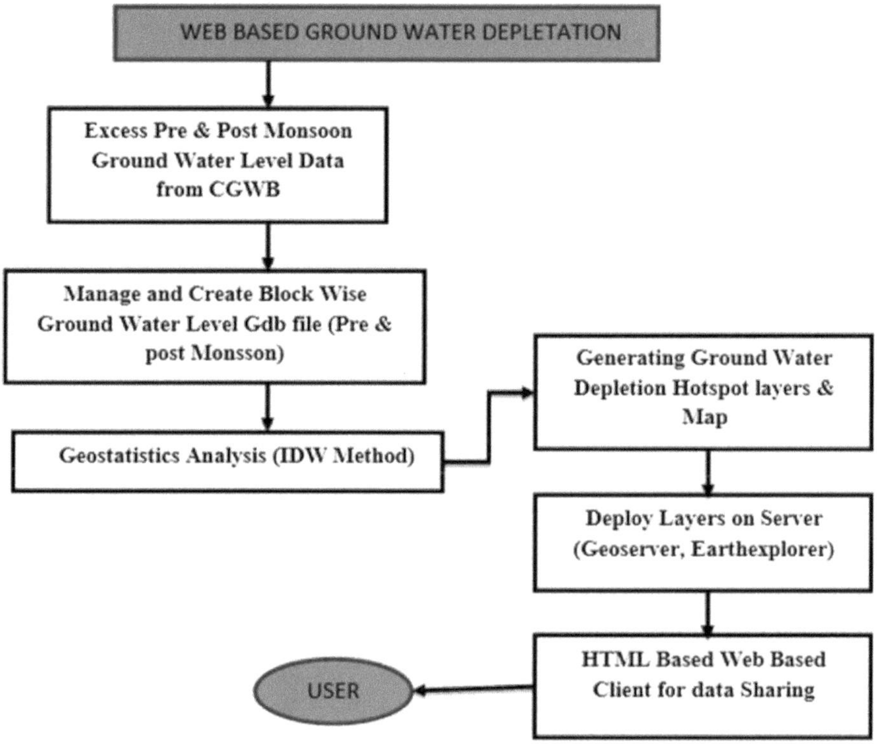

FIGURE 8.2 Method for sharing spatial information on the Web.

denudation hills, alluvial plains, and aeolian plains have been identified as three major geomorphic units. Nearly 95% of the area is covered by quaternary alluvium and aeolian sediments, while the rest is made up of proterozoic rocks from the Delhi Supergroup and Tertiary rocks. Delhi is 1483 sq. km in size and is bordered on the north, west, and south by Haryana, and on the east by Uttar Pradesh. It has mostly flat terrain, with the exception of an NNESSW-trending Delhi Ridge in the southeast. Rajasthan is home to Achaean, Metamorphites, and modern alluvium and wind-blown sand, as represented by the Bhilwara Super Group (over 2500 million years old) (Chopra, 1990). Rajasthan's western and northwestern regions are covered in Thar desert (Marusthal) unconsolidated deposits. Precambrian metamorphic schist, quartzite, marbles, and gneisses, along with acid and basic intrusives Sedimentary rock includes the Mature Granulite series, which includes rocks from the Aravalli Super Group, Delhi Super Group, and upper Precambrian Vindhyan Super Group, as well as rocks from the Cambrian to Jurassic,

Cretaceous, and Tertiary ages. The state's southeasternmost region is dominated by lava flows from the Cretaceous Deccan Traps, which formed the modern-day city of Hyderabad, India. Along with the aforementioned rock units, several economic mineral deposits can be found.

8.5 DATA AND METHOD

During January, April-May, August, and November, the level of groundwater is measured by the Central Groundwater Board four times a year. The surveillance program was first established in 1969 by the Central Groundwater Board. Fourteen thousand six hundred seventy-four observation wells are dug wells, while 6024 are piezometers. There have been approximately 14,391 wells to monitor and analyze during January 2014. In April and May of each year, groundwater samples are obtained from observation wells to obtain data on groundwater quality change on a regional scale. The database generated is the basis for developing the groundwater management policies. We have analyzed the data from 2005 to 2014 premonsoon and postmonsoon data that was available on the CGWB website; now, the data can be accessed from Water Resources Information System (India-WRIS) (*India—Water Resources Information System*, n.d.). The principle of geostatistical analysis is to create (interpolate) a surface from sample points taken at various locations throughout a landscape (ESRI, 2012). It is much easier to make a continuous surface or map using the geostatistical analysis to take measurements stored in a point feature layer, raster layer, or polygon centroids (Figure 8.3a and b). Measures such as elevation, depth to the water table, or pollution levels could be used as sample points.

During this work, data were downloaded by the CGWB (Central Ground Water Board) in the Excel sheet form, and for sharing data online, there is a two-way method applied for online data sharing; first and foremost, Geostatistics analysis is in ArcGIS software. Second is the Web-based approach for sharing the information online Geoserver and Earthexplorer (Figures 8.4 and 8.5), open-source software were used during the development of Web client or user interface (Osorio and Carlos, 2011).

8.6 RESULTS

The groundwater depletion has been estimated using the data obtained from the Central Ground Water Board. The results so obtained have been shown

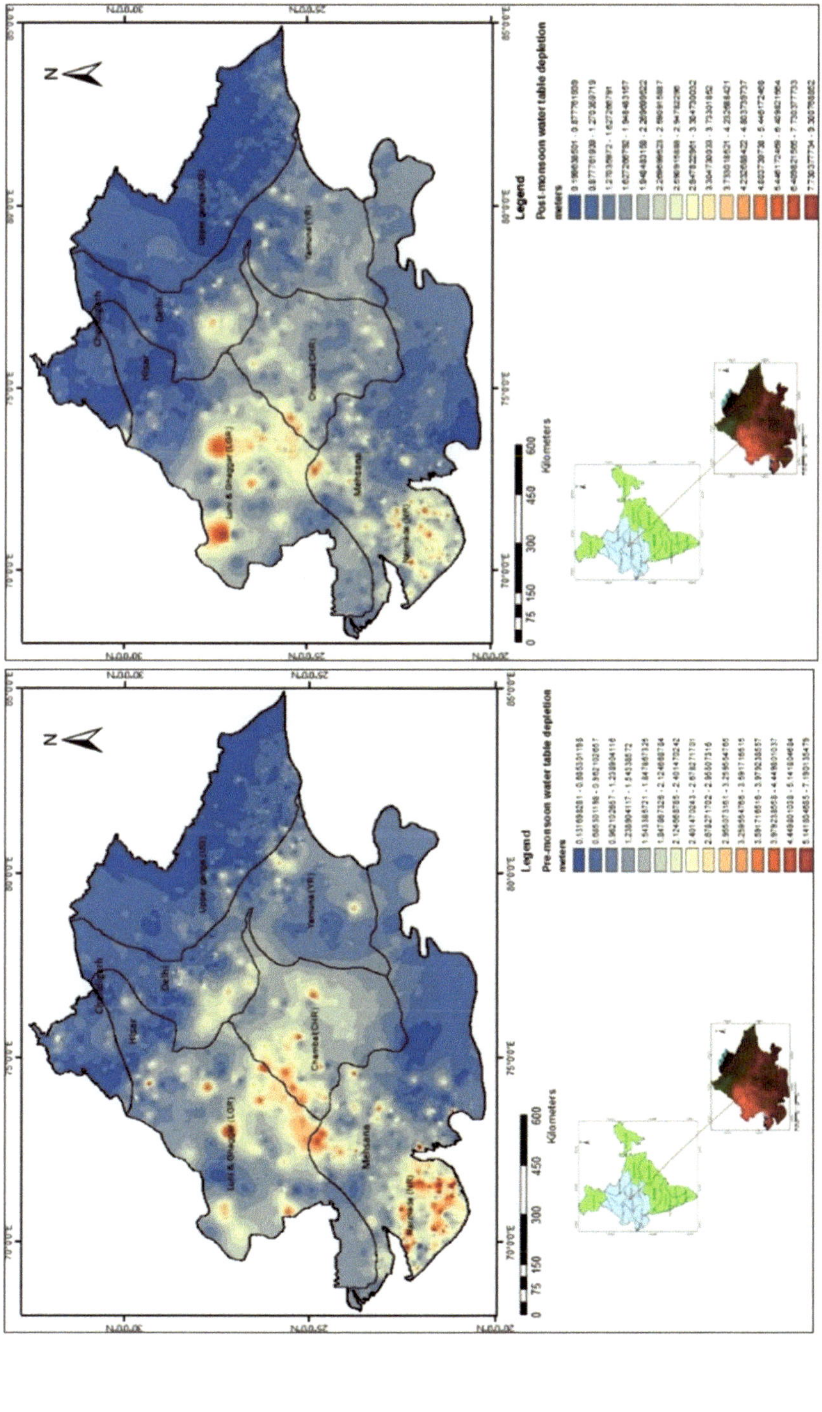

FIGURE 8.3 (a) Premonsoon groundwater and (b) postmonsoon groundwater depletion map for water table data.

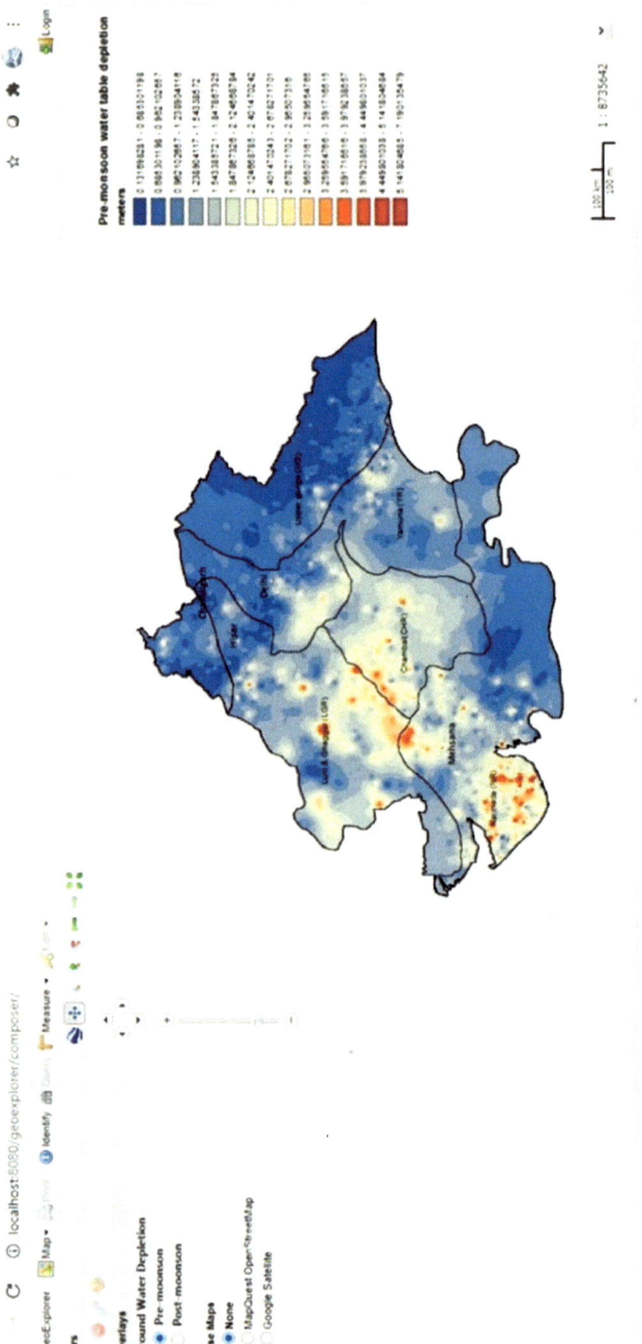

FIGURE 8.4 Premonsoon groundwater depletion layer visualization over the Internet.

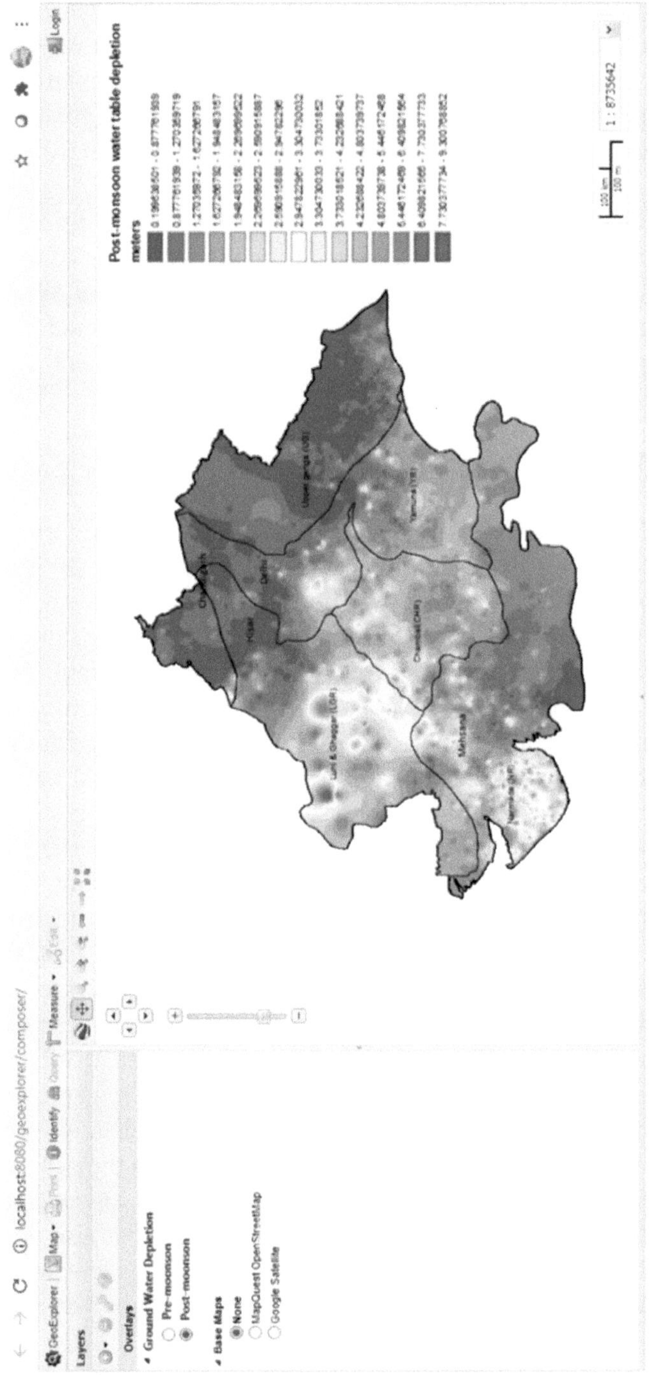

FIGURE 8.5 Post monsoon groundwater depletion layer visualization over the internet.

below. Pre- and postmonsoon depletion maps of groundwater have been prepared to visualize the status of groundwater level in northwest India. The groundwater fluctuation map depicts that the maximum depletion has been happening in the northern Gujarat, Rajasthan, Delhi NCT, Haryana, and Uttar Pradesh. The complete depletion can be perceived in Gujarat, Rajasthan, and Delhi.

As per this study, our main objective was to share this scientific output on the Web using open source software, after the derived out of groundwater depletion, pre- and postmonsoon layer was examined and shared on the Open-GIS to visualize the layers. Many tasks can be performed in the online platform, that is, overlapping, comparison, print, measurement, etc.

8.7 CONCLUSION AND DISCUSSION

In this study, Web-based GIS's capability for detecting and monitoring depletion of groundwater was demonstrated. The spatial analysis of groundwater fluctuation was done by Geostatistical analysis. It gave a good matching between groundwater depletion for the study area. Some of the areas with high groundwater levels decline. As per the water supply view, groundwater is the major source in urban and rural areas, and assessment of it involves identifying the hazard associated with it (Smith, 2001). The Web-based approach is to sharing this geospatial data awareness of groundwater depletion and its effect. After collecting all the geoinformation using satellite-based information and field research in various application types such as disaster management, snow and glacier, geology, wildfire monitoring, defense and logistics, urban and rural information system, hydrology, agriculture, and forestry. The extracted spatial information or results in maps, table data, raster or vector, and representation of geospatial data frozen in time. Geospatial data over the Web allow the user to access the scientific output associated with spatial and nonspatial RDBMS technologies (Singh et al., 2012). Geoserver over the Web represented the situation when the data were gathered (Yang et al., 2002). Though useful in its way, it is limited in its application scope unless the features and results represent on the Web for sharing the geoinformation. The integration of geospatial information with internet environment online maps and information has created a consumer market that also demands the user, government, nongovernment officials, decision-makers who interface with the data.

KEYWORDS

- **geospatial**
- **groundwater**
- **web-based GIS**
- **aquifers soil**
- **excel sheets**

REFERENCES

CGWB. *Dynamic Groundwater Resources of India*; Central Groundwater Board, Ministry of Water Resources, Government of India, 2009.

Chopra, S. A Geological Cum Geomorphological Framework of Haryana and Adjoining Areas for Landuse Appraisal Using Landsat Imagery. *J. Indian Soc. Remote Sens.* **1990,** *18* (1–2), 15–22. doi:10.1007/BF03017813.

Foerster, T.; Stoter, J. Establishing an OGC Web Processing Service for Generalization Processes. In *Workshop of the ICA Commission on Map Generalisation and Multiple Representation*, 2006.

Government of India. *Aquifer Systems of India*; Central Groundwater Board, Ministry of Water Resources, 2012.

India—Water Resources Information System, n.d. https://indiawris.gov.in/downloads/wris-Brochure.pdf

Michaelis, C. D.; Daniel, P. A. Evaluation and Implementation of the OGC Web Processing Service for Use in Client-side GIS. *Geoinformatica* **2009,** *13*, 109–120.

Open Geospatial Consortium. Open Geospatial Consortium, n.d. [Online]. Available: http://semanticommunity.info/A_QuintCross_Information_Sharing_and_Integration/Open_Geospatial_Consortium.

Osorio, M.; Carlos, A. *Parallelization of Web Processing Services on Cloud Computing: A Case Study of Geostatistical Methods*; Germany, 2011.

Singh, P. D.; Chutia, D.; Sudhakar, S. Development of a Web Based GIS Application for Spatial Natural Resources Information System Using Effective Open Source Software and Standards. *J. Geogr. Inf. Syst.* **2012,** *4* (3), 261–266. doi:10.4236/jgis.2012.43031.

Smith, K. *Environment Hazards: Assessing Risk and Reducing Disaster* (3rd ed.); London: Routlege, 2001; p 324.

Yang, B.; Tian, Y.; Zhu, Z. *A Study on Applications of Future GIS Technologies. Microcomputer Applications*, Department of Automation, Shanghai Jiaotong University Shanghai, 2002.

CHAPTER 9

Assessing the Risk of Watersheds to Soil Erosion Using GIScience

UMAIR ALI[1*], UMMAR AHAD[2], SYED AHMAD ALI[3], MUKEEM AHMAD[3], ABAAS AHMAD MIR[4], and TANZEEL KHATOON[3]

[1]*Department of Earth Sciences, University of Kashmir, Srinagar, Kashmir, India*

[2]*Centre of Central Asian Studies, University of Kashmir, Srinagar, Kashmir, India*

[3]*Department of Geology, Aligarh Muslim University, Aligarh, UP, India*

[4]*ICSSR, New Delhi, India*

[*]*Corresponding author. E-mail: umairgeo121@gmail.com*

ABSTRACT

The calculations of terrain aspects are a crucial tool to estimate the soil erosion risk of any watershed. The Lolab watershed was separated into 12 subwatersheds (LSB1 to LSB12) to evaluate subwatershed-wise soil erosion behavior using digital elevation model in Arc GIS 10.2 and Global Mapper software. The Lolab watershed is a sixth-order basin with varied drainage pattern from dendritic to subdendritic. Linear, shape, and relief aspects were calculated and given a rank based on the value/relationship with erodibility, to delineate more erosion prone areas. The area was categorized into four classes, namely, low, medium, high, and very high based on alarm of soil conservation and management. Ranking based on compound value exposes that subwatersheds LSB8 & LSB9 fall in very high priority with respect to

GIScience for the Sustainable Management of Water Resources. Gowhar Meraj, Shruti Kanga, Majid Farooq, Suraj Kumar Singh, & Sudhanshu (Eds.)
© 2023 Apple Academic Press, Inc. Co-published with CRC Press (Taylor & Francis)

soil erodibility. The soil erosion chances in these high priority subwatersheds are more because of steep slopes, deep valleys, and high altitudinal, topography which increase upper loose soil cover erosion underlain by highly structured lithology. The low priority LSB1, LSB5, and LSB10 may be covered by dense forest, hence are less prone to soil erosion. The morphometric calculations and other terrain aspects showed their significance in the measurement of soil erosion behavior suggested for initiating soil conservation and landslide mitigation activities, and resultantly water conservation. The present work is mainly focused to recognize vulnerable areas to soil erosion necessary to begin defensive procedures to minimize the ill-effects originated due to soil erosion.

9.1 INTRODUCTION

Soil and water conservation is a key issue in watershed management based on delineation of watersheds. The watershed management and conservation is performed by linking different factors like geomorphology, land use, soils, slope, uplands, and low lands. Morphometry is the measurement and mathematical analysis of the configuration of the earth's surface, shape, and dimension of its landforms (Agarwal, 1998; Obi Reddy et al., 2002). It is well established that analysis of morphometric parameters is very important in characterizing the landform processes, soil physical properties, and erosional characteristics. Morphometric analysis is performed by computation of linear, aerial, and relief parameter of the basin and slope contribution (Nag and Chakraborty, 2003).

Presently, geospatial system is applied to describe diverse terrain and morphometric parameters of the basins. The investigation of geospatial data highlighting critical areas boosted by ground truths makes it possible to recognize and delineate different features, such as geological structures, geomorphic features, and their hydraulic characters (Ali et al., 2003; Ali and Ali, 2017). Prioritization of watershed, that is, the ranking in terms of soil erosion was carried out in order to take conservation and mitigation measures. Consideration of watershed conservation and management work, it is not helpful to analyze the whole drainage basin at once (Altaf et al., 2013, 2014). Therefore, the watershed has been divided into subwatersheds in order to describe the characteristics of drainage system. In the area, categorization of subwatersheds is performed by applying geospatial techniques that could be utilized for the conservation and management strategies in terms of soil erosion risk in the area (Meraj et al., 2015, 2016).

9.2 GEOLOGICAL SETUP

The Lolab valley is subvalley in Himalaya situated in the Kashmir Valley in NW corner. Lolab valley is surrounded in the south, north, and west by the Kashmir valley, the Neelum valley, and Nagmarg meadows from Bandipora, respectively. The Lolab valley lies between 34°43′30″–34°24′0″ N latitudes and 74°15′0″–74°39′0″ E longitudes and formed from east to west of Lahwal (Lalkul) river (Figure 9.1). The length of trunk stream flowing in a westerly direction is about 30 km and drains an area of 447 km^2. Three distinct physiographic units can be observed, that is, High Karewa Plateau lands, sharp hard rock ridges, and intervening valleys and river terraces. The mountain ridges have a NW to SE direction and the master slope is toward the southeast. The area shows the presence of conical ridges with very steep to steep slope distribution. The lithological variation is observed in the form Zewan beds/Gondawana formations, Dogra Slates, Panjal Traps, Karewa formations, and Quaternary alluvium ranging in age from Cambro–Silurian to recent, respectively (Raza et al., 1978).

9.3 DATABASE AND METHODOLOGY

To assess the morphometric conditions, satellite imagery and SRTM digital elevation model (DEM) were used supported by the existing geological map. The ERDAS Imagine 9.1 and Arc GIS 10.2software were applied in georegistration as well as for georectification, image processing of geospatial data. Arc GIS 10.2 software was used for digitization, computation of spatial as well as attribute database for the drainage system characteristics, lithology, and lineament analysis. Different thematic maps like slope, aspect, elevation, and contour map are generated from DEM that was integrated by overlay technique in GIS environment for updating the stream network and analyzing other terrain characteristics. The parameters like basin area and perimeter, length and number of streams were computed automatically in GIS environment (Singh and Pandey, 2014; Bhatt et al., 2017; Sharma and Kanga, 2020). The other parameters are computed using the formula suggested by Horton (1945), Strahler (1964), Schumm (1956), Nooka Ratnam et al. (2005), and Miller (1953) (Table 9.1).

Finally, each individual parameter is given knowledge-based weightage by interpreting its influence on soil erosion and land degradation. Based on the compound factor of subwatersheds, the highest priority is given to the subwatershed having highest compound value. Thus, the priority values obtained were categorized into four classes, that is, very high, high, medium, and low priority, which was correlated with the lineament map and 3D-terrain map for confirmation of results. The high priority indicates critical subwatersheds that need soil conservation and mitigation measures.

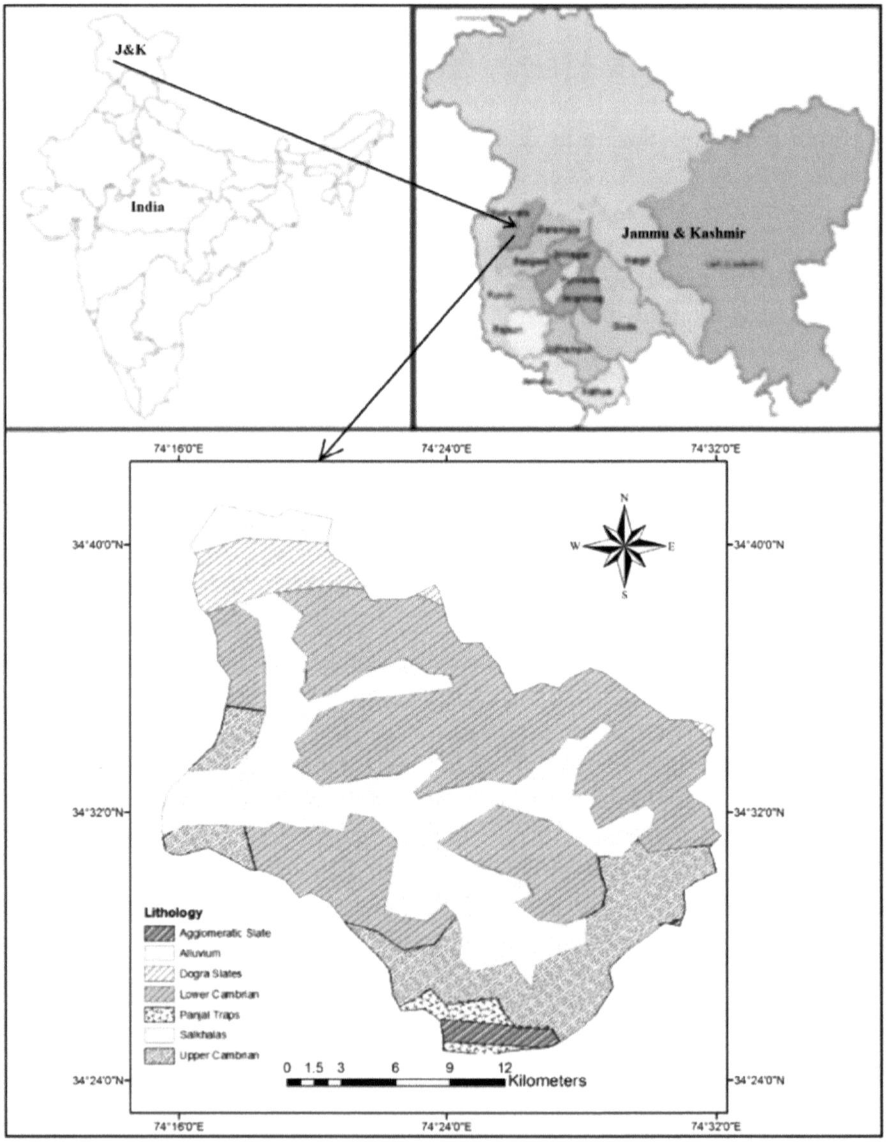

FIGURE 9.1 Lolab valley location map with lithological variation.

TABLE 9.1 Morphometric Parameters and Their Mathematical Expressions.

S. No.	Morphometric parameters	Formula	References
1	Stream order	Nu	Strahler (1964)
2	Stream length	Lu	Horton (1945)
3	Bifurcation ratio (Rb)	Rb = Nu/Nu + 1	Horton (1932)
4	Stream frequency (Fs)	Fs = Nu/A	Horton (1945)
5	Drainage density (Dd)	Dd = Lu/A	Horton (1945)
6	Texture ratio (T)	T = Nu/P	Smith (1950)
7	Form factor (Ff)	Ff = A/LSB2	Horton (1945)
8	Relief ratio (Rh)	Rh = H/L	Schumm (1963)
9	Ruggedness number (Rn)	Rn = Bh × Dd	Schumm (1956)
10	Elongation ratio (Re)	$Re = 2/LSB) * (A/Pi)^{0.5}$	Schumm (1956)
11	Circularity ratio (Rc)	$Rc = 4 * Pi * A/P^2$	Miller (1953)
12	Compactness coefficient (Cc)	Cc = 0.2821P/A0.5	Horton (1945)
13	Length of overland flow (Lo)	Lo = 1/Dd * 2	Horton (1945)
14	Shape factor (Bs)	$Bs = LSB^2/A$	Horton (1956)

9.4 RESULTS AND DISCUSSION

The present study was carried out in three sections: the first was delineation and updation of stream numbers, stream order, and stream lengths using SRTM DEM (30 m) in GIS environment. The second was the computation of various linear, shape, and relief parameters and the third was the categorization of area based on value of linear, shape, and relief aspects.

9.4.1 STREAM ORDER (U)

The basin drains an area of 447 km^2 with sixth-order stream. Among 12 subwatersheds, LSB1, LSB2, LSB3, LSB6, LSB9, LSB10, and LSB12 are fifth-order streams, whereas as LSB4, LSB5, LSB8, and LSB11 are having fourth-order streams and LSB7 is having third-order stream. The structural and physiographic scenario of the area largely affects the stream order and size of subwatersheds (Figure 9.2).

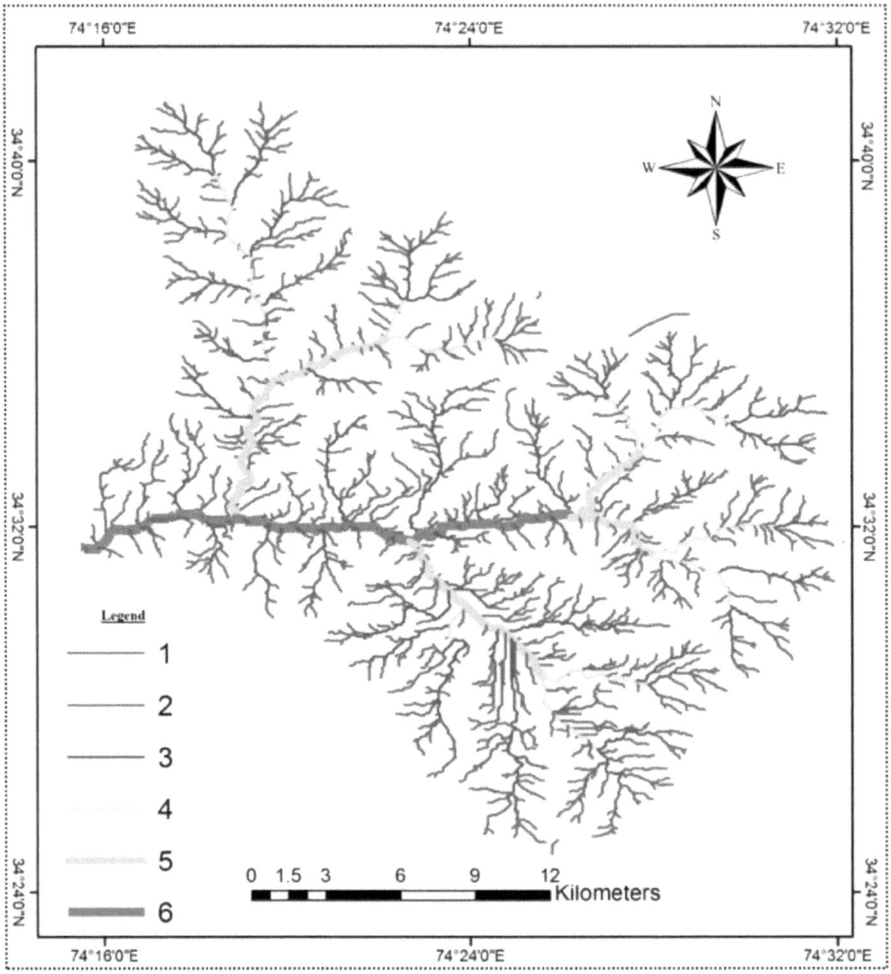

FIGURE 9.2 Subwatersheds of Lolab valley and their drainage network.

9.4.2 *STREAM LENGTH (LU)*

The stream length was calculated directly in ArcGIS. The length of stream follows Horton's law in all 12 subwatersheds. As a rule, the total length of stream segments decrease as the stream order increase. The LSB1 shows the highest Lu, and LSB8 shows the lowest Lu. The reason for this variation can be result of flowing of streams from high altitude, lithological variations, and moderately steep slopes (Singh and Singh, 1997).

9.4.3 STREAM FREQUENCY (FS)

Normally, high Fs is attributed to impermeable subsurface material, sparse vegetation, high relief conditions, and low infiltration capacity (Reddy et al., 2004). The Fs value showed variation from 3.04 to 6.32 per km^2 for 12 subwatersheds. Among 12 subwatersheds, LSB11 shows highest Fs followed by LSB1, LSB10, LSB5, LSB2, and LSB3, respectively, all having Fs > 4. Higher Fs reflects greater surface run-off and a steeper ground surface. Subwatershed LSB1, LSB10, LSB5, LSB2, and LSB3 have high Fs, indicating that these subwatersheds generate more water flow at the time of high-intensity rainfall. High-stream frequency associated with high relief conditions and more water flow makes these subwatersheds prone to soil erosion.

9.4.4 DRAINAGE DENSITY (DD)

It is a key indicator of linear scale of landform elements in a stream-eroded topography (Horton, 1945). Drainage density (Dd) indicates the closeness of spacing of channels, thus determines the time of travel by water (Horton, 1932). In the area, the Dd values range from 2.35 to 3.8 km/km^2. The LSB1 shows highest Dd followed by LSB11, LSB10, LSB2, LSB3, and LSB6, respectively, the reason is mountainous relief and impermeable subsurface material, thus these subwatersheds will offer more runoff in a short period of time, which ultimately cause flooding in lower areas (Pandey et al., 2010, 2013). Therefore, quicker runoff process of mountainous subwatersheds indicates more chances of soil erosion.

9.4.5 DRAINAGE TEXTURE (T)

It is a product of Dd and frequency. Smith (1950) classifies *T* as coarse (<4 per km), intermediate (4–10 per km), fine (10–15 per km), and ultrafine (>15 per km). In the area, texture shows variation from 7.42 to 22.82 signifying intermediate to ultrafine texture, respectively. The subwatersheds having fine texture with high relief may have more chances of erosion.

9.4.6 LENGTH OF OVERLAND FLOW (LO)

It is taken as half the reciprocal of Dd (Horton, 1945). Length of overland flow (Lo) relates inversely to the average channel slope, thus it is an important independent parameter that has great influence on hydrological and physiographical development of the area. The subwatersheds of Lolab valley showed Lo range from 0.52 for LSB1 to 0.84 for LSB8. The larger Lo for LSB8 infers slower runoff process and shorter Lo for LSB1 infers the quicker runoff process. The higher length of overland flow and slower runoff give more time for mud to settle brought during flooding from high altitude subwatersheds.

9.4.7 FORM FACTOR (FF)

Form factor (Ff) represents the shapes and outlines of a basin (Horton 1945; Gregory and Walling, 1973). The value of Ff would always be greater than 0.78 for a perfectly circular basin, and the basin will be more elongated if the value turns to be smaller. The Ff value for 12 subwatersheds ranges from 0.28 for LSB8 to 1.15 for LSB11. Except LSB11, all the subwatersheds have Ff valueless than 0.78 depicting their elongated shape. This elongated shape of subwatersheds with lower Ff value highlights that they have a flatter peak flow for longer duration, thus exposing more area to soil erosion for a longer time.

9.4.8 ELONGATION RATIO (RE)

It is a measure of the shape of the basin and it depends on the climatic and geologic types. A circular basin is more efficient than an elongated basin in runoff discharge (Singh and Singh, 1997). The elongation ratio (Re) normally ranges from 0.6 to 1.0. The basins which show ratio variation of 0.6–0.8 are attributed with strong relief and steep ground slopes, whereas the ratio range of 0.8–1.0 is typical of regions with low relief (Strahler, 1964). The Re value varies from 0.60 for LSB8 to 1.21 for LSB11. The LSB2, LSB3, LSB4, LSB5, LSB6, and LSB12 have Re value range between 0.8 and 1, thus indicating low relief that results in lower run-off. Therefore, these low relief and low lying subwatersheds show less soil erosion, whereas the Re value for LSB1, LSB7, LSB8, LSB9, andLSB10 ranges between 0.6 and 0.8, thus showing high relief and steep ground slopes that could be the cause of soil erosion during harsh weather. The area showing lower Re ratio is described by high susceptibility to erosion and sediment load (Reddy et al., 2004).

9.4.9 COMPACTNESS COEFFICIENT

It can be represented as basin perimeter divided by the circumference of a circle to the same area of the basin. This parameter is directly related to erosion risk estimation, that is, lower compactness value indicates less vulnerability for risk factors, while higher values show great vulnerability, require mitigation, and preservative actions. In this watershed, 0.15 is highest while 0.07 is lowest that indicates that LSB5 and LSB7 is more elongated as compared to LSB1.

9.4.10 BASIN RELIEF (H)

It is defined as difference in elevation between the reference points, that is, the highest (divide) and lowest (outlet) elevation in the basin (Figures 9.3 and 9.4). The relief ranges from 700 to 1900 m for subwatersheds. The high-relief subwatersheds are more in number than the low relief. The high relief shows low gravity of water flow as well as infiltration and high runoff conditions as well as sediment down the slope.

9.4.11 RELIEF RATIO (RH)

Relief ratio (Rh) is a measure of overall steepness of the area. Rh is an important indicator to measure the intensity of water flows from a slope in a catchment. The Rh ranges from 0.6 for LSB9 to 0.22 for LSB5 and LSB11. The higher the Rh, the higher is the surface runoff. In general, the Rh is the controlling factor for surface runoff. The LSB2, LSB4, LSB5, LSB10, LSB11, and LSB12 show high Rh that reveals steep slopes in the area. The subwatersheds with highest Rh have a higher peak flows with greater velocities (Altin and Altin, 2011). These subwatersheds contribute more water in a short time period and are more prone to soil erosion.

9.4.12 RUGGEDNESS NUMBER

It is the product of Dd and relief of the basin in the same unit (Strahler, 1957; Melton, 1958). The highest value was observed in LSB1, LSB2, LSB3, LSB10, and LSB11 in which high values of basin relief and Dd are computed, that is, these subwatersheds have very steep slope associated with its slope length. Thus, the above subwatersheds showing high slopes in upper reach give lower time of concentration of overland flow and increase the soil erosion potential.

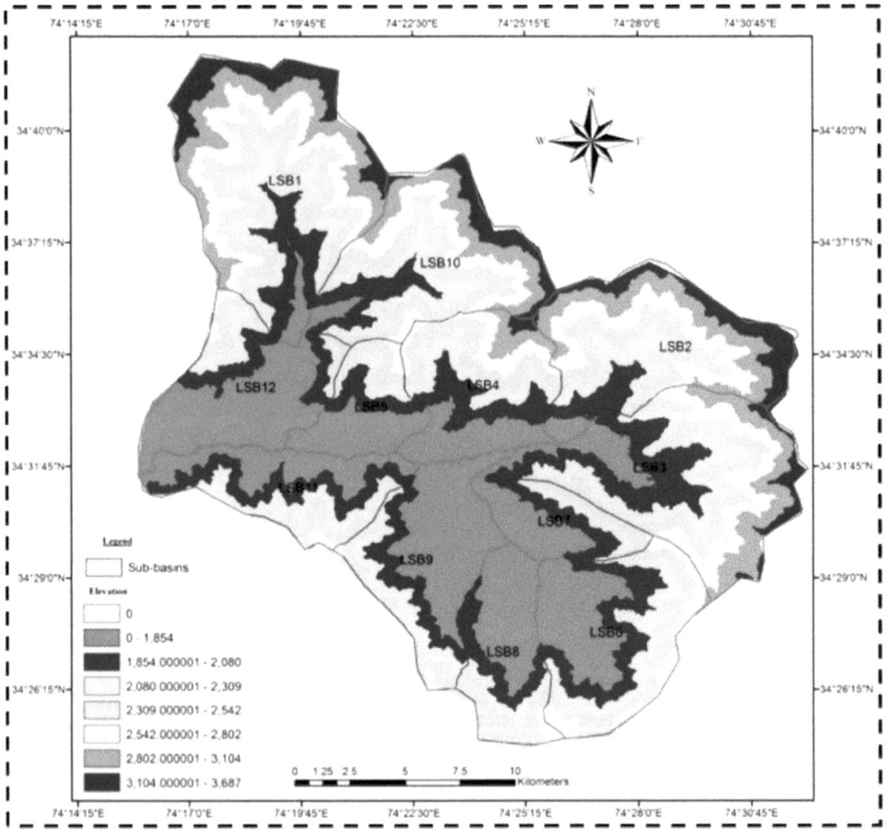

FIGURE 9.3 Variation of relief in Lolab valley overlain by subwatershed map.

9.4.13 RANKING OF MORPHOMETRIC PARAMETERS

The morphometric parameters computed are applied in erosion risk estimation and prioritization (Biswas et al., 1999). The parameters like stream frequency, Dd, drainage texture, Lo, and mean bifurcation ratio have a direct link with erodibility, whereas parameters like circularity ratio, Re, compactness coefficient, Ff, and basin shape have an inverse link with erodibility (Nooka Ratnam et al., 2005). Therefore, categorization was done with highest priority/rank to highest value of linear parameters and lowest value of shape parameters. The linear parameter with highest value was ranked 1, the second-highest as 2, and so on. The shape parameters have converse relation which means lower the value, higher is the erodability. Hence, the shape parameter with lowest value was given rank 1 and the second lowest as 2, and so on. Consequently, the addition of linear as well as shape parameters and then dividing by number of parameters were applied to calculate compound factor (Table 9.2). The subwatersheds were ranked, with the highest priority rank given to the subwatershed with lowest compound factor and so on (Figure 9.5).

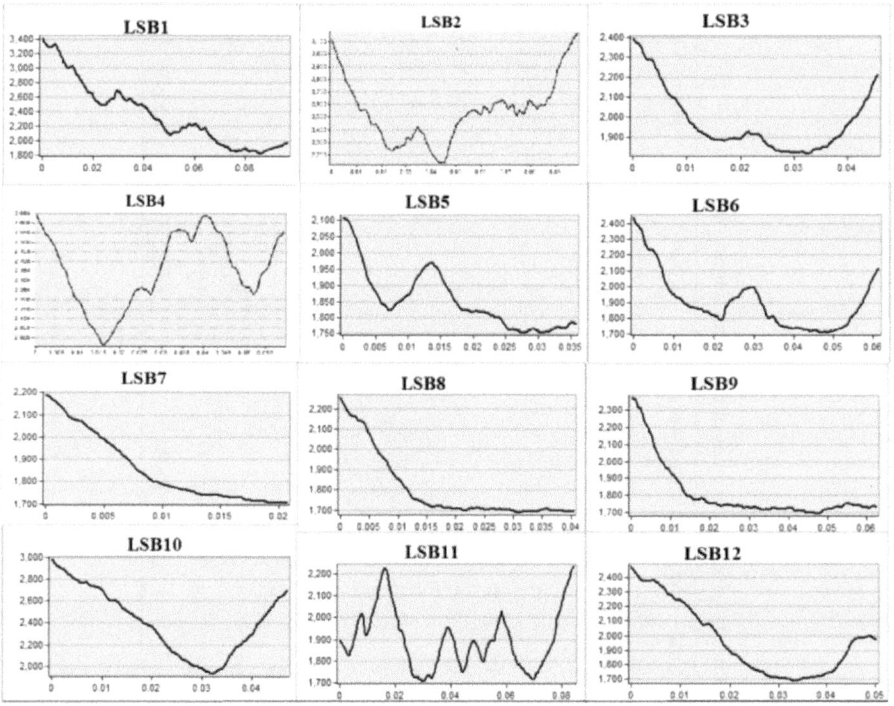

FIGURE 9.4 The topographic profiles plotted across all the subwatershed, as shown in Figure 9.3.

9.5 CONCLUSION

Computation of compound factor by the analysis of morphometric parameters in GIS environment provided considerable information for evaluating

TABLE 9.2 Priority Rank and Priority of Subwatersheds of the Lolab Valley.

	Priority	Low	Medium	Medium	High	Low	High	High	Very high	Very high	Low	Very low	High
Ordering of subwatersheds	Priority rank	8	5	5	4	7	4	4	2	1	6	9	3
	Compound factor	7.8	6	6	5.88	7.44	5.88	5.88	3.88	3.33	6.77	8.11	5
	Compactness constant (Cc)	1	3	2	4	8	3	8	7	5	3	6	5
	Elongation ratio (Re)	4	6	7	6	8	7	2	1	3	5	10	9
	Circulatory ratio (Rc)	11	6	5	9	8	12	10	4	1	7	2	3
	Shape factor (Fs)	8	6	4	6	3	5	10	11	9	7	1	2
	Mean bifurcation ratio (Rbm)	9	4	6	10	8	2	12	7	3	5	11	1
	Texture ratio (*T*)	12	8	9	4	5	7	2	1	3	11	10	6
	Form factor (Ff)	4	6	8	6	9	7	2	1	3	5	11	10
	Stream frequency (Fs)	10	7	6	3	8	4	4	2	1	9	11	5
	Drainage density (Dd)	12	8	7	5	10	6	3	1	2	9	11	4
	Subwatersheds	LSB1	LSB2	LSB3	LSB4	LSB5	LSB6	LSB7	LSB8	LSB9	LSB10	LSB11	LSB12

watershed characteristics as well prioritization and highlighted their relative characteristic with respect to hydrologic response. These morphometric parameters calculated using DEM helped in highlighting the hydrological behavior of a Lahwal (lalkul) river that influences the erosion-related activities in the Lolab basin.

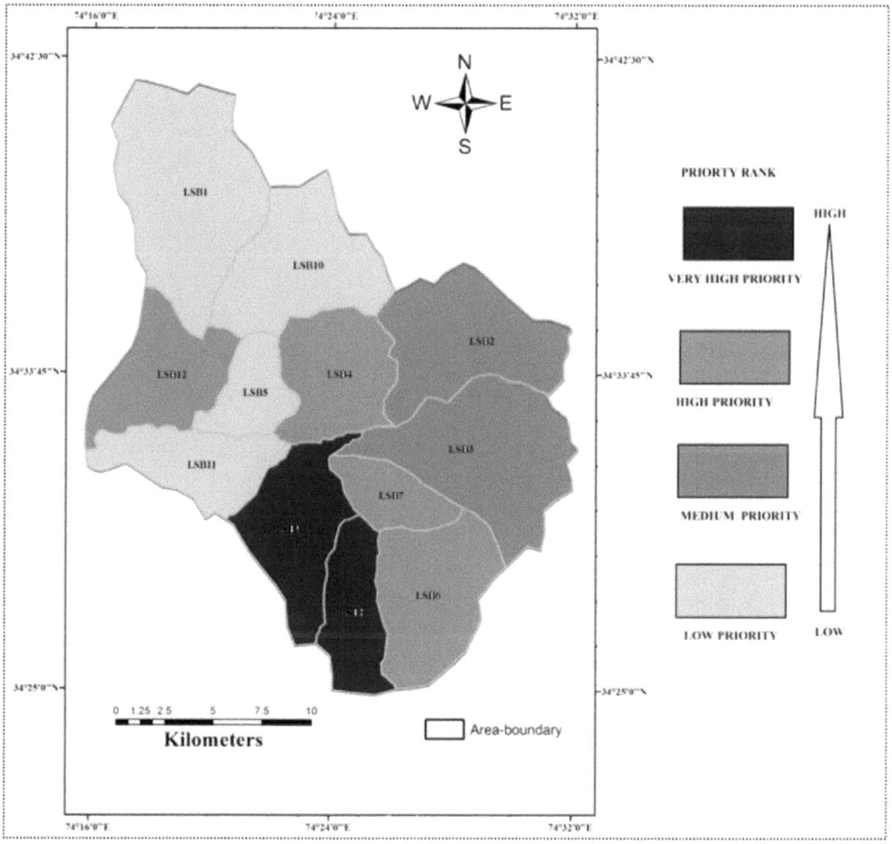

FIGURE 9.5 Prioritization classes and the area under different priority zones.

Results of prioritization reveal that LSB8 and LSB9 subwatersheds have been kept in very high priority zone with respect to soil erosion susceptibility. Thus, immediate attention should be given to preserve the land from further erosion and to reduce natural hazard. The LSB4, LSB6, LSB7, and LSB12 subwatersheds fall under high priority and thus need to be taken care of, to prevent any kind of soil erosion. The chances of land sliding increases with erosion of upper loose cover due to the presence of weak planes (structural features), and seepage of water through these weak planes in hard rock terrain act as lubricant to enhance the movement. On the other hand, siltation (mud) problems are more in lower plainer subwatersheds as it is faced in every flood event. Thus, the examination of various morphometric parameters can successfully be applied to highlight watershed behavior and prioritization liked to natural hazards like soil erosion. In order to prevent any kind of soil erosion, mitigation measures should be taken into consideration to protect the agricultural land, crops, and human settlements in Lolab valley.

KEYWORDS

- **morphometric analysis**
- **compound parameter**
- **geospatial environment**
- **soil erosion**

REFERENCES

Agarwal, C. S. Study of Drainage Pattern through Aerial Data in Naugarh Area of Varanasi District, U. P. *J. Indian Soc. Remote Sens.* **1998,** *26*, 169–175.

Ali, S. A.; Rangzen, K.; Pirasteh, S. Use of Digital Elevation Model for Study of Drainage Morphometry and Identification of Stability and Saturation Zones in Relations to Landslide Assessments in Parts of Shahbazan Area, Zagros Belt, SW Iran. *Cartography* **2003,** *32* (2), 162–169.

Altaf, F.; Gowhar, M.; Shakil, A. R. Morphometric Analysis to Infer Hydrological Behaviour of Lidder Watershed, Western Himalaya, India. *Geogr. J.* **2013**. https://doi.org/10.1155/2013/178021.

Altaf, S.; Gowhar, M.; Shakil, A. R. Morphometry and Land Cover Based Multi-criteria Analysis for Assessing the Soil Erosion Susceptibility of the Western Himalayan Watershed. *Environ. Monit. Assess.* **2014,** *186* (12), 8391–8412.

Altin, T. B.; Altin, B. N. Drainage Morphometry and Its Influence on Landforms in Volcanic Terrain, Central Anatolia, Turkey. *Proc. Soc. Behav. Sci.* **2011,** *19*, 732–740.

Bhatt, C. M.; et al. Satellite-based Assessment of the Catastrophic Jhelum Floods of September 2014, Jammu and Kashmir, India. *Geomat. Nat. Hazards Risk* **2017,** *8* (2), 309–327.

Biswas, S.; Sudhakar, S.; Desai, V. R. Prioritisation of Sub-watersheds Based on Morphometric Analysis of Drainage Basin: A Remote Sensing and GIS Approach. *J. Indian Soc. Remote Sens.* **1999,** *27* (3), 155–166.

Faniran, A. The Index of Drainage Intensity—A Provisional New Drainage Factor. *Aust. J. Sci.* **1968,** *31*, 328–330.

Government of Jammu and Kashmir. *A Satellite Based Rapid Assessment on Floods in Jammu and Kashmir—September, 2014*; Dept. of Ecology, Environment and Remote Sensing: Bemina, Srinagar; National Remote Sensing Centre Indian Space Research Organization: Hyderabad, 2014.

Horton, R. E. Drainage Basin Characteristics. *Trans. Am. Geophys. Union* **1932,** *13*, 350–361.

Horton, R. E. Erosional Development of Streams and Their Drainage Basins: A Hydrophysical Approach to Quantitative Morphology. *Geol. Soc. Am. Bull.* **1945,** *56* (3), 275–370.

Melton, M. A. Correlations Structure of Morphometric Properties of Drainage Systems and Their Controlling Agents. *J. Geol.* **1958,** *66*, 442–460.

Meraj, G.; et al. Assessing the Influence of Watershed Characteristics on the Flood Vulnerability of Jhelum Basin in Kashmir Himalaya. *Nat. Hazards* **2015,** *77* (1), 153–175.

Meraj, G.; Shakil, A. R.; Sadaf, A. Inferring Land Surface Processes from Watershed Characterization. In *Geostatistical and Geospatial Approaches for the Characterization of Natural Resources in the Environment*; Springer: Cham, 2016; pp 741–744.

Nag, S. K. Morphometric Analysis Using Remote Sensing Techniques in the Chaka Sub-basin, Purulia District, West Bengal. *J. Indian Soc. Remote Sens.* **1998,** *26*, 69–76.

Nag, S. K.; Chakraborty, S. Influence of Rock Types and Structures in the Development of Drainage Network in Hard Rock Area. *J. Indian Soc. Remote Sens.* **2003,** *31*, 25–35.

Nooka Ratnam, K.; Srivastava, Y. K.; Venkateswara, R. V.; Amminedu, E.; Murthy, K. S. R. Check Dam Positioning by Prioritization of Microwatersheds Using SYI Model and Morphometric Analysis—Remote Sensing and GIS Perspective. *J. Indian Soc. Remote Sens.* **2005,** *33* (1), 25–38.

Obi Reddy, G. E.; Maji, A. K.; Gajbhiye, K. S. GIS for Morphometric Analysis of Drainage Basins. *GIS India* **2002,** *4* (11), 9–14.

Pandey, A. C.; Singh, S. K.; Nathawat, M. S. Waterlogging and Flood Hazards Vulnerability and Risk Assessment in Indo-Gangetic Plain. *Nat. Hazards* **2010,** *55* (2), 273–289.

Pandey, A. C.; Singh, S. K.; Nathawat, M. S.; Saha, D. Assessment of Surface and Subsurface Waterlogging, Water Level Fluctuations, and Lithological Variations for Evaluating Groundwater Resources in Ganga Plains. *Int. J. Digit. Earth* **2013,** *6* (3), 276–296.

Raza, M.; Ahmad, A.; Mohammad, A. *The Valley of Kashmir: A Geographical Interpretation, Vol. 1: The Land*; Vikas Publishing House Pvt. Ltd.: New Delhi, 1978; pp 1–59.

Reddy, O. G. P.; Maji, A. K.; Gajbhiye, S. K. Drainage Morphometry and Its Influence on Landform Characteristics in a Basaltic Terrain, Central India—A Remote Sensing and GIS Approach. *Int. J. Appl. Earth Obs. Geoinf.* **2004,** *6*, 1–16.

Schumm, S. A. Evolution of Drainage Systems and Slopes in Badlands at Perth Amboy, New Jersey. *Geol. Soc. Am. Bull.* **1956,** *67*, 597–646.

Sharma, A.; Kanga, S. *Surface Runoff Estimation of Sind River Basin Using SCS-CN Method and GIS Technology*, 2020.

Singh, S.; Singh, M. C. Morphometric Analysis of Kanhar River Basin. *Nat. Geogr. J. India* **1997,** *1*, 31–43.

Singh, S. K.; Pandey, A. C. Geomorphology and the Controls of Geohydrology on Waterlogging in Gangetic Plains, North Bihar, India. *Environ. Earth Sci.* **2014,** *71* (4), 1561–1579.

Smith, K. G. Standards for Grading Textures of Erosional Topography. *Am. J. Sci.* **1950,** *248*, 655–668.

Strahler, A. N. Quantitative Analysis of Watershed Geomorphology. *Trans. Am. Geophys. Union* **1957,** *38*, 913–920.

Strahler, A. N. Quantitative Geomorphology of Drainage Basins and Channel Networks. In *Handbook of Applied Hydrology*, Chow, V. T. (Ed.); McGraw Hill Book Company: New York, 1964; pp 4–11.

Yildiz, O. An Investigation of the Effect of Drainage Density on Hydrologic Response. *Turk. J. Eng. Environ. Sci.* **2004,** *28*, 85–94.

CHAPTER 10

Sentinel-2 Red-Edge Band as Virus Indicators for Sustainable Wetland Management Assessed During COVID-19 Lockdown

RAJASHREE NAIK and LAXMI KANT SHARMA*

Department of Environmental Science, School of Earth Sciences, Central University of Rajasthan, Bandarsindri 305817, Ajmer, Rajasthan, India

**Corresponding author. E-mail: laxmikant_evs@curaj.ac.in*

ABSTRACT

The emergence of COVID-19 has emphasized that the destruction of the ecosystem leads to destruction of human. Ecosystem that protects us has been altered allowing pathogens including corona virus. Wetlands are diverse ecosystems in which slight alteration puts enormous impact on socioecological, physiochemical, and economical aspects. India had longest lockdown from March 24, 2020 to May 30, 2020. Current studies attempt to study India's largest inland saline wetland Sambhar salt lake in terms of pre-COVID-19 and ongoing pandemic situation using remote sensing technology, in the absence of on-field observations. Satellite images of January, March, and May for the years 2019 and 2020 have been taken. Normalized difference water index has been calculated to change in spatiotemporal extent classifying them as water, saline soil, salt crust, barren land, and vegetation. The red-edge band (705 nm) was used for algal sheet dynamics as a qualitative measure. It was found that the water level in all the images of 2020 is much higher in 2019. Based on index values, two new classes as shallow water and deep water have come

GIScience for the Sustainable Management of Water Resources. Gowhar Meraj, Shruti Kanga, Majid Farooq, Suraj Kumar Singh, & Sudhanshu (Eds.)
© 2023 Apple Academic Press, Inc. Co-published with CRC Press (Taylor & Francis)

up indicating the self-restoration capabilities of the lake. Viral correlations are established with the algal sheets from the literature for identifying potential substrates. The study prompted that geospatial approach can be used to identify and quantify indicators for viral diversities and abundance even in the absence of field observations to bridge research gaps important for wetlandscape management.

10.1 INTRODUCTION

Saline wetlands are among the most important productive wetland ecosystems mostly located in arid and semiarid regions. Though temporary and endorheic such wetlands play vital role in groundwater recharge, improving surface water quality, mitigating flash floods, providing edible salt, and critical habitat for halophytes, halophiles, and migratory birds. The complex hydrology of saline wetlands controls spatiotemporal distribution of salts movements subsequently influencing the distribution of haloalkaliphiles and faunal breeding and feeding grounds (Kaplan and Avdan, 2017). The decline of wetlands has gained intense attention over the past few decades; however, inland saline wetlandscapes are most neglected by more than 50% since 1900 (WWAP, 2003).

The water quality parameters of a lake are vital information for proper management. Collection of samples, analysis in laboratory from a wetland is a labor-intensive and tedious task. Remote sensing technology has unlimited potential utility for complex ecosystems since the launch of the Landsat series in the 1970s (Klemas et al., 1971). Every object on the earth has different behavior while interacting with electromagnetic radiation. They are identified in the satellite images as per their reflections at a particular wavelength range. A minimum change in any of their properties (physical or chemical) is detected through their change in spectral signatures. Wetlands, both permanent and temporary are very dynamic in nature. Wetlandscape signatures vary with the change in season, water-level, depth and extent, suspended matter, chlorophyll (both floating, emergent, and submerged) content, turbidity, concentration of atmospheric constituent, Sun-elevation angle (Moore, 1980). On scrutinizing the change in reflectance of the water class, the existence and concentration of any component can be both qualitatively or quantitatively be evaluated using remote sensing. It is important to emphasize the wetland parameters that affect its optical. Therefore, the research works are mainly focused on wetland boundary delineation, wetlandscape patches and corridors, landscape heterogeneity, turbidity, and chlorophyll content, sedimentation, colored dissolved organic matter concentration, floating, and submerged vegetation delineation (Lim and Choi, 2015; Luis et al., 2019).

In most of the researches, different indexes have been used for the identification of wetlands from other land use land covers. Landsat satellite data have been used from 1985 to 2009 for wetland dynamics in northern Virginia using the normalized difference vegetation index (NDVI) (Kayastha et al., 2012). (Dong et al., 2014) used NDVI and land surface water index (LSWI) for mapping wetlands of China. Other indexes as soil and atmosphere resistant vegetation index, atmospherically resistant vegetation index; normalized difference water index (NDWI) (Dvorett et al., 2016). Other datasets such as LiDAR (Huang et al., 2014), Digital Elevation Model (Li and Chen, 2005), hyperspectral, and microwave satellite datasets (Morandeira et al., 2016) have also been used for wetlands identification and monitoring. However, least studies have been conducted for inland saline wetlands using geospatial technology.

Novel Coronavirus (COVID-19) has occurred as the worldwide deadliest disease (Chauhan and Singh, 2020; Meraj et al., 2020a, 2020b; Kanga et al., 2020a, 2020b), which has confirmed around 23,739,184 by August 23, 2020, out of which 814,797 death cases (World meters, 2020) in 213 countries of the world. Looking at its characteristics, World Health Organization confirmed it as a pandemic situation (Saadat et al., 2020). To protect people from this pandemic, each country declared nation-wide lockdown in phase-wise manner (Chauhan and Singh 2020; Ranga et al., 2020a, 2020b). India announced its nation-wide longest lockdown to fight the pandemic in three phases from March 24, 2020 to May 30, 2020.

Researchers report that air and water conditions have visibly improved by reduction in pollution in the world as the result of lockdown (Braga et al., 2020; Lal et al., 2020; Li et al., 2020; Muhammad et al., 2020). India also experienced an environmental boost up through improved air and water quality index during lockdown (CPCB 2020; Yunus et al., 2020; Kan et al., 2020). In the current study, an attempt has been made to verify it using remote sensing data for India's largest inland saline wetland, Sambhar salt lake, which was on the verge of extinction due to illegal saltpan encroachment. The temporal changes in red-edge bands of the Sentinel-2A/B datasets and NDWI have been studied for pre- and post-nationwide lockdown. Since no field visits could be done during this phase; the results are showcased only in qualitative terms.

10.2 MATERIALS AND METHODS

10.2.1 STUDY AREA

Sambhar salt lake (SSL) is in Rajasthan, India, (Figure 10.1) with 26°52′ to 27°02′ N; 74°54′–75°14′ E latitude and longitude, respectively. It is in ENE–WSW direction. It is elliptical (Yadav et al., 2013). It is the seventh declared Ramsar site of India satisfying criteria A. It has Ramsar site No. 464 and Important Bird Area No. IN073 (Bhat et al., 2015). SSL is 230 km^2 with 22.5 km and 3–1 km length and width, respectively (Baid, 1975). It comes under tropical climatic conditions. It is shallow downs to 60 cm during dry to 3 m after monsoon season appearing a blackish muddy wetland (Gupta et al., 2015). It has 24.4°C–40.7°C in summer during March–June and 11°C in winter during November–February (Kumar, 2005). Except for main reservoir and legal saltpans, the lake dries up exposing salt flakes (Sanadhya et al., 2013). It receives around 500 mm of rainfall and 250–300 sunny days (Yadav et al., 2013). Ephemeral rivers (Mendha, Rupnagar, Khandel, Kharian) form a catchment of 5520 km^2, 5.16 km long dam divides the lake into two unequal parts (77 km^2 toward east as reservoir and rest 113 km^2 is wetland) (Gaur, 2015). Migratory birds through Central Asian, East Asian Australasian, and Asian East African flyways and amphibians, crustaceans, and other invertebrates come during monsoon (Baid, 1975). It is habitat of 37 herbs, 14 shrub, 14 trees, 15 grass, 6 Chlorophycea, 25 Cyanophyceae, and 7 Bacilariophyceae species (Charan and Sharma, 2016).

10.2.2 DATA USED

To map the water extent and vegetation (halophiles and halophytes) in the study area, the Sentinel-2A/B dataset of both 2019 and 2020 (January, March, and May) has been used. The Sentinel-2 A/B was launched on June 23, 2015 and March 7, 2017, respectively. Its multispectral instrument has 13 spectral bands with VNIR (visible-NIR), short wave infrared (SWIR) with four special vegetation red-edge band. It has swath 290 km with spatial resolution of 10 m for VNIR, 20 m for SWIR, and red and 60 m for atmospheric correction bands (Table 10.1).

The specification of each tile downloaded and taken into use are given in Table 10.2. The cloud-free data were downloaded from the USGS Earth Explorer (https://earthexplorer.usgs.gov/). For the study, the Level-1C data, which are radiometrically and geometrically corrected, were used to retain the uniformity on the basis of atmospheric correction.

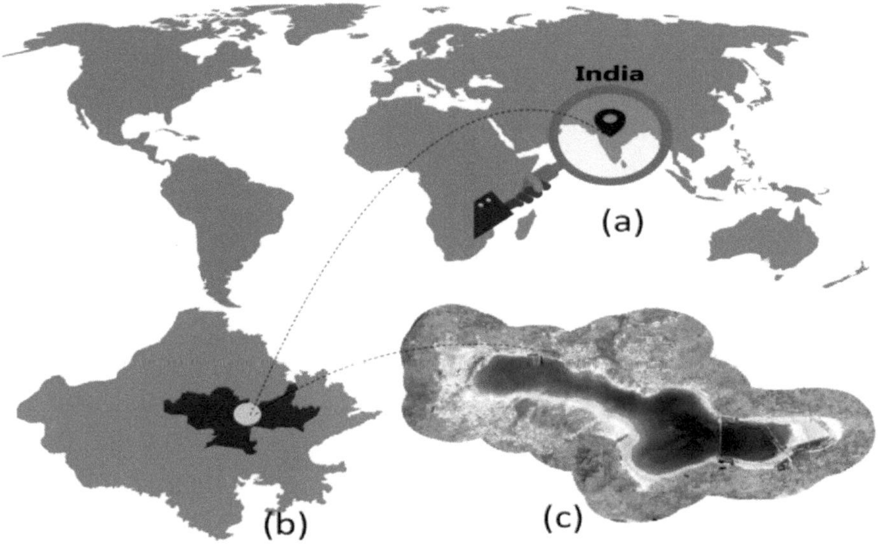

FIGURE 10.1 Study area where (a) is global map highlighting India, (b) is Rajasthan state highlight convergence of three districts Ajmer, Jaipur, and Nagaur, (c) is FCC of Sambhar lake, Jan. 2020.

TABLE 10.1 Bands of Sentinel Satellite.

Sr. no.	Band no.	Band region	Central wavelength (nm)	Spatial resolution (m)
1	B1	Coastal aerosol	443	60
2	B2	Blue	490	10
3	B3	Green	560	10
4	B4	Red	665	10
5	B5	Vegetation red edge	705	20
6	B6	Vegetation red edge	740	20
7	B7	Vegetation red edge	783	20
8	B8	NIR	842	10
9	B8A	Vegetation red edge	865	20
10	B9	Water vapor	945	60
11	B10	SWIR (cirrus)	1.375	60
12	B11	SWIR	1.610	20
13	B12	SWIR	2.190	20

TABLE 10.2 Details of Satellite Data Used.

S.no.	Sensor	Tile no.	Acquisition date
1	Sentinel-2B	T43RDK	January 28, 2019
2	Sentinel-2A	T43REK	January 30, 2019
3	Sentinel-2B	T43RDK	March 29, 2019
4	Sentinel-2B	T43REK	March 29, 2019
5	Sentinel-2A	T43RDK	May 23, 2019
6	Sentinel-2B	T43REK	May 28, 2019
7	Sentinel-2B	T43RDK	January 3, 2020
8	Sentinel-2B	T43REK	January 3, 2020
9	Sentinel-2A	T43RDK	March 8, 2020
10	Sentinel-2A	T43REK	March 8, 2020
11	Sentinel-2A	T43RDK	May 7, 2020
12	Sentinel-2A	T43REK	May 7, 2020

10.2.3 METHODOLOGY

Due to the current pandemic, India announced a shutdown in four phases from March 25 to April 14, 2020, April 15 to May 3, 2020, May 4 to May 31, country-wide, and till June 30, 2020 for containment zones. Due to lockdown and shut down, industrial and commercial activities except medical and essential services were shut, and people were ordered to remain inside their houses. This step resulted in the least environmental disturbance. There was no salt production from the lake, no extraction of underground water, least anthropogenic pressure, and other activities along its course. In the current study, remote sensing data have been used as there are no other supporting data or scientific investigations available to interrelate. The overall methodology is given in Figure 10.2.

10.2.3.1 QUALITATIVE ESTIMATION OF WATER PIXELS THROUGH NDWI

To monitor and compare the lake within 3 km buffer area from the official boundary of lake, optical data of Sentinel-2A/B satellite) of pre- and post-shutdown time for 2019 and 2020 were used (Table 10.2). The data were collected for January, March, and May of both the years to establish a balanced comparative approach for the lake during the pandemic situation. This satellite provides datasets from visible to SWIR range at 10, 20, and 60 m resolution. NDWI set by McFeeters (1996) was used to estimate water pixels and classify the lake accordingly.

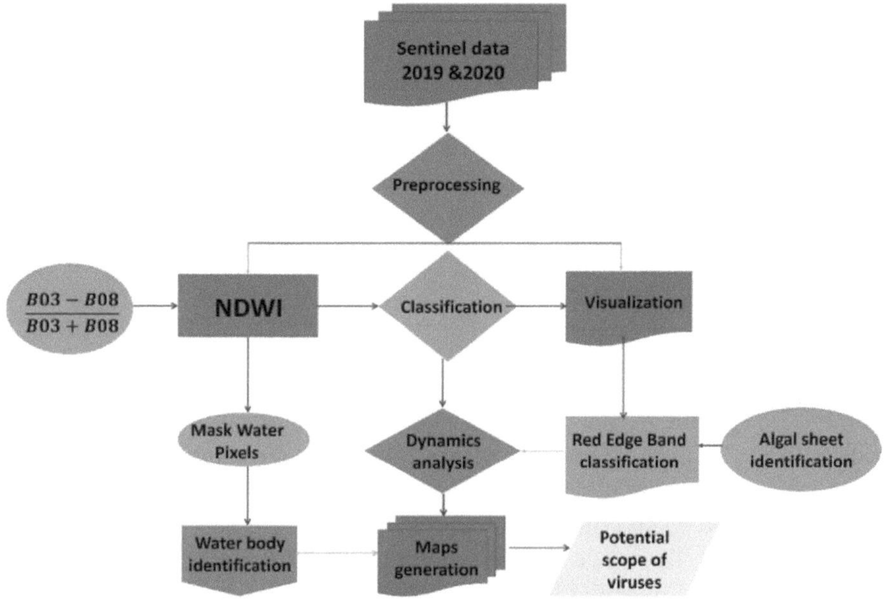

FIGURE 10.2 Methodology workflow chart followed in the study.

$$\text{NDWI} = \frac{B03 - B08}{B03 + B08}$$

Equation 1: Normalized Difference Water Index formula

where B03 is green band and B08 is NIR band, respectively. As water reflects highest in the visible region, absorbing completely in NIR wavelength, these two bands were selected for index-based classification. The range of NDWI lies within −1 to +1, where theoretically values above 0 are considered as waterbodies in NDWI classified image and below zero as nonwater body classes. However, it is always advisable to go for trial and error methods before deciding threshold for water pixels. Finally compared the indexed images of 2019 and 2020 as per values obtained. Based on the values obtained, the study area has been classified into five classes. Water, saline soil, salt crust, barren land, and vegetation, however, in 2020 due to improved condition of the lake, two new classes as deep and shallow water came into the picture.

10.2.3.2 CHANGE IN REFLECTANCE IN RED EDGE SPECTRAL BAND OF SENTINEL-2A/B

Literature state the efficacy of red edge bands for vegetation identification. Red edge band of Sentinel-2A/B has been used for mapping mangroves (Castillo et al., 2017), agricultural management (Zhang et al., 2017), rubber boom (Xiao et al., 2020), postfire detection using red edge based NDVI (Evangelides and Nobajas, 2020). As the study area is a saline habitat, it is abode of halophytes and halophiles surrounded by xerophytes. Band 05 (705 nm) was used for visual color ramp to extract the sheets of floating vegetation within the lake and other terrestrial vegetation.

10.3 RESULT

Sentinel-2 data have been used for visualization and comparison of pandemic situations. The high spatial resolution data have an enormous capacity for earth observation and understanding. True color composite (blue, green, and red bands) and false color composite (FCC) (green, red NIR bands) have been generated, as shown in Figures 10.3 and 10.4, which clearly state the improved condition of the lake. Vegetation red edge band has been used for algal sheet observation. The water index has been used to classify and map the status of India's largest inland saline wetland and source of edible salt. The surface water body and its extent have been analyzed comparatively between the first six months of 2019 and 2020 for pre–post-COVID-19 lock-down situation. In the present study, optical data of Sentinel-2A/B have been applied for all three phases of lockdown. The results obtained are discussed in the following sections.

10.3.1 QUALITATIVE ESTIMATION OF WATER PIXELS THROUGH NDWI

The study area was classified using NDWI into five classes like water, saline soil, salt crust, barren land, and vegetation with two especially emerging classes as deep and shallow water during January and March 2020. The generated classified maps are presented in Figure 10.5. The values of nonwater classes were obtained as negative values. Vegetation remained with −0.8 to −0.3, barren land between −0.3 and −0.2, salt crust between −0.2 and −0.1, saline between −0.1 and −0.5 constantly in all the maps for both the years. Analyzing water pixel values of 2019, it ranges from 0.0 to 0.5 in January, 0.0 to 0.4 in March, and −0.05 to 0.21 in May as single classes. In the case of January and March 2020, the pixels emerged into two new water classes as shallow water (between 0.1 and 0.3) and deep water (0.3–0.5) and May 2020 map is the same as other period-generated maps. From Figure 10.5a, c and f, it is very distinct that lake is devoid of water even during the winter (January) and Spring (March) season. Due to a few days of rainfall in May, there is a little amount of water in the lake. However, their respective counter images 5b, 5d, and 5f depict lake full of water. The reflectance values of water are very high in the indexed maps during COVID-19 situation. This could be possible due to the halt of illegal collection of lake water using electrical pumps.

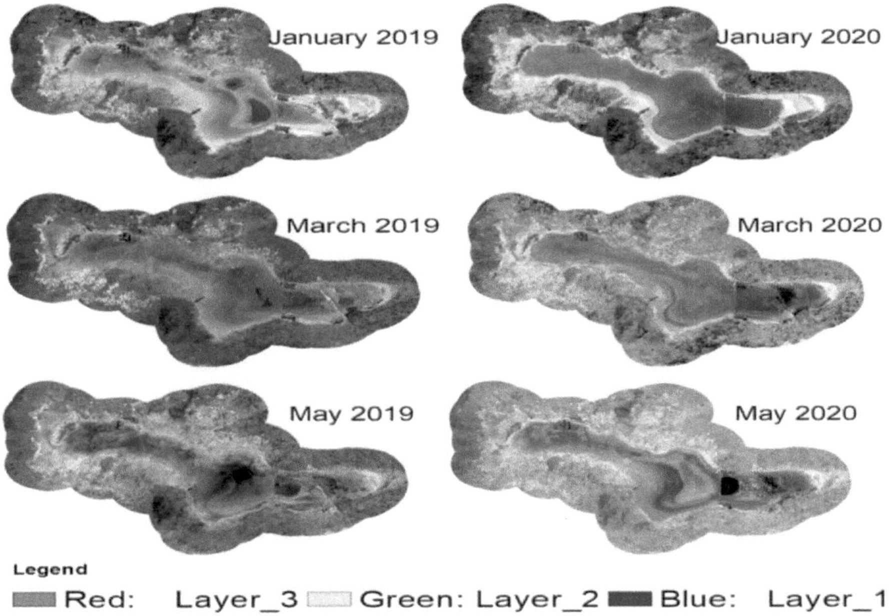

FIGURE 10.3 True color composite from sentinel-2 for visual comparison.

10.3.2 CHANGE IN REFLECTANCE IN RED EDGE BAND OF SENTINEL-2

In this study, research was conducted through remote sensing and GIS using datasets of Sentinel. Initially, the change in lake surface water, change was investigated through NDWI using green and NIR bands. However, when

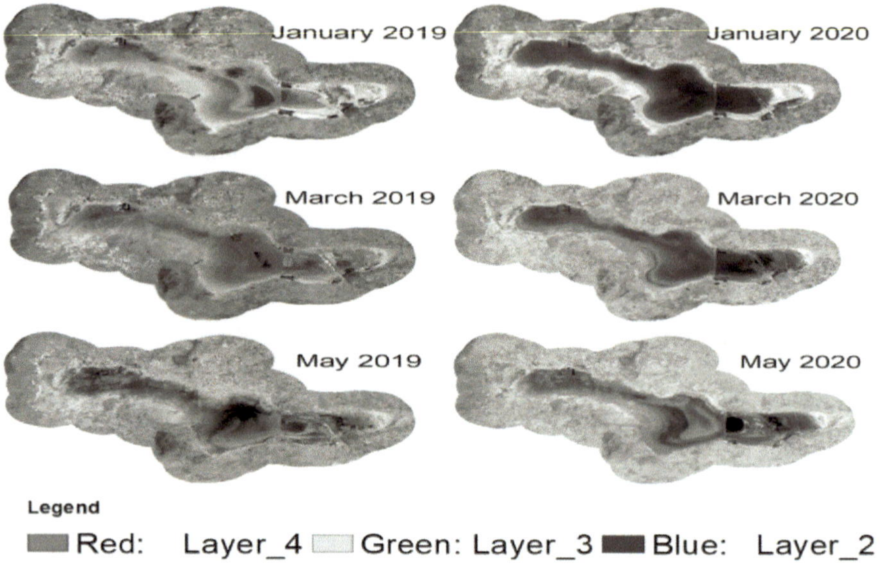

FIGURE 10.4 False color composite of sentinel-2 for visual comparison.

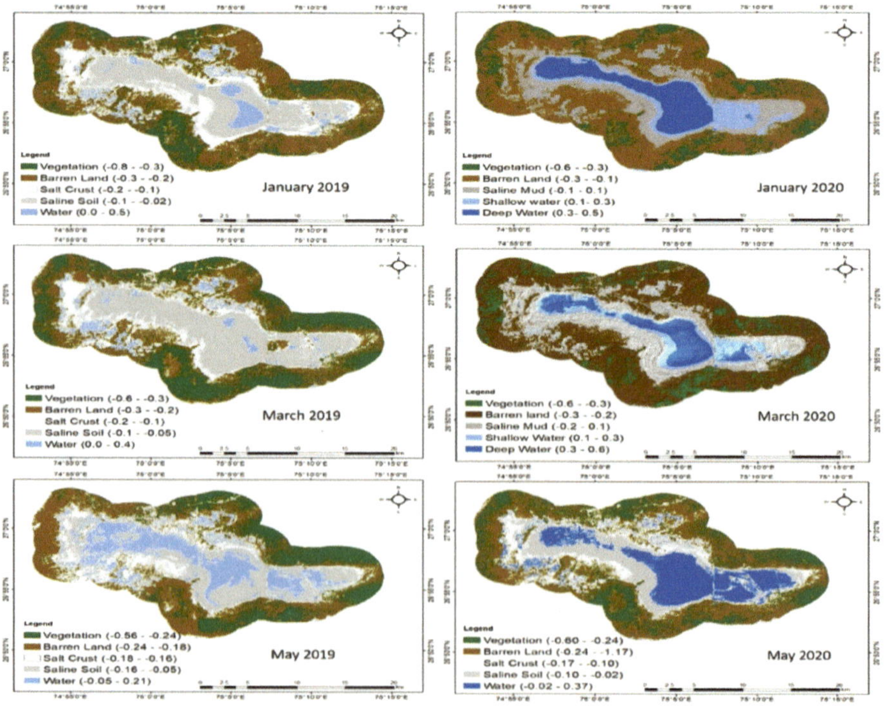

FIGURE 10.5 NDWI generated maps of the lake.

observed through FCC comparing 2019 and 2020 images, there is a change in algal sheets also. To obtain accurate information, red edge band (705 nm) has been taken into use, as the minute changes in reflectance would be observed in this wavelength region. Using this Band-05, three major classes could be visualized (salt, barren land, and vegetation), as shown in Figure 10.6. Vegetation represents both aquatic and terrestrial forms. Terrestrial vegetation comprises of xerophytes and halophytes. Aquatic vegetation comprises of floating algal sheets. The other classes in yellowish color could be saline soil, saline mud, shallow water without algal sheets, which could be identified using this wavelength range. From Figure 10.6a–f, visualization indicates the increase in algal sheets in all the images of 2020 compared to 2019. In Figure 10.4a, the lake had only some water patches and hence least algal sheet. On the other hand, in Figure 4b, there is a continuous sheet of algae in the whole lake as well as in the reservoir. The deplorable condition had remained the same Figure 6c and 6e; however, there is improvement in their counterimage 6d and 6f. Sentinel red edge band has provided very encouraging results for an inland saline wetland ecosystem. Study area had been under the constant pressure of saltpan encroachment, extensive groundwater extraction. These illegal activities had led to wetland shrinkage, habitat destruction. The improved condition during COVID-19 lockdown may be due to the least groundwater extraction and salt production. The whole rainfall may have remained in the lake naturally encouraging sufficient algal growth. These algae are the attraction point of migratory birds. Lakhs and lakhs of flamingoes, pelicans come for feeding, breeding, resting, and roosting. Due to deteriorating conditions, these birds had started shifting their habitat toward the wetlands of Haryana and Delhi, India. Due to improvement in the condition during this pandemic situation, it is expected to welcome a large group of migratory birds in the subsequent years.

10.4 DISCUSSION

Sambhar lake has a very rich historical past, ecological and economic importance at global platform. During 1817, it was the only source of edible salts for the whole country along with welcoming lakhs of migratory birds special flamingoes. However, in the last few decades, illegal saltpan encroachment, extensive underground water extraction, and pollution have almost converted this wetland to a wasteland. The analysis of water and vegetation (halophiles, halophytes, and xerophytes) has been studied completely using optical datasets following the given methodology of the previous section.

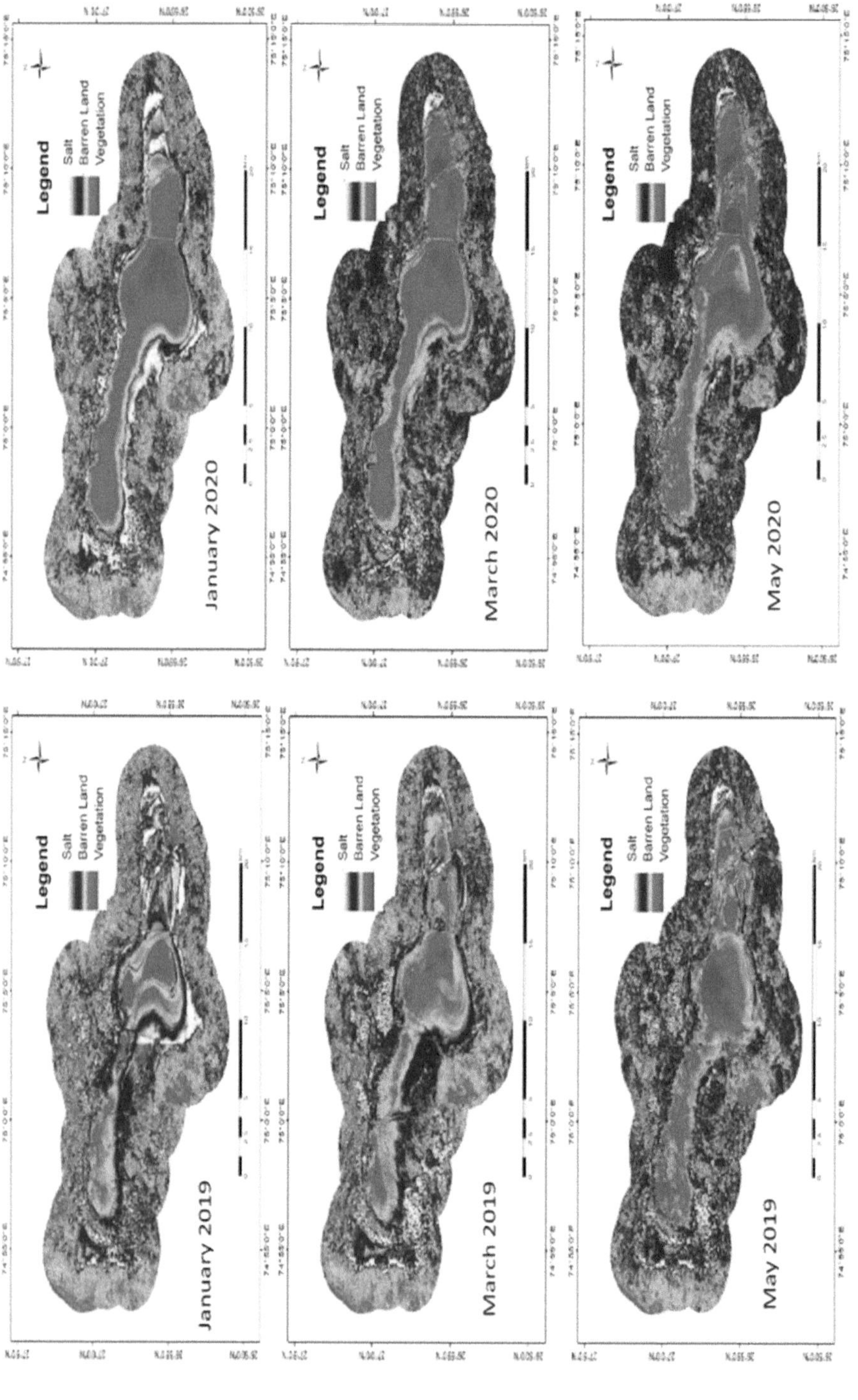

FIGURE 10.6 Red edge band maps of the study area.

10.4.1 VIRUSES IN NATURAL WETLAND

This pandemic situation has turned around the world toward viruses. It is the need of the hour to be prepared for future such situations. Viruses from the wetland have been least studied except for wastewater treatment. The study area has four major classes of algae as Chlorophyceae, Cyanophyceae, Bacilariophyceae, and Eugleanophyceae; however, no studies have been conducted for viruses in saline wetlands. This section describes wetland virus and the research scope.

Virus behave differently in different wetlands. Generally, they are causal factors for bacterial mortality, cell lysis diverting particulate organic matter to dissolved organic matter, releasing nutrient usually in organic forms changing nutrient availability for other organisms, viruses that host in algae or cyanobacteria decline phytoplankton blooms, affect microbial food web, phages could also facilitate genetic exchange (Kitajima et al., 2018). Most of the virus studies for wetlands have been done for sediments as substrates providing a stable physical environment for viral growth than fluctuating water column; however, diversity changes with changing climatic conditions (Rosario et al., 2009). The presence of organic matter encourages transition between aerobic and anaerobic environments (Nicaise, 2014). Abundance of vegetation provides a surface for viral abundance (Mehle and Ravnikar, 2012). Studies for the freshwater state the presence of bacterial biomass as biofilms are indicators for wetland virus mostly for bacteriophages (Chen et al., 2019). Viral abundance and viral densities are high on surface of sediments and submerged macrophytes as compared to deep water columns that are important concepts for viral ecology (Payet and Suttle, 2013). Viral dynamics are highly linked to dynamics and growth rates of host mostly bacteria and algae (Kitajima et al., 2018). Studies state eutrophication status encourage more virus (Cantalupo et al., 2011). Less water, shallow wetlands with high temperatures encourage more viruses (Mehle and Ravnikar, 2012). 16SrRNA and pulse-field gel electrophoresis are two popular methods applied for viral studies of wetlands (Bacnik et al., 2020). Limitations of methods have also restricted knowledge toward this aspect. Studies state that these marine viruses infect organisms from microorganisms to large mammals (Rosario et al., 2009). Most of the studies have been done for constructed wetlands for wastewater treatment for indicator viruses (e.g., coliphages) pathogen removal efficiency tests; however, viruses for natural wetlands are limited to primarily marine wetlands, sediment substrates, paddy field, flood plains, and oxbow lakes out of 18 types of wetlands as per Ramsar convention (Nicaise, 2014).

Depending on flooding time, wetness times, ecotonal, and nonecotonal wetlands, wetland formation, climate, topography, proximity to a water source, soils and substrates, soil composition, waterlogging effects on soil, soil color and flooding impacts on soil development, factors affecting soil color, organic and hydric soils, aerobic and anaerobic soil, vegetation, wetland hydrology, water budget, environmental changes due to waterlogging and flooding, they morphologically adapt to hypertrophied stems, shallow root systems, pneumatophores, looping roots adventitious roots, aerenchyma hypertrophied lenticels, buttressed trunks, stem hypertrophy, hollow stems, adventitious water roots, shallow root systems, pneumatophores, loosely packed root nodules, swelling of root, aerial root tips, succulent roots, viviparous seeds, buoyant seeds, hydrochory, succulent leaves and heterophylly. These are indicators of hydrology, pollution and disturbance, water sources, water chemistry, soil type, water stress, floristic quality and wetland condition, and microbial trophic structure. Microbial community of wetlands mediates vital biogeochemical cycles like carbon, nitrogen, sulfur, phosphorous, and iron. Studies suggest bacteria are responsible redox reactions in anoxic hydric soils; however, the right time has come to step beyond toward the role played by viral communities in these diverse ecosystems.

10.4.2 RESEARCH SCOPE

Viruses play vital role in the ecosystem that they have ignored in wetland ecosystem. The research gaps include (1) studies on wetland viral communities, diversities, abundance, distribution of viruses in different wetlands, (2) search for substrates than other sediments, (3) impact of virus on benthic microbial process, (4) seasonal behavior of virus and structural and functional components of wetland, (5) relationship among wetland organisms like algae, bacteria, archaea, fungi, diatoms, macrophytes, and wetland virus, (6) relation of viral communities with different sediment size, (7) viral composition, diversity, and abundance at different spatio-temporal scale, (8) variation of viral diversity and abundance at landscape level, (9) diversities between ecotonal regions and surroundings aquatic and terrestrial ecosystems, (10) environmental factors controlling viral growth rate and diversity, (11) transmission mode of virus among wetland hydric soil, aquatic plants in fluctuating water level, (12) diversities during wetland flooding and drought seasons, (13) different hosts for wetland virus should also be explored, (14) methanogen-infecting phages of wetlands, (15) role of viruses in wetland biogeochemical cycle and nutrient cycling, carbon cycling, and global climate change is unclear (Jackson and Jackson, 2008).

10.5 CONCLUSION

The presence of water is very essential for a wetland. It is reported that environmental conditions have improved during pandemic lockdown phases in India. In the current study, purely Sentinel-2A/B satellite data of 2019 and 2020 for January, March, and May has been used for spatio-temporal study of Sambhar salt lake, the largest inland saline wetland of India. Investigation methodology broadly involves qualitative estimation using geospatial indices and red edge band (705 nm). The study lacks field validations due to nationwide lockdown. Initially, NDWI was used to delineate surface water, extract water boundary, and compare water values for both the years using green and NIR wavelength regions. Results clearly show the increase in water extent of the lake after COVID-19 complete lockdown for March–May. This is justified by the increased indices value. Since neither true color (blue, green, and red band composite) composite nor false-color composite (green, red, and NIR band composite) could give proper information about the vegetation status of the lake, red edge band 05 was used. This band has an enormous capacity to map the heterogeneity of vegetation and spatial scale as chlorophyll content is best reflected in this wavelength region. The analysis confirmed that there is increase in water level and extent and subsequent increase in algal sheets during the lockdown. It was realized that ground sampling and lab analysis of soil and water sample from the lake would have taken much time to get the status of the lake even at point location. However, the use of remote sensing data enabled to derive information, compare at spatio-temporal scale at landscape level more accurately; however, the results are qualitative. Therefore, the retrieved information through geospatial approach requires field validation, which could be conducted once lockdown is over and laboratories start again. Further, the obtained information may improve with very high, spatial, spectral, and temporal resolution satellite data for such a small but important study area. The results obtained can be directly be referred for the lake management and conservation practices as it was on the verge of extinction. After so long, it is restoring naturally, it should be maintained to prevent from Montreux record.

KEYWORDS

- **COVID-19**
- **lockdown**
- **red edge band**
- **saline wetland**
- **sentinel-2**
- **virus**

REFERENCES

Bacnik, K.; Kutnjak, D.; Pecman, A.; Mehle, N.; Žnidarič, M. T.; Aguirre, I. G.; Ravnikar, M. Viromics and Infectivity Analysis Reveal the Release of Infective Plant Viruses from Wastewater into the Environment. *Water Res.* **2020,** 115628.

Baid, I. C. Description of a New Species of *Branchinella sayce* from Sambhar Lake, India (Crustacea: Branchiopoda: Anostraca). *J. Bombay Nat. History Soc.* **1975,** *72* (3), 786–793.

Bhat, A. H.; Sharma K. C.; Banday. U. J. Impact of Climatic Variability on Salt Production in Sambhar Lake, a Ramsar Wetland of Rajasthan, India. *Middle-East J. Sci. Res.* **2015,** *23* (9), 2060–2065. https://doi.org/10.5829/idosi.mejsr.2015.23.09.95224

Braga, F.; Scarpa, G. M.; Brando, V. E.; Manfe, G.; Zaggia, L. COVID-19 Lockdown Measures reveal Human Impact on Water Transparency in the Venice Lagoon. *Sci. Total Environ.* **2020,** *736*, 139612.

Cantalupo, P. G.; Calgua, B.; Zhao, G.; Hundesa, A.; Wier, A. D.; Katz, J. P.; Grabe, M.; Hendrix, R. W.; Girones, R.; Wang, D.; Pipas, J. M. Raw Sewage Harbors Diverse Viral Populations. *MBio* **2011,** *2*, 1–11. https://doi.org/10.1128/mBio.00180-11

Castillo, J. A. A.; Apan, A. A.; Maraseni, T. N.; Salmo, S. G. Estimation and Mapping of Above-Ground Biomass of Mangrove Forests and Their Replacement Land Uses in the Philippines Using Sentinel Imagery. *ISPRS J. Photogrammetry Remote Sens.* **2017**. https://doi.org/10.1016/j.isprsjprs.2017.10.016

Charan, P. D.; Sharma, K. C. Floral Diversity of Thar Desert of Western Rajasthan, India. *J. Phytol. Res.* **2016,** *29* (1) 55–71.

Chauhan, A.; Singh, R. P. Decline in PM2.5 Concentrations over Major Cities around the World Associated with COVID-19. *Environ. Res.* **2020,** *187*, 109634.

Chen, X.; Wei, W.; Wang, J.; Li, H.; Sun, J.; Ma, R.; Zhang, R. Tide Driven Microbial Dynamics through Virus-Host Interactions in the Estuarine Ecosystem. *Water Res.* **2019,** *160*, 118–129.

CPCB (Central Pollution Control Board). Impact of Lockdown on Water Quality of River Ganga. CPCB, Ministry of Environment, Forest and Climate Change, Govt. of India, New Delhi, 2020. https://cpcb.nic.in/openpdffile.php?id=TGF0ZXN0RmlsZS8yOTNfMTU4Nzk3ODU3MV9tZWRpYXBob3RvMTY3MDYucGRm (accessed May 12, 2020).

Dong, Z. Y.; Wang, Z. M.; Liu, D. W.; Song, K. S.; Li, L.; Jia, M. M. Mapping Wetland Areas Using Landsat- Derived NDVI and LSWI: A Case Study of West Songnen Plain, Northeast China. *J. Indian Soc. Remote Sens.*, **2014,** *42* (3), 569–576.

Dvorett, D.; Davis, C.; Papes, M. Mapping and Hydrologic Attribution of Temporary Wetlands Using Recurrent Landsat Imagery. *Wetlands* **2016,** *36* (3), 431–443.

Evangelides, C.; Nobajas, A. Red-Edge Normalised Difference Vegetation Index (NDVI705) from Sentinel-2 Imagery to Assess Post-Fire Regeneration. *Remote Sens. App.: Soc. Environ.* **2020,** 17. https://doi.org/10.1016/j.rsase.2019.100283

Gaur, A. Isolation and Characterization of Halotolerant *Bacillus sp.* with Extra Cellular a-Amylase Production Potential from Sambhar Salt Lake, India **2015,** *4* (6), 7–10.

Gupta, M.; Aggarwal, S.; Navani, N. K.; Choudhury, B. Isolation and Characterization of a Protease-Producing Novel Haloalkaliphilic Bacterium *Halobiforma sp.* Strain BNMIITR from Sambhar Lake in Rajasthan, India. *Ann. Microbiol.* **2015,** *65* (2), 677–686. https://doi.org/10.1007/s13213-014-0906-z

https://earthexplorer.usgs.gov/

Huang, C.; Peng, Y.; Lang, M.; Yeo, I. Y.; McCarty, G. Wetland Inundation Mapping and Change Monitoring Using Landsat and Airborne LiDAR Data. *Remote Sens. Environ.* **2014,** *141*, 231–242.

Kanga, S.; Sudhanshu, Meraj, G.; Farooq, M.; Nathawat, M. S.; Singh, S. K. Reporting the Management of COVID-19 Threat in India Using Remote Sensing and GIS Based Approach. *Geocarto Int.* **2020a,** 1–8.

Kanga, S.; Meraj, G.; Farooq, M.; Nathawat, M. S.; Singh, S. K. Risk Assessment to Curb COVID-19 Contagion: A Preliminary Study Using Remote Sensing and GIS, 2020b.

Kaplan, G.; Avdan, U. Mapping and Monitoring Wetlands Using Sentinel-2 Satellite Imagery. In *ISPRS Annals of the Photogrammetry, Remote Sensing and Spatial Information Sciences*, Volume IV-4/W4, 2017 *4th International GeoAdvances Workshop, 14–15* October 2017, Safranbolu, Karabuk, Turkey, 2017; pp 1–7. https://doi.org/10.5194/isprs-annals-IV-4-W4-271-2017.

Kayastha, N.; Thomas, V.; Galbraith, J.; Banskota, A. Monitoring Wetland Change Using Inter-Annual Landsat Time- Series Data. *Wetlands* ***2012,*** *32* (6), 1149–1162.

Kitajima, M.; Sassi, H. P.; Torrey, J. R. Pepper Mild Mottle Virus as a Water Quality Indicator. *NPJ Clean Water* **2018,** *1*, 1–19. https://doi.org/10.1038/s41545-018-0019-5

Klemas, V.; Borchardt, J. F.; Treasure, W. M. Suspended Sediment Observations from ERTS-1. *Remote Sens. Environ.* **1971,** *2*, 205–221.

Kumar, S. Fauna of Sambhar Lake (Rajasthan). *Wetland Ecosyst. Ser.* **2005,** *6*, 1–200.

Lal, P.; Kumar, A.; Kumar, S.; Kumari, S.; Saikia, P.; Dayanandan, A.; Adhikari, D.; Khan, M. L. The Dark Cloud with a Silver Lining: Assessing the Impact of the SARS COVID-19 Pandemic on the Global Environment. *Sci. Total Environ.* **2020,** *732*, 139297.

Li, J. H.; Chen, W. J. A Rule-Based Method for Mapping Canada's Wetlands Using Optical, Radar and DEM Data. *Int. J. Remote Sens.* **2005,** *26* (22), 5051–5069.

Li, L.; Li, Q.; Huang, L.; Wang, Q.; Zhu, A.; Xu, J.; Liu, Z.; Li, H.; Shi, L.; Li, R. Air Quality Changes during the COVID-19 Lockdown Over the Yangtze River Delta Region: An Insight into the Impact of Human Activity Pattern Changes on Air Pollution Variation. *Sci. Total Environ.* **2020,** *732*, 139282.

Lim, J.; Choi, M. Assessment of Water Quality Based on Landsat 8 Operational Land Imager Associated with Human Activities in Korea. *Environ. Monit. Assess.* **2015,** *187* (6), 384.

Luis, K. M. A.; Rheuban, J. E.; Kavanaugh, M. T.; Glover, D. M.; Wei, J.; Lee, Z.; Doney, S. C. Capturing Coastal Water Clarity Variability with Landsat 8. *Mar. Pollut. Bull.* **2019,** *145*, 96–104.

Mehle, N.; Ravnikar, M. Plant Viruses in Aqueous Environment—Survival, Water Mediated Transmission and Detection. *Water Res.* **2012,** *46*, 4902–4917. https://doi.org/10.1016/j.watres.2012.07.027

Meraj, G.; Farooq, M.; Singh, S. K.; Romshoo, S. A.; Nathawat, M. S.; Kanga, S. Coronavirus Pandemic versus Temperature in the Context of Indian Subcontinent: A Preliminary Statistical Analysis. *Environ. Dev. Sustain.* **2020a,** 1–11.

Meraj, G.; Farooq, M.; Singh, S. K.; Romshoo, S. A.; Nathawat, M. S.; Kanga, S. Statistical Analysis for Understanding the Relationship between Coronavirus Pandemic and Temperature in the Context of Indian Subcontinent. Climatological, Meteorological, and Environmental Factors in the COVID-19 Pandemic, 2020b.

Moore, G. K. Satellite Remote Sensing of Water Turbidity. *Hydrol. Sci. J.* **1980,** *25* (4), 407–421.

Morandeira, N. S.; Grings, F.; Facchinetti, C.; Kandus, P. Mapping Plant Functional Types in Floodplain Wetlands: An Analysis of C-Band Polarimetric SAR Data from RADARSAT-2. *Remote Sens.* **2016,** *8* (3).

Muhammad, S.; Long, X.; Salman, M. COVID-19 Pandemic and Environmental Pollution: A Blessing in Disguise? *Sci. Total Environ.* **2020,** *728*, 138820.

Nicaise, V. Crop Immunity against Viruses: Outcomes and Future Challenges. *Front. Plant Sci.* **2014,** *5*, 1–18. https://doi.org/10.3389/fpls.2014.00660

Payet, J. P.; Suttle, C. A. To Kill or Not to Kill: The Balance between Lytic and Lysogenic Viral Infection Is Driven by Trophic Status. *Limnol. Oceanogr.* **2013,** *58*, 465e474. https://doi.org/10.4319/lo.2013.58.2.0465

Ranga, V.; Pani, P.; Kanga, S.; Meraj, G.; Singh, S. K.; Farooq, M.; Nathawat, M. S. Health GIS-A Long lasting Solution for the Effective Pandemic Management in India. Climatological, Meteorological, and Environmental Factors in the COVID-19 Pandemic, 2020a.

Ranga, V.; Pani, P.; Kanga, S.; Meraj, G.; Farooq, M.; Nathawat, M. S.; Singh, S. K. National Health-GIS Portal-A Conceptual Framework for Effective Epidemic Management and Control in India, 2020b.

Rosario, K.; Nilsson, C.; Lim, Y. W.; Ruan, Y.; Breitbart, M. Metagenomic Analysis of Viruses in Reclaimed Water. *Environ. Microbiol.* **2009,** *11*, 2806–2820. https://doi.org/10.1111/j.1462-2920.2009.01964.x

Saadat, S.; Rawtani, D.; Hussain, C. M. Environmental Perspective of COVID-19. *Sci. Total Environ.* **2020,** *728*, 138870.

Sanadhya, S.; Nagarajappa, R.; Sharda, A. J.; Asawa, K.; Tak, M.; Batra, M.; Daryani, H. The Oral Health Status and the Treatment Needs of Salt Workers at Sambhar Lake, Jaipur, India. *J. Clin. Diagnostic Res.* **2013,** *7* (8), 1782. DOI: 10.7860/JCDR/2013/5887.3275

WorldMeters.https://www.worldometers.info/coronavirus/?utm_campaign=homeAdUOA?Si (accessed Aug 25, 2020).

WWAP, U. UN World Water Development Report: Water for People. Water for Life, 2003.

Xiao, C.; Li, P.; Feng, Z.; Liu, Y.; Zhang, X. Sentinel-2 Red-Edge Spectral Indices (RESI) Suitability for Mapping Rubber Boom in Luang Namtha Province, Northern Lao PDR. *Int. J. Appl. Earth Observ. Geoinfo.* **2020,** *93* (C), 102176. https://doi.org/10.1016/j.jag.2020.102176

Yadav, A. K.; Vardhan, S.; Kashyap, S.; Yandigeri, M.; Arora, D. K. Actinomycetes Diversity among rRNA Gene Clones and Cellular Isolates from Sambhar Salt Lake, India. *Sci. World J.* **2013**. https://doi.org/10.1155/2013/781301

Yunus, A. P.; Masago, Y.; Hijioka, Y. COVID-19 and Surface Water Quality: Improved Lake Water Quality during Lockdown. *Sci. Total Environ.* **2020,** *731*, 139012.

Zhang, T.; Su, J.; Liu, C.; Chen, W. H.; Liu, H.; Liu, G. Band Selection in Sentinel-2 Satellite for Agriculture Applications. *ICAC 2017—2017 23rd IEEE International Conference on Automation and Computing: Addressing Global Challenges through Automation and Computing*, 2017. https://doi.org/10.23919/IConAC.2017.8081990

PART II

GIScience for Evaluating the Change Dynamics of Earth System Processes

CHAPTER 11

Assessing Crop Production Dynamics Using Geospatial Patterns of Soil Moisture, Aridity, and Changing Climate

P. DHANYA* and V. GEETHALAKSHMI

Tamil Nadu Agricultural University, Coimbatore, Tamil Nadu, India

**Corresponding author. E-mail: dhanya.eptri@gmail.com*

ABSTRACT

Geospatial technologies provide huge potential for land degradation and climate change impact assessments. The endeavor was to analyze the long term trends in the moisture and aridity conditions prevailing in the Kancheepuram district, South India and project decadal changes in peanut yield using Decision Support System for Agro Technology Transfer (DSSAT) under medium emission scenarios RCP 4.5. Precipitations (P) and potential evaporation (PE) data were used to create Moisture Index and to assess the temporal and spatial variations in aridity status. Gridded data analysis of the mean annual moisture conditions during the period 1951–2008 showed no significant trend for any of the stations. The annual mean moisture index for Chittamur, Madhuranthakam, Thirukazhikundram, Kancheepuram, Wallajabad, St. Thomas Mount were −35.8, −35.5 −28.6, −42.4, −36.2, −28.4, respectively. Among the stations, Kancheepuram recorded the lowest moisture levels, indicating its critical semiarid conditions throughout 57-year period. Similarly GIS analysis showed that aridity may be increased in the coastal areas. PNUTGRO crop simulation model output revealed that the productivity of the peanut crops may fall toward the end of the 21st century. Yield of groundnut indicated a strong declining trend in the mid-century. The decadal mean yield under the

GIScience for the Sustainable Management of Water Resources. Gowhar Meraj, Shruti Kanga, Majid Farooq, Suraj Kumar Singh, & Sudhanshu (Eds.)
© 2023 Apple Academic Press, Inc. Co-published with CRC Press (Taylor & Francis)

control simulation over 127 years was 1732.4 kg/ha, whereas the decadal mean groundnut yield under elevated carbon dioxide condition was 1846.03 kg/ha. The yield is likely to be reduced by −12.7% for the decade 2091–2098 due to climate change. This spatiotemporal assessment threw light on observed hydroclimatic conditions and crop yield changes under climate change and the urgent need for water and land management to cope with climate change impacts.

11.1 INTRODUCTION

Cartography is a significant tool to understand spatiotemporal changes that happens over land areas. Geospatial tools, in general, and maps in particular are used widely to communicate spatial variations in climate change and land degradations all over the world (Gujree et al., 2017; Pall et al., 2019; Romshoo et al., 2020; Carolyn S, 2020). Thirty percent of the world's terrestrial area is located in semiarid regions which are found in the Americas, Oceania, Asia, and Africa (Rossato et al., 2017). There is a high likelihood that global warming will increase the rate of desertification across the globe (IPCC, 2012). Desertification is hurting the livelihoods of more than 1 billion people and an additional 12 million hectares of arable land is lost each year because of droughts (UNCCD, 2019). It is a known fact that any changes in hydrological cycle may force far reaching consequences on the rainfall availability, soil moisture conditions, and land quality in an area. Not only the changes in rainfall patterns but evaporation and transpiration also impacts crop yield positively or negatively depending on its intensity (Brunetti et al., 2002). Alterations in rainfall intensity could increase the risk of floods and droughts in coastal wetlands occupied mostly by farming (IPCC, 2013, 2014; Singh et al., 2017b; Kanga et al., 2020a, 2020b). Unexpected climatic stresses compel the farmers to adopt adjustments in the timing of farm operations, crop diversification, alternative livelihood options, and agroforestry practices (Kumar et al., 2011). Decrease in soil moisture, increased aridity, increased salinity and surface and groundwater depletion can cause tremendous reduction in the arable land. Adopt any measure is the only way to change in response to current conditions (IFAD, 2014, Oyiga et al., 2011, Bastiaanssen and Steduto, 2016).

Anthropogenic global warming may pose significant human forcing of earth systems and serious implications on geomorphologic processes (Lane, 2019). Climate-change-induced global warming may have adverse impacts of greater magnitude on current species diversity, ecosystem services, and landscapes (Farooq and Muslim, 2014; Nathawat et al., 2010; Kumar et al., 2018; Joy et al., 2019; UNCCD, 2019). Aridification is a serious consequence of climate change. Semiarid regions in the world is faced with recurrent droughts and related socioeconomic distress (Barbosa and Lakshmi Kumar, 2016; Marengo et al., 2016). Reports from FAO (2012) and WWAP (2012) indicate the severity of water crisis around the world. Climatic stressors play a major role in shaping up the geomorphology of a place. Physical scarcity of water is faced by almost one-fifth of the world's population (1.2 billion people) and around one-quarter (1.6 billion people) of the world's population, probably faces economic water shortage (where countries lack the necessary infrastructure to take water from rivers and aquifers). Deforestations and reduction in the available moisture level in the top soil create less humus content in the soil and thus soil compactions and land erosion and landslides. It has four main social implications, and they can be seen as four different ways according to Paredes et al. (2015), as stated by Paredes et al. (2015), these are meteorological, hydrological, agricultural, and socioeconomic (Marengo et al., 2016, Fagariba et al., 2018).

India covers a total length of 7500 km and accounts for 16.7% of the world's population. Furthermore, it has a fast population growth rate, which puts the country in a vulnerable position when confronted with environmental issues (De et al., 2005). India has encountered greater and increasing vulnerability to the socioeconomic impacts of extreme weather events like floods, droughts, cyclones, heat waves, etc., apart from that Indian coastal districts have a share of 17% of our national population, making this areas more and more sensitive (Usha et al., 2014; Rossato et al., 2017; Saied et al., 2017; Pandey et al., 2010; Pandey et al., 2013; Singh and Pandey, 2014; Meraj et al., 2015a, 2015b, 2016; Bhatt et al., 2017; Sharma and Kanga 2020). The geographical setting of Tamil Nadu makes the state vulnerable to natural disasters such as cyclones, droughts, floods, and earthquake-induced tsunami. Many researchers have suggested to have a detailed local level studies to quantify the impacts and to deepen our understanding of crop vulnerability, society's responses to climate change (Knight and Ager 2009; Meraj et al., 2013, 2018a, 2018b).

An estimated 77% of the global population growth over the next two decades will occur in developing countries, with the result that global demand for food and water will rise in urban areas (Meraj 2020; Meraj and Kumar 2021; Meraj et al., 2021). Cities in lower income countries will account for most of the future growth in global population. It is essential to have regional and local specific research studies to get the pulse of a region. Cartography plays a significant role to capture the impact of rising population on climate, quality of water available, agriculture yield, forests, etc. Geospatial tools support postprocessing the modeled climate data (Kanga et al., 2017a, 2017b; Rather et al., 2018; Hassanin et al., 2020; Quach and Bernhard, 2020; Kanga et al., 2021). As an increasing number of people move to cities, agriculture has the capacity to support a larger population but it must have water to sustain crop and livestock production. Being a semiarid state, groundnut as a dry land oil seed crop is cultivated in 6.19 lakh hectares with a production of 10.98 lakh tones. In this, 70% is under rain-fed cultivation and remaining 30% under irrigated conditions. Hence in this study, an emphasis has been given the groundnut crop as it commonly and majorly cultivated in these areas.

11.2 MATERIALS AND METHOD

11.2.1 STUDY AREA

Kancheepuram district is situated in the northeast Agro Climatic Zone (ACZ) of Tamil Nadu, South India. It lies in the tropics between the latitude 11° 00′ to 12° 00′ north and 77° 28′ to 78° 50′ east longitude (Figure 11.1). It receives rainfall mainly from northeast (retrieval) and southwest monsoon with 54% and 36% contribution each to the total annual rainfall, respectively. The major river that runs through the district is Palar which is nonperennial, active only during the months of October, November, and December. The majority of the people who live in rural areas depend on agriculture and fishing as their main source of income. Paddy, groundnut, sugarcane, cholam, and cumbu are the major crops grown in these areas. This district has eight administrative divisions known as "talukas." In that, Cheyyur, Thirukalikundram, Chengalpattu, and Tambaram talukas share the coastal boundaries of the state, whereas Sriperumbudur, Kancheepuram, Uthiramerur, and Maduranthakam talukas are in the interior parts of the district. The ancestors of this region constructed several water tanks across the river basins, and they were able to rely on the water collected in these tanks for irrigation. Groundwater depletion is occurring at a high rate due to the rapid depletion of groundwater resources to feed the growing population of Chennai, India's largest city. Additionally, bore wells can go dry during the off-season. There are fewer lakes, reservoir tanks, and ponds in these areas because of the reduction in rainfall. The rivers that flow through Kancheepuram's Palar, Cheyyur, Killiyar, and Vegavathi drainages are the Palar, Cheyyur, Killiyar, and Vegavathi. A major drainage that runs from north to south and ends into the sea is the Buckingham canal, which merges with the Palar river to the south. These rivers have a seasonal cycle, and during the monsoon season, they carry a substantial amount of water. The area's main water sources are reservoirs, wells, and tube wells. As far as climate is concerned, the maximum day time temperature rarely exceeds 43°C and minimum night time temperatures seldom falls below 18°C. The annual mean rainfall of this area is 1202 mm with a potential evapotranspiration (PE) of 1750 mm. The rainfall in this region is showing remarkable annual variability. Lathur, Uthiramerur of the Kancheepuram district comes under over-exploited category as far as ground water is concerned.

FIGURE 11.1 Location of the study area.

As the crop production statistics are concerned the net sown area covers 58.3% of the total geographic area of the district (Seasonal Crop Report 2009–2010). Paddy, groundnuts, sugarcane, and pulses are the major crops cultivated in this region. The field studies have revealed that in the recent past, some patches are converted into mango orchards, vegetables cultivation, and floriculture crops especially different varieties of Jasmine, tube roses (Champaki), chrysanthemum, etc. (Dhanya et al., 2013). The density of population of Kancheepuram is 1049 km^2 and it has increased almost three times in the study area. Preserving crop lands in the fringe areas of fast expanding real-estate markets in the city like Chennai is actually a serious cause of concern.

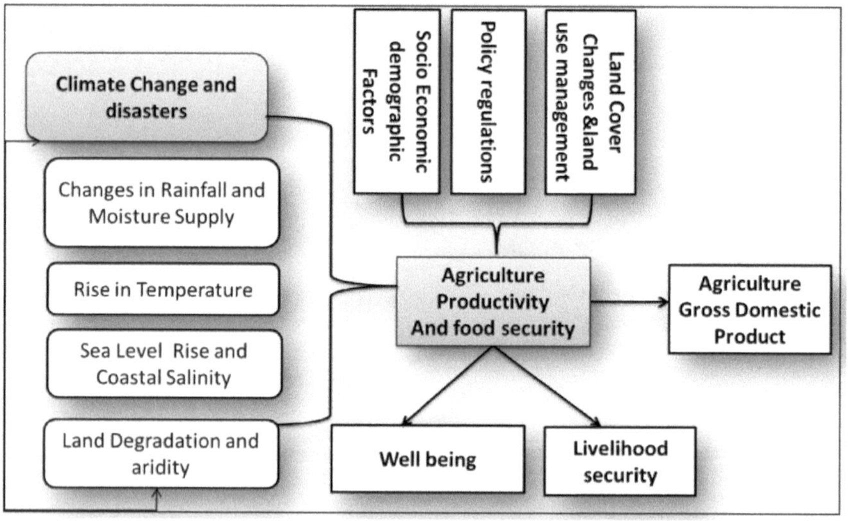

FIGURE 11.2 Conceptual framework on the methodology. Cartographic method to understand complex mechanisms which affect agriculture productivity and food security in the coastal areas.

11.2.2 THE PERCENTAGE DEVIATION (DI)

It is derived through the below-mentioned formula: years (1951–2010) temperature and precipitation is used to calculate the PE for various stations of Kancheepuram were obtained from India Meteorological Department

$$[(P-Pi/Pi) \times 100]............ \quad (1)$$

where P is the annual rainfall in the year and Pi is the long term annual mean rainfall.

11.2.3 MOISTURE INDEX (IM)

Moisture index was developed by Raju et al. (2013) and Roa et al. (1971) and is being used here to classify the monthly and seasonal and climatic conditions prevailing in this area. The moisture index is an indicator of the supply of water in an area relative to the demand under prevailing climatic condition, Mather (1978). It uses precipitation (P) and PE to classify an area into different climatic zones. The PE data of 31 districts of Tamil Nadu except Chennai were collected from the published report of Rao et al. (1971). The seasonal moisture index (Im) was computed based on the model of Thornthwaite and Mather (1955) as given below

$$\text{Moisture Index (MI)} = MI = \frac{(P - \text{PE})}{(\text{PE})} \times 100 \qquad (2)$$

TABLE 11.1 Seasonal Moisture Index Values of Different Climate Types.

Sr. No	Category	MI Values	Seasons
1	Arid	>66.7 n, m	NE Monsson, Winter
2	Semiarid	–66.6 to 33.3	NE Monsson, Winter
3	Dry subhumid	–33.3 to 0	Summer
4	Moist subhumid	0 to + 20	
5	Humid	20.1–99.9	Southwest monsoon
6	Per humid	>100	

11.2.4 TREND ANALYSIS OF MOISTURE STATUS

In a Mann–Kendall test is carried out to understand the trend in the time series data. Under this test, there are two hypotheses, H0 indicates there is no trend in the series and Ha indicates a significant trend in the series. Trends in the observed moisture levels to the baseline period were assessed by running linear regressions on the MI index formula suggested by Raju et al. (2013). Trends for the observed dataset have been calculated for monthly, annual, and all four seasonal datasets like southwest monsoon season (June–September) and northeast monsoon season (October–December). Winter (January, February), summer (March–May). The magnitude and direction of trends and its statistical significance can be tested by time series trend tests. Traditionally, methods like trend analyses have been widely used to identify changes in the climate around the world. For the same reason, several climatologists have provided extensive methodology (Dhorde and Gadgil, 2009; Jain and Kumar, 2012). This test examines the null hypothesis that there is no upward or downward trend in the time series data, against the alternative hypothesis that the trend in the data exists.

The endeavors of the present study were made to analyze the long term trends in the moisture conditions prevailing in the Kancheepuram district, South India and project decadal changes in groundnut crop yield using Decision Support System for Agro Technology Transfer (DSSAT) under the IPCC medium emission scenarios RCP 4.5 trajectories. The meteorological parameters precipitations (P) and PE were used to create Moisture Index and to assess the temporal and spatial variations in aridity status. Analysis of the mean annual moisture conditions was carried out for six stations of the Kancheepuram district, Tamil Nadu during the period 1951–2008.

11.2.5 CLIMATE DATA ANALYSIS AND DSSAT MODELS

Dynamical downscaling of regional climate model, Regional Climate Model-RegCM 4.4 rc22, with HadGEM 2-Earth System model is employed to generate future likely climate change (Hoogenboom et al., 2017; Olajire et al., 2020). The latest available emission trajectory, Representative Concentration Pathway 4.5 (RCP 4.5), proposed by the AR5 report of IPCC has been used for the study. Data were culled from the 14 grid points covering the entire study area with the Arc GIS 10.2. GIS data extraction and analysis tools utilize ArcGIS geo-processing capabilities (Rashid et al., 2013; Singh et al., 2017a). To analyze the 88-year simulation data of near century (2010–2040), mid-century (2041–2070), and end century (2071–2098) climate scenarios and the previous 30-year base period (1970–2000), 88 years of data pertaining to the simulations were analyzed. For agricultural impact assessment, the major weather variables (daily maximum and minimum temperatures, solar radiation, and rainfall) were extracted from the outputs of the Regional Climate Models (RCMs). Mean changes for the 30-year period and decadal estimations were made from the daily data outputs. Use of simulation model, Decision Support System for Agrotechnology Transfer, was utilized for the assessment of crop yield impact (DSSAT-V4.5). To represent future changes in yield, CROP-GRO was utilized as part of the legume modules. Data obtained from on-the-ground research and experimental work performed at Tamil Nadu Agricultural University's research centers was critical in formulating the simulation models.

11.3 RESULTS AND DISCUSSION

11.3.1 TRENDS IN THE DAY AND NIGHT TEMPERATURE AND RAINFALL DURING 1950S AND 2010

It appeared as though the mean maximum temperature increased, while the mean minimum temperature stayed the same, and the precipitation levels dropped over the studied area. Both day and night time temperature significantly increased over the course of 64 years, beginning in 1951 and continuing until 2014, as shown in Figures 11.2 and 11.3. In general, the minimum temperature trended higher compared to the maximum temperature. There was no clear trend in the amount of rainfall, as the computed *p*-value (*P*-value) was greater than the significance level alpha (significance level of 0.05) and one null hypothesis H0 could not be rejected.

TABLE 11.2 Mann-Kendal Test Statistical of Annual Mean Temperature and Rainfall.

	Kendall tau value	*P* value	SD	Min	Max	Mean	Sen's slope value
Annual M T max	0.645	< 0.0001	0.501	31.656	34.008	32.614	0.02
Annual M T min	0.670	< 0.0001	0.768	22.175	25.125	23.286	0.03
Annual T rainfall	0.115	0.184	237.218	763.293	1934.200	1189.387	2.208

11.3.2 PERCENTAGE DEVIATION OF RAINFALL FROM THE NORMAL

Assessing the characteristics of drought in an area would help in understanding the prevailing situations. As it is recurring climatic extreme event, analysis of the frequency and extent to which it appears also requires special attention. The analysis revealed that Kancheepuram experienced Moderate drought years in 1974, 1980, 1982, 1999 with −35.78%, −30.09%, −30.54%, −27.30% deviation from the normal rainfall, respectively. The percentage deviation of rainfall for the years 1973, 1991, and 2002 were −20.35%, −20.39%, (−22.2%), respectively. The results of the analysis reveal that the mild drought years were 1970, 1971, 1986, 1987, 1989, 1990, 1994, 1995, 1998, 2000, 2001, 2004, 2006, and 2009, etc. (Figure 11.3).

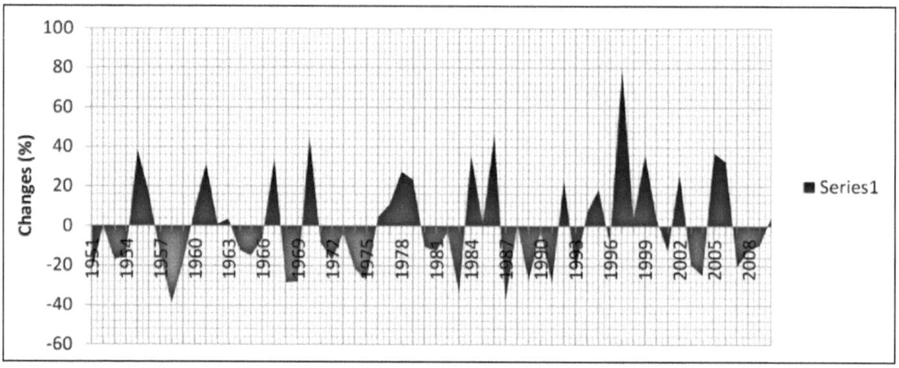

FIGURE 11.3 Deviations in rainfall 1951–2010 in Kancheepuram.

11.3.3 MONTH AND SEASON WISE TIME SERIES ANALYSIS OF MOISTURE CONDITIONS

In January and February months, the trend in moisture availability in showing a significant declining trend the analysis for the period 1951–2008 shows that 1961 (144.8), 1967 (159.4), 1970, the moisture index has a value above 119 from 1970s, onward it shows a declining trend. During the years 1953, 1958, and in 1997, 2001, 2003, and 2004 the moisture index values were around −95 experiencing an arid climate. Similarly the month of February also showed a negative trend 1954 and 1962 and 2004 recorded −95 and −97, −92.4 MI, respectively. In March, moisture conditions during the period 1951–2008 showed an ordinary trend following a median way throughout, and March month in 1978 and 1968 marked par humid conditions with MI values 358.1 and 118.1, respectively. Till 1971, the month of April experienced conditions very similar to semidry, however after that till 1983, this area experienced excess moisture conditions and positive trend with the year 1973, 1977, 1980, 1983 recording MI of 103, 134.6, 238.4, 124.5, respectively. After 1984, there is a declining trend in moisture conditions in April. In May on the contrary moisture availability improved during these years, especially from 1978 with MI of 209.1 and 341.9 during 1979 and 1986, etc. After 1987, the values tend to follow a decreasing trend, indicating moisture stress conditions during summer till 2008.

In June, moisture conditions showed a significant increasing trend especially from 1983 onward. The years 1984, 1992, 1995, and 1999 have showed recording of 154.3, 383, 208.7, and 192.5, respectively. In 1996 onward it showed a declining trend. Especially in 2004 it went down to −76.9.

Similar to June, July also followed a similar positive trend in 1998 and 2002 recording the 145.2 and 143.7 MI values. Similar to June, July–August also had shown increase in the moisture levels since 1995. The year 2006 and 1997 had recorded an MI of 337.6 and 319.7, showcasing mere par humid conditions. In September, onward on contrary to September showed a declining trend in 1960s–1999 thereafter it started rising with peaks during 2002 and 2006 was 39.3, 167.9 MI. On the contrary, October, November, and December showed a significant declining trend with an R^2 value of 0.4. During the years 1980s, 2003, 2005, October MI was around −100, depicting a semiarid climate. However, maximum moisture deficit was noted in November month with an R^2 value of 0.63. During 1952, 1968, 1984, 1999, and 2007, MI were −99.2, −86.2, −99.1, −100, and −100, respectively. In December, 1991 and 2008, the MI was −98 and −100, respectively.

11.3.4 ANNUAL MEAN MOISTURE CONDITIONS

Analysis of the annual mean moisture conditions was carried out using precipitation and PE data of six station data in the district during the period 1951–2008. As far as annual mean moisture index was concerned, none of the stations showed any significant trend. The annual mean moisture index for Chittamur, Madhuranthakam, Thirukazhikundram, Kancheepuram, Wallajabad, St. Thomas Mount were −35.8 ($R^2 = 0.01$), −35.5 ($R^2 = 0.02$), −28.6 ($R^2 = 0.05$), −42.4 ($R^2 = 0.01$), −36.2 ($R^2 = 0.05$), −28.4 ($R^2 = 0.10$), respectively. Among the stations, Kancheepuram recorded the lowest moisture levels, indicating its chronic dry conditions throughout 57-year period.

Delay in the onset of monsoon rains affects groundnut seed germination and extreme heat and dryness in the upper soil during the growing period causes flower dropping and less pod setting in groundnut. Dry matter production may be substantially reduced due to dry spells and droughts. Water-deficit conditions in the dry spells in the fruiting zone result in unfilled pods and less, calcium concentrations in the hull and seed. As reported by Olayide et al. (2016), climate smart irrigation systems especially, drip irrigation and mulching with groundnut crop stubbles are the options to sustain yields in dry land systems.

The droughts of 2002, 2003, and 2012 and floods during the years 1984, 1999, and 2005 also affected their standing crops and lost their valuable traditional seeds that we have been preserving for years. Pandey and Ramasastri (2002) and Drought network news (1998) have reported recurrent droughts conditions of the whole of Tamil Nadu in an elaborate way. Nearly

TABLE 11.3 Grid Wise Trends in Moisture Levels at Various Locations.

	Kendall Z value	P value	Min	Max	Mean	SD	Median	25th	75th	Skew	Kurtosis	S value
A Im-Madhuranthakam	*1.153*	*0.248*	−68.6	15.2	−35.46	19.20	−39.36	−48.6	−21.2	0.65	−0.11	173
A Im-Thirukalikundram	*1.770*	*0.076*	−67.5	28.39	−28.6	21.7	−31.61	−44.5	−15.9	0.62	−0.02	265
A Im- Wallajabad	*1.81*	*0.08*	−63.6	11.46	−36.19	17.87	−39.56	−49.9	−23.2	0.67	−0.02	267
A Im- St. Thomas Mount	*2.388*	*0.016*	−64.1	27.42	−28.35	21.64	−31.61	−44.4	−16.9	0.68	0.23	357
A Im- Kancheepuram	*0.8521*	*0.3941*	−65.5	5.5	−42.43	15.38	−45.4	−53.1	−34.8	0.86	0.37	128
A Im- Chittamur	0.724	0.468	−69.1	5.12	−35.7	19.5	−38.07	−50.5	−24.3	0.56	−0.38	109

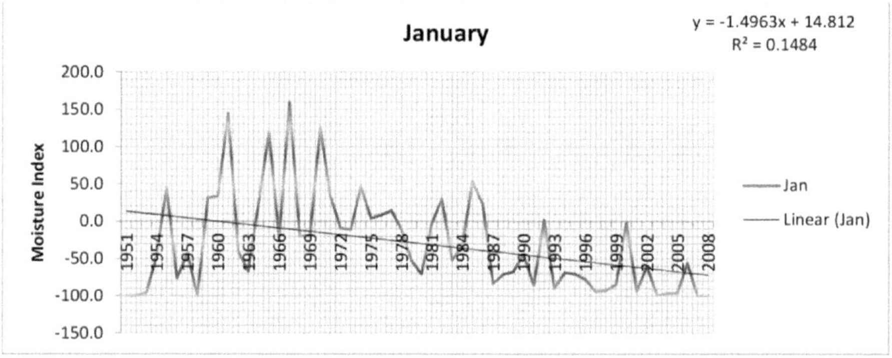

FIGURE 11.4 Graph showing the average moisture index values for the month of January from 1951 to 2008.

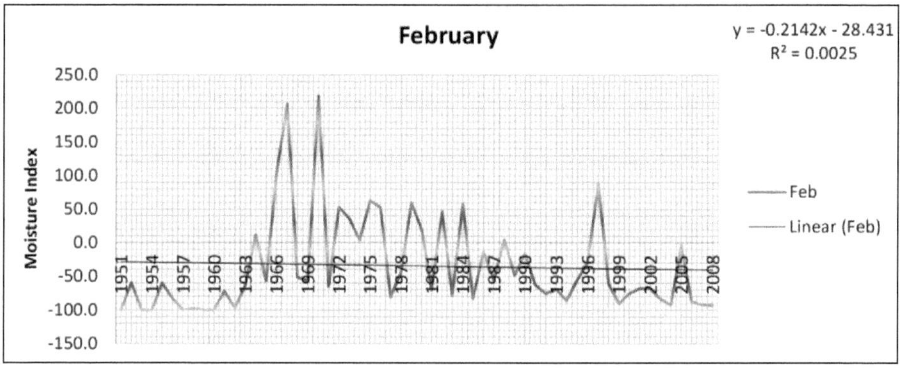

FIGURE 11.5 Graph showing the average moisture index values for the month of February from 1951 to 2008.

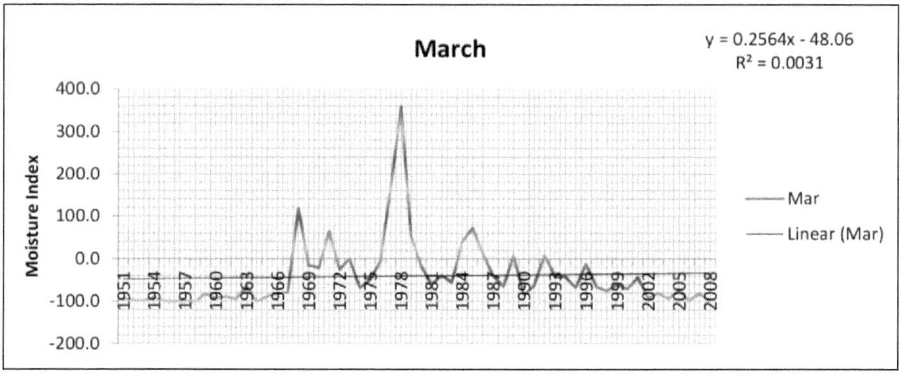

FIGURE 11.6 Graph showing the average moisture index values for the month of March from 1951 to 2008.

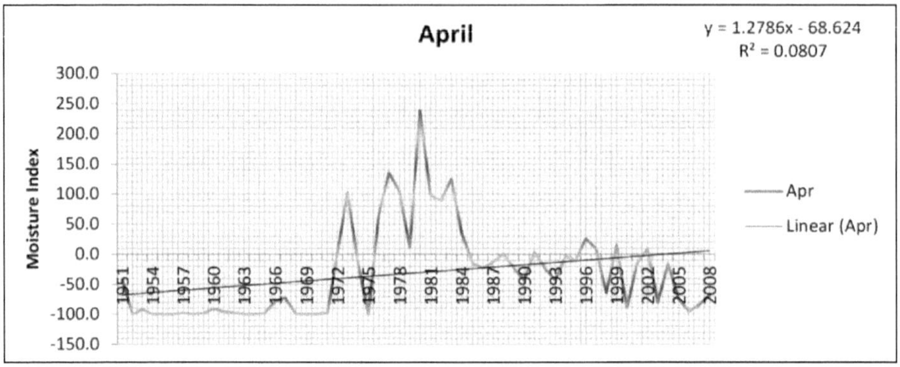

FIGURE 11.7 Graph showing the average moisture index values for the month of April from 1951 to 2008.

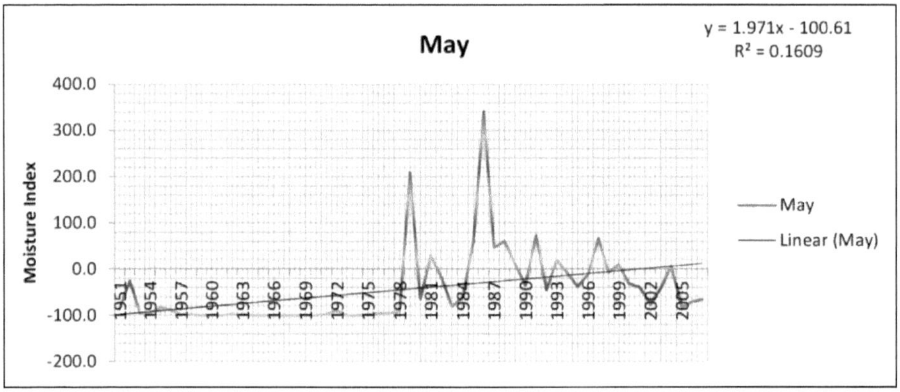

FIGURE 11.8 Graph showing the average moisture index values for the month of May from 1951 to 2008.

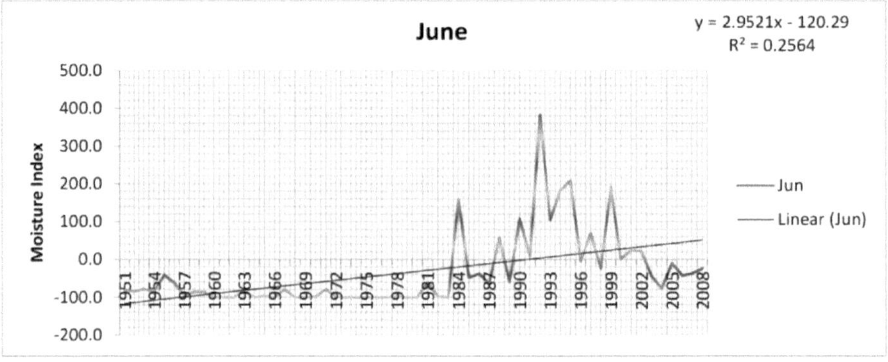

FIGURE 11.9 Graph showing the average moisture index values for the month of June from 1951 to 2008.

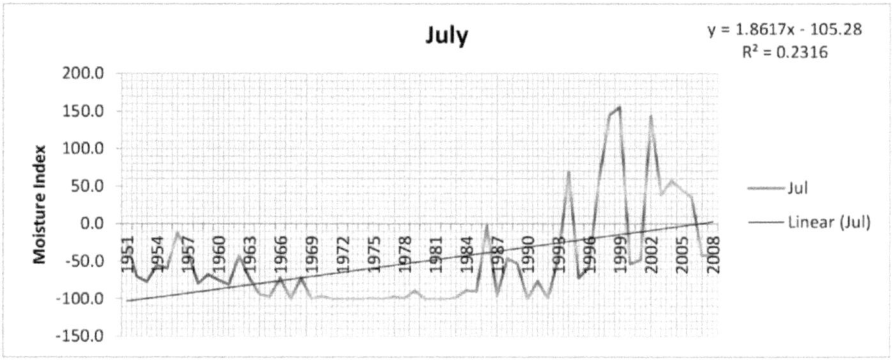

FIGURE 11.10 Graph showing the average moisture index values for the month of July from 1951 to 2008.

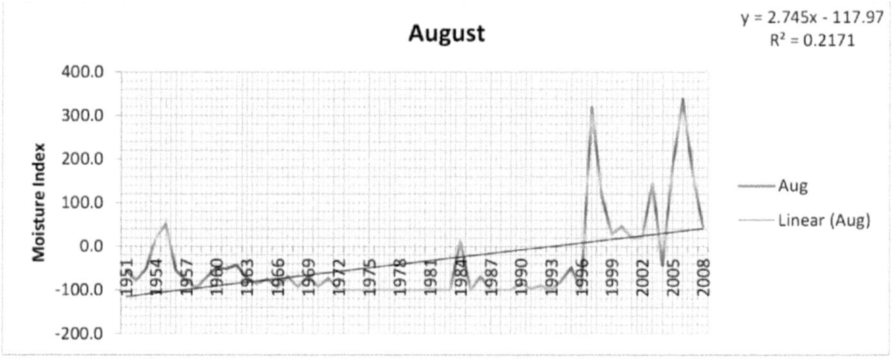

FIGURE 11.11 Graph showing the average moisture index values for the month of August from 1951 to 2008.

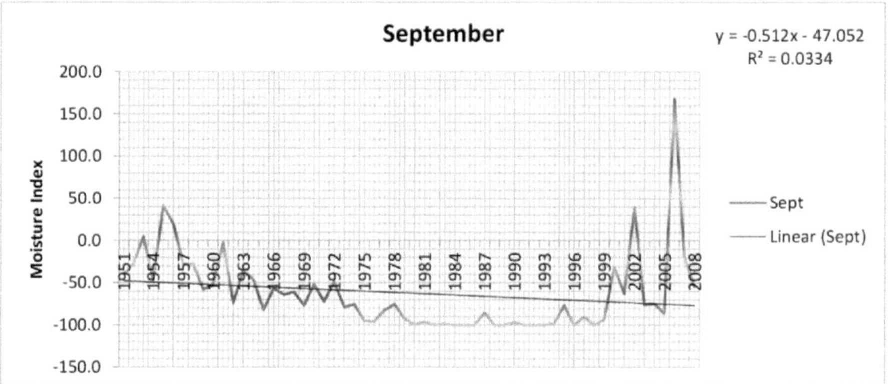

FIGURE 11.12 Graph showing the average moisture index values for the month of September from 1951 to 2008.

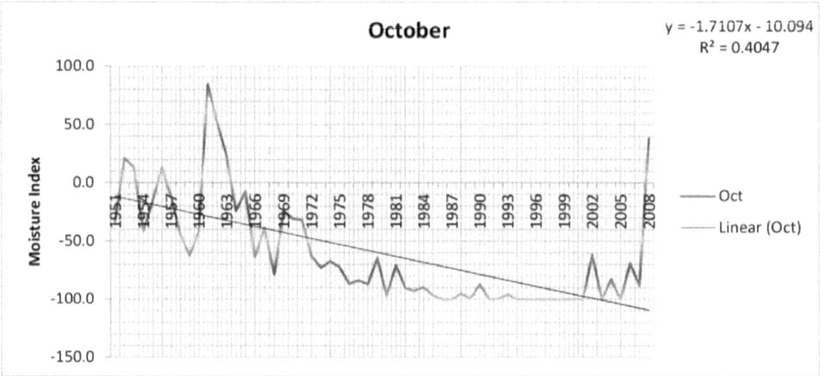

FIGURE 11.13 Graph showing the average moisture index values for the month of October from 1951 to 2008.

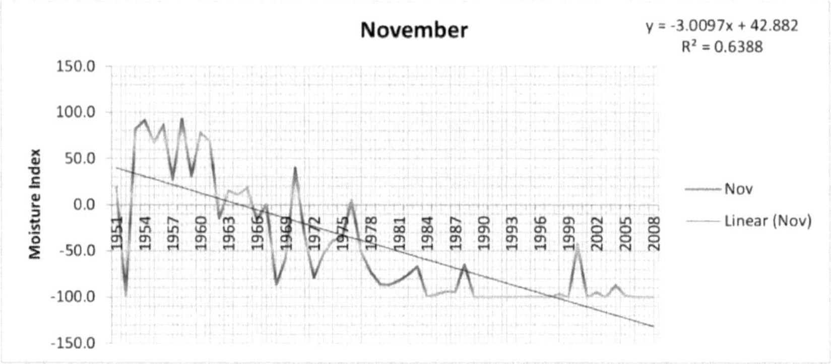

FIGURE 11.14 Graph showing the average moisture index values for the month of November from 1951 to 2008.

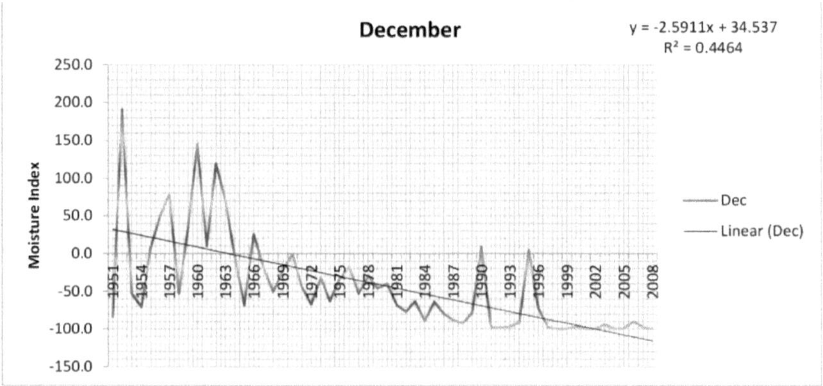

FIGURE 11.15 Graph showing the average moisture index values for the month of December from 1951 to 2008.

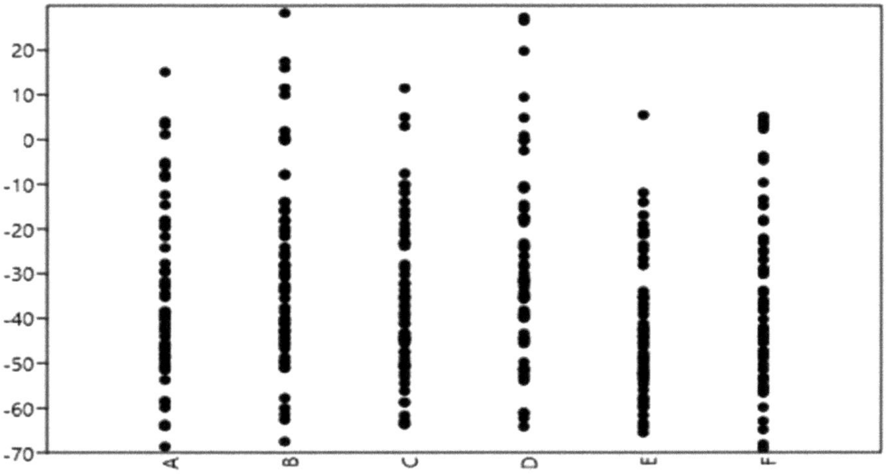

FIGURE 11.16 Jitter plot of annual mean moisture index for all the grid points for the Kancheepuram district, Tamil Nadu, South India during the period from 1951 to 2010.

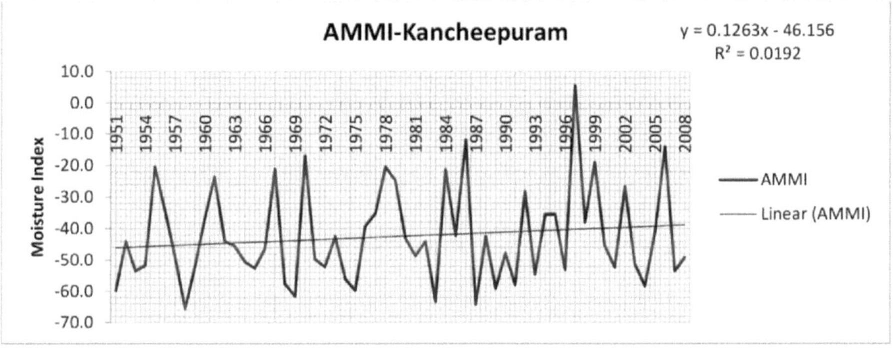

FIGURE 11.17 Annual mean monthly moisture index.

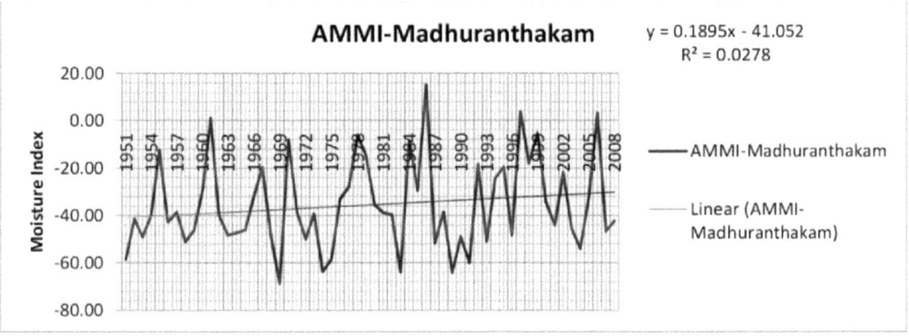

FIGURE 11.18

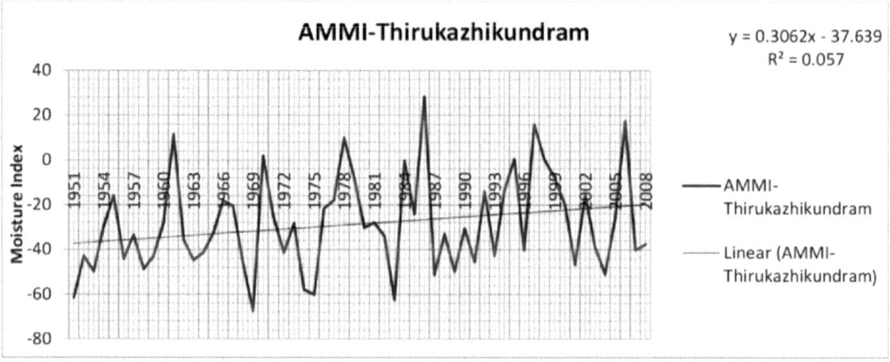

FIGURE 11.19

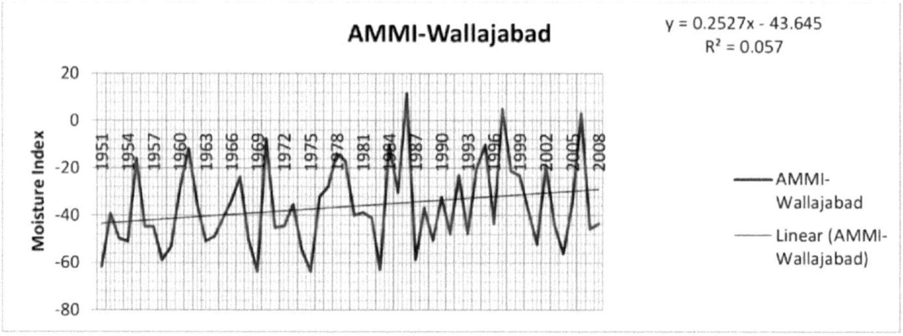

FIGURE 11.20

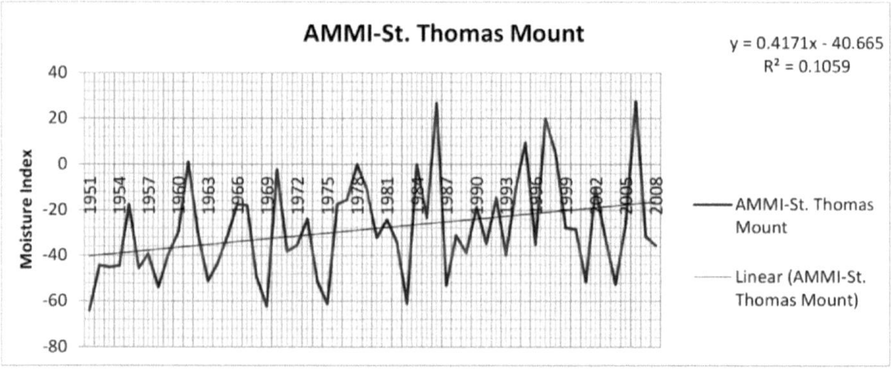

FIGURE 11.21

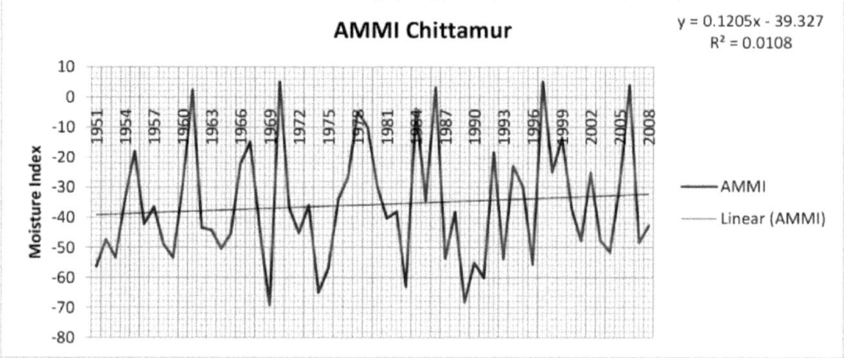

FIGURE 11.22

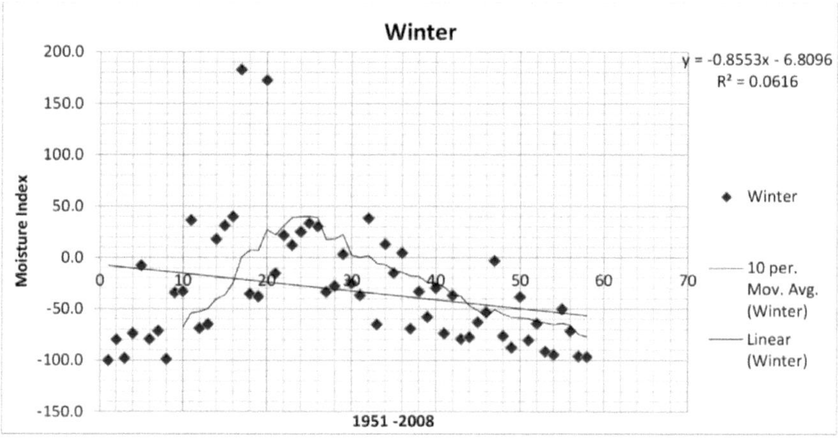

FIGURE 11.23

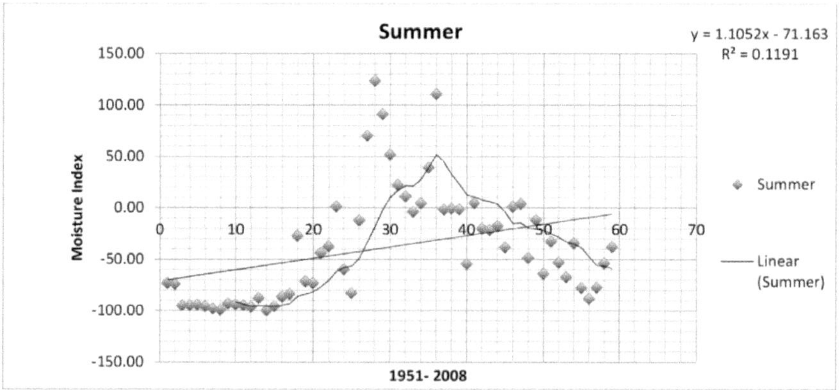

FIGURE 11.24

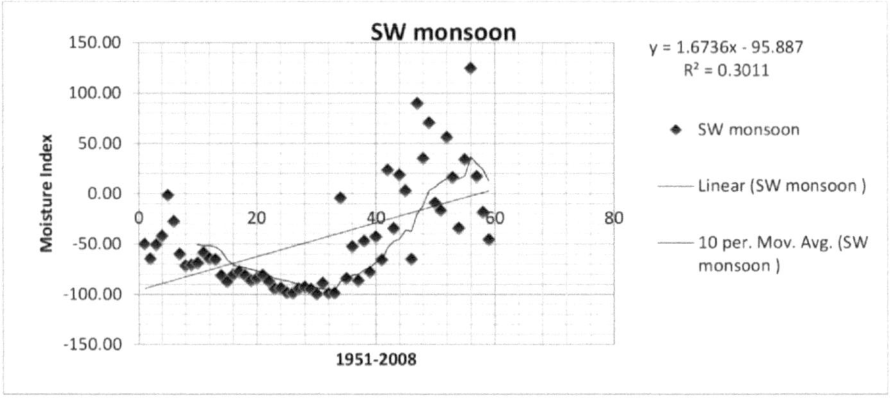

FIGURE 11.25

50% of the districts in the state are drought prone. The state receives nearly 80% of its annual rainfall during the northeast monsoon season. The state as a whole faces acute water crisis almost every 2.5 years (Nathan et al., 1998). The northeast zone of the state, comprising of the present study area Chengelpet districts, experiences cyclonic storms and associated flash floods during the northeast monsoon seasons (October–December) adding more climatic and land degradation vulnerability (Dhanya and Ramachandran, 2016).

11.3.5 SIMULATED CLIMATE CHANGE OVER THE STUDY AREA

Figures 11.26–11.28 show the spatial changes in mean annual temperature, rainfall, and aridity over the study area. The annual mean temperature is anticipated to change in the range of 0.8°C–1°C in the near century period and 1.7°C–2°C in the mid-century and 2.1°C–2. 6°C in the end-century period as per Reg CM model as per IPCC medium emission trajectory RCP 4.5.

There is a slight wetting trend noted in the precipitation distribution toward the end of the 21st century. However, the aridity values tend to decline in future especially for the mid-century period toward the coastal regions. Maps helped to understand the visual interpretation that even though there is wetting trend, due to future warming, aridity may be pronounced even in the coastal tracts of the study area.

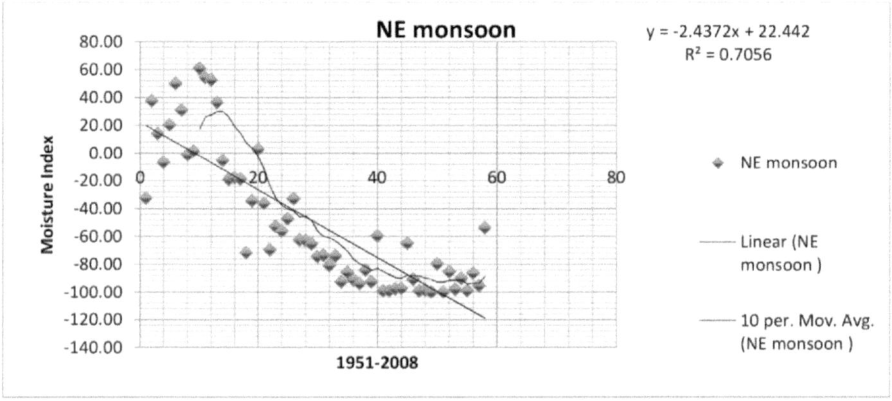

FIGURE 11.26

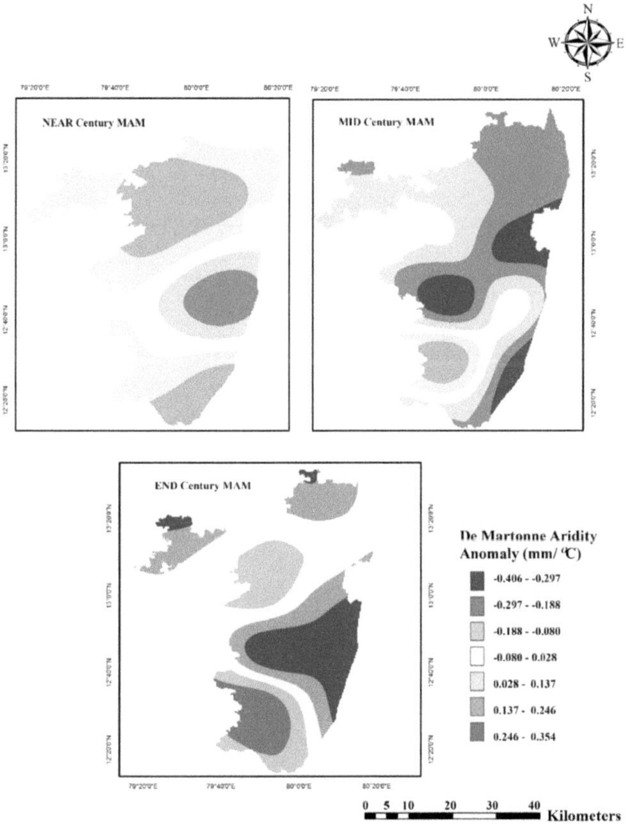

FIGURE 11.27 Spatial changes in mean annual temperature over the study area for the 21st century.

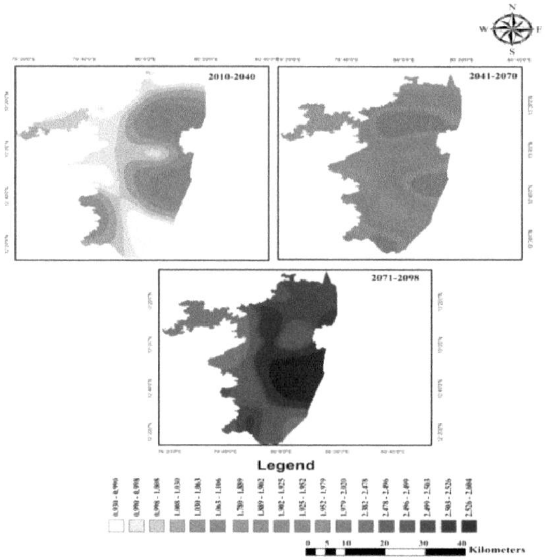

FIGURE 11.28 Spatial pattern of rainfall changes for the 21st century.

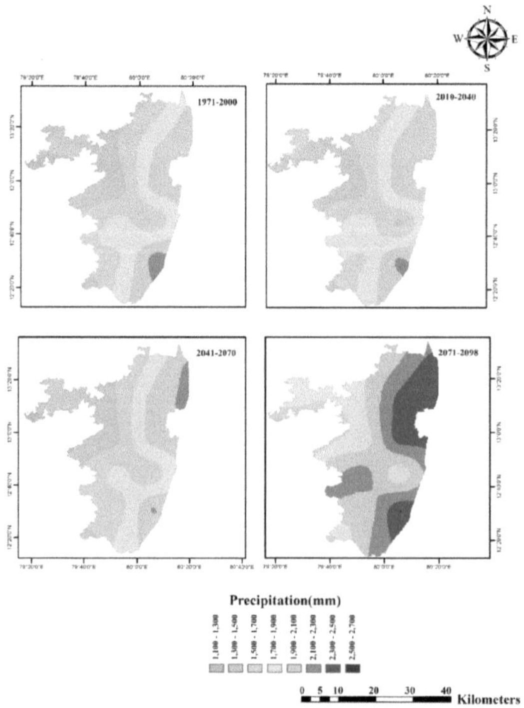

FIGURE 11.29 Anticipated changes in the annual mean temperature levels in the study area.

11.3.6 SIMULATED CROP YIELD FOR THE STUDY AREA

The decadal yield projections of day- and night-time temperature and groundnut crop yield are given in Figure 11.30.

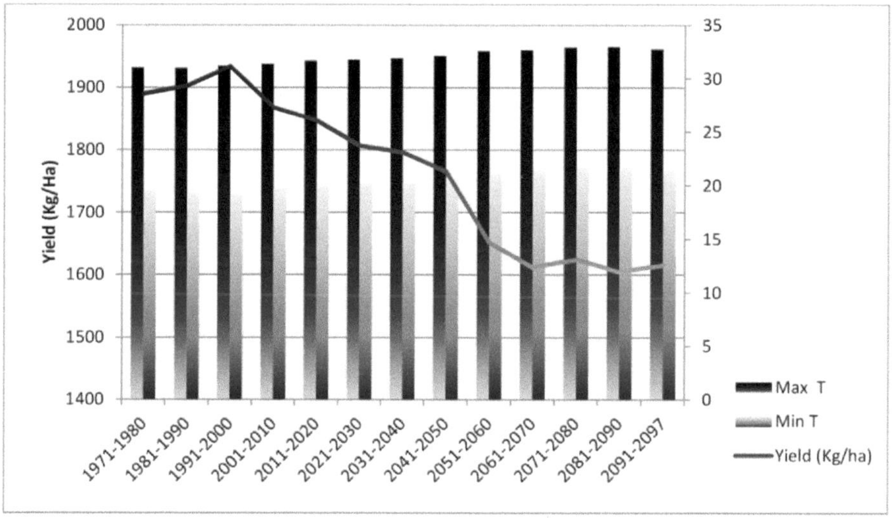

FIGURE 11.30 Decadal changes in Groundnut Crop Yield from 1970–1980 to 2091–2100 in Chengalpet areas.

PNUTGRO crop simulation model output revealed that the productivity of the peanut crops may fall toward the end of the 21st century. Yield of groundnut indicated a strong declining trend ($R^2 = 0.61$) between mid-century period, 2050s–2060s under control simulations. The mean yield under the control simulation over these 127 years is 1732.4 kg/ha, whereas the decadal mean groundnut yield under elevated CO_2 conditions is 1846.03 kg/ha. Differences in the decadal groundnut yield under control and elevated CO_2 simulation is more (−12.7%) from the baseline during 2091–2098 of the end century period (Figure 11.31). The yield is likely to be reduced by −11.8%, −12.7% for the decade 2061–2070 and 2091–2098 under CO_2-enriched scenario, respectively. The Spatio-temporal assessment helped to gain an understanding on the dynamics of the observed hydro climatic conditions and crop yield changes under climate change.

Maps and graphs are vital cartographic tools to understand the spatial temporal variations. Understanding level and severity of land degradation at a place, combined with the increasing adverse climate change and its impacts on crop productivity requires immediate and assertive national, state and district level actions. Degradation of land resources cause enhanced greenhouse gas emission from these areas. Land degradation neutrality (LDN) is an effective tool for sustaining healthy natural resources and restoring the degraded ones (UNCCD, 2019). It has the potential to act as a powerful

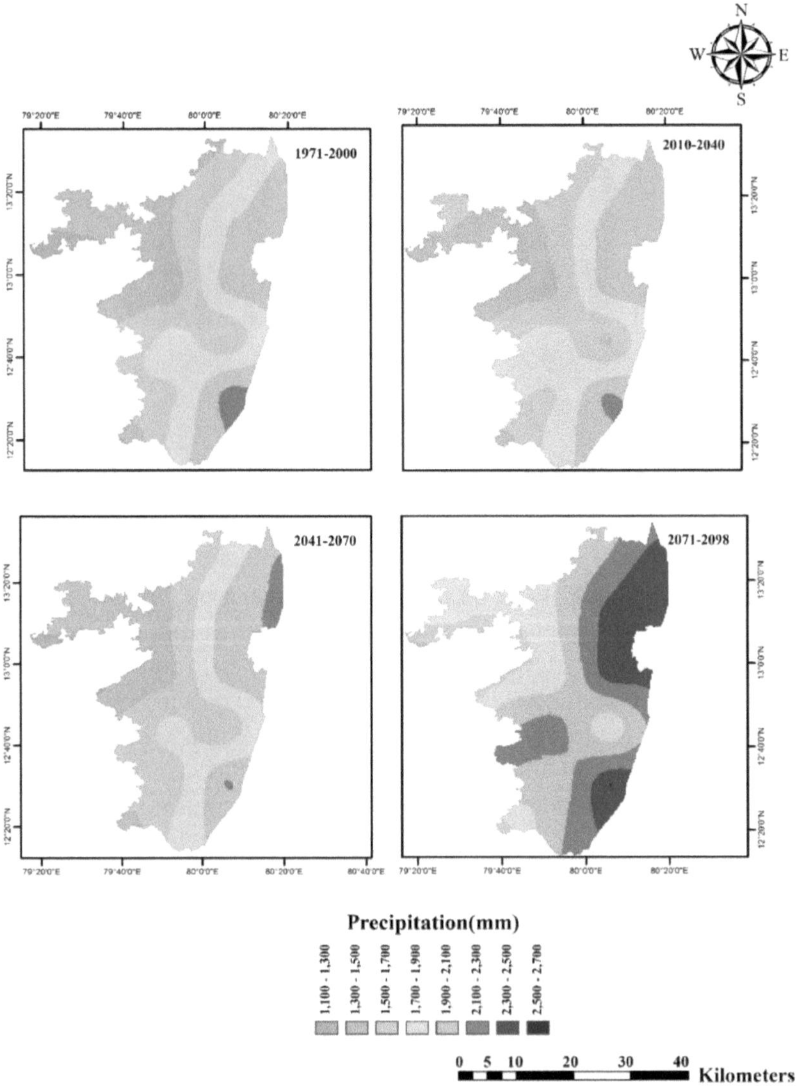

FIGURE 11.31

climate change mitigation and adaptation tool by increasing carbon stocks in soils and vegetation and build better ecosystem services in the study area and resilience to local communities.

11.4 CONCLUSION

This chapter also indicates the need to have robust and resilient community-based adaptations to tackle the combined challenge to meet the demands of the exponentially growing population, and achieve India's Nationally Determined Contributions and Sustainable Developmental Goals. Through this chapter, long-term local specific data analysis has been carried out which is a valuable information to monitor and understand the potential rate of climatic induced geomorphological changes in the study area. Geospatial analysis was helpful in identifying the NITI Aayog, India, in issuing guidelines to the planning departments of the state governments to undertake identifying SDG and INDC focal points in every other department to have inclusive sustainable development of all sectors including agriculture. There should be a link between state action plans on climate change (SAPCC, TN) and state level SDGs implementations that are associated with implementation of INDCs at state level. This paper identifies that inter departmental coordination is the need of the hour. Identify credible NGOs working with in the district who can aid in collecting evidences of the field realities. Areas that are frequently hit by multiple hazards like dry spells and droughts and cyclones should have a separate budget allocation for disaster-risk reduction. Government has to seek possibilities and scopes for interlinking of departments like TWAD board, rural development, Municipal Corporation, groundwater board, agriculture and allied sectors, and fisheries to have better moisture management plans. Planning at the smallest administrative level is most crucial for creating solutions at district and block level. Adaptive measures have to be taken to save crops during distress and more resilient crop varieties have to be made available to the farmers. This geospatial analysis indicated the need for location and context specific moisture management strategies have to be introduced to Kancheepuram areas. The efficient use of the available scientific outcomes is still at an infancy level in our country. Understanding the location specific trends will help in timely management of risks. It is also important to have risk communication to the concerned authorities and so that coordinated actions are possible at the block level.

KEYWORDS

- **moisture**
- **aridity**
- **geospatial technologies**
- **climate change**
- **RCP 4.5**
- **adaptations**

REFERENCES

Barbosa, H. A.; Lakshmi Kumar, T. V. Influence of Rainfall Variability on the Vegetation Dynamics over Northeastern Brazil. *J. Arid. Environ.* **2016,** *124*, 377–387. DOI: 10.1016/j.jaridenv.2015.08.015

Bastiaanssen, W. G. M.; Steduto, P. The Water Productivity Score (WPS) at Global and Regional Level: Methodology and First Results from Remote Sensing Measurements of…. *Sci. Total Environ.* **2016,** http://dx.doi.org/10.1016/j.scitotenv.2016.09.032

Bhatt, C. M. et al. Satellite-Based Assessment of the Catastrophic Jhelum Floods of September **2014,** Jammu & Kashmir, India. *Geomat. Nat. Hazards Risk* **2017,** *8* (2), 309–327.

Brunetti, M.; Maugeri, M.; Nanni, T.; Navarra, A. Droughts and Extreme Events in Regional Daily Italian Precipitation Series. *Int. J. Climatol.* **2002,** *22*, 543–558 .

Carolyn, S. F.. Cartographic Content Analysis of Compelling Climate Change Communication. *Cartogr. Geog. Info. Sci.* **2020,** *47* (6), 492–507. DOI: 10.1080/15230406.2020.1774421

Ch. Srinivasa, R.; Gopinath, K. A.; Prasad, J. V. N. S.; Prasannakuma, R.; Singh, A. K. Chapter Four—Climate Resilient Villages for Sustainable Food Security in Tropical India: Concept, Process, Technologies, Institutions, and Impacts. *Adv. Agron.* **2016,** 140, 101–214.

Dhanya, P.; Ramachandran, A.; Bal, P. K.; Thirumurugan, P. Recent and Future Weather and Climate Trends of Kancheepuram (TN) and Its Possible Impacts. *J. Agrometeorol.* **2013,** *2*, (2), 176–182.

Dhanya, P.; Ramachandran, A. The Urgent Need to Have Land Degradation Vulnerability. *AIMS Agric.* **2016**.

Dhorde, A.; Gadgil, A. S. Long-Term Temperature Trends at Four Largest Cities of India during the Twentieth Century. *J. Ind. Geophys. Union* **2009,** *13* (2), 85–97.

Fagariba, C. J.; Shaoxian Song, S.; Baoro, S. K. G. S. Climate Change Adaptation Strategies and Constraints in Northern Ghana: Evidence of Farmers in Sissala West District. *Sustainability* **2018,** *10*, 1484. https://doi.org/10.3390/su10051484

FAO. Climate-Smart Agriculture Sourcebook. Executive Summary, Food and Agriculture Organization of the United Nations, Italy, 2013. http://www.faoorg/3/a-i3325epdf (accessed Apr 28, 2015).

Farooq, M.; Muslim, M. Dynamics and Forecasting of Population Growth and Urban Expansion in Srinagar City-a Geospatial Approach. *Int. Arch. Photogrammetry Remote Sens. Spatial Inf. Sci.* **2014,** *40* (8), 709.

Food and Agriculture Organization of the United Nations (FAO). *Crop Yield Response to Water*; Steduto, P.; Hsiao, T. C.; Fereres, E.; Raes, D.; Eds.; Food and Agriculture Organization of the United Nations: Rome, Italy, 2012.

Gadgil, A.; Dhorde, A. Temperature Trends in Twentieth Century at Pune, India. *Atmos. Environ.* **2005,** *35* (39), 6550–6556.

Gujree, I et al. Evaluating the Variability and Trends in Extreme Climate Events in the Kashmir Valley Using PRECIS RCM Simulations. *Model. Earth Syst. Environ.* **2017,** *3* (4), 1647–1662.

Hassanin, M.; Kanga, S.; Farooq, M.; Singh, S. K. Mapping of Trees Outside Forest (ToF) From Sentinel-2 MSI Satellite Data using Object-Based Image Analysis. *Gujarat Agric. Univ. Res. J.* **2020,** 207.

Hoogenboom, G.; Porter, C. H.; Shelia, V.; Boote, K. J.; Singh, U.; White, J. W.; Hunt, L. A.; Ogoshi, R.; Lizaso, J. I.; Koo, J.; Asseng, S.; Singels, A.; Moreno, L. P.; Jones, J. W. Decision Support System for Agrotechnology Transfer (DSSAT) Version 4.7. DSSAT Foundation, Gainesville, 2017. www.DSSAT.net (accessed Dec 25, 2018).

IFAD. *The Adaptation Advantage: The Economic Benefits of Preparing Small-Scale Farmers for Climate Change*; International Fund for Agricultural Development: Rome, Italy, 2014.

IPCC. Terrestrial and Inland Water Systems, Final Draft, 2013; Chapter 4. http://ipccwg2.gov/AR5/images/uploads/WGIIAR5-Chap4_FGDall.pdf (accessed on Oct 21, 2015).

IPCC. Climate Change 2014: Impacts, Adaptation, and Vulnerability. Part A: Global and Sectoral Aspects. Contribution of Working Group II to the Fifth Assessment Report of the International Panel of Climate Change, 2014.

IPCC. Managing the Risks of Extreme Events and Disasters to Advance Climate Change Adaptation. A Special Report of Working Groups I and II of the Intergovernmental Panel on Climate Change; Field, C. B., Barros, V., Stocker, T. F., Qin, D., Dokken, D. J., Ebi, K. L., Mastrandrea, M. D., Mach, K. J., Plattner, G.-K., Allen, S. K., Tignor, M., Midgley, P. M., Eds.; Cambridge University Press: Cambridge, UK; New York, NY, USA, 2012; 582 pp.

IPCC. Summary for Policymakers. In Climate Change 2007: The Physical Science Basis; Contribution of Working Group I to the Fourth Assessment Report of the Intergovernmental Panel on Climate Change; Cambridge University Press: Cambridge, UK, 2007; pp 1–18.

Jacobs, J. M.; Mohanty, B. P.; Hsu, E.; Miller, D. SMEX02: Field Scale Variability, Time Stability and Similarity of Soil Moisture. *Remote Sens. Environ.* **2004,** *92*, 436–446. DOI: 10.1016/j.rse.2004.02.017

Jain, S. K.; Kumar, V. Trend Analysis of Rainfall and Temperature Data for India. *Curr. Sci.* Jan 10, **2012,** *102* (1), 37–49.

Joy, J.; Kanga, S.; Singh, S. K. Kerala Flood 2018: Flood Mapping by Participatory GIS Approach, Meloor Panchayat. *Int. J. Emerg. Technol.* **2019,** *10* (1), 197–205.

Kanga, S.; Rather, M. A.; Farooq, M.; Singh, S. K. GIS Based Forest Fire Vulnerability Assessment and Its Validation Using Field and MODIS Data: A Case Study of Bhaderwah Forest Division, Jammu and Kashmir (India). *Indian Forester* **2021,** *147* (2), 120–136.

Kanga, S.; Singh S. K.; Sudhanshu. Delineation of Urban Built-Up and Change Detection Analysis Using Multi-Temporal Satellite Images. *Int. J. Recent Res. Aspects* **2017a,** *4* (3), 1–9.

Kanga, S.; Kumar, S.; Singh, S. K. Climate Induced Variation in Forest Fire Using Remote Sensing and GIS in Bilaspur District of Himachal Pradesh. *Int. J. Eng. Comput. Sci.* **2017b,** *6* (6), 21695–21702.

Krishna Kumar, K.; Patwardhan, S. K.; Kulkarni, A.; Kamala, K.; Rao, K.; Jones, R. Simulated Projections for Summer Monsoon Climate over India by a High-Resolution Regional Climate Model (PRECIS). *Curr. Sci.* Aug 10, **2011,** *101* (3), 312–326.

Kumar, N. et al. Impacts of Climate Change on Indian Agriculture. *Curr. Sci.* **2012.**

Kumar, S. et al. Delineation of Shoreline Change Along Chilika Lagoon (Odisha), East Coast of India Using Geospatial Technique, 2018.

Marengo, J. A.; Torres, R. R.; Alves, L. M. Drought in Northeast Brazil—Past, Present, and Future. *Theor. App. Clim.* **2016,** *20*, 1–12. DOI: 10.1007/s00704-016-1840-8

Mather, J. R. *The Climatic Water Budget in Environmental Analysis*; Lexington Books: Lexington, MA, 1978; p 239.

Meraj, G. Ecosystem Service Provisioning—Underlying Principles and Techniques. *SGVU J. Clim. Change Water Res.* **2020,** *7*, 56–64.

Meraj, G.; Kumar, S. Economics of the Natural Capital and the Way Forward. Preprints 2021, 2021010083. DOI: 10.20944/preprints202101.0083.v1.

Meraj, G. et al. Assessing the Influence of Watershed Characteristics on the Flood Vulnerability of Jhelum Basin in Kashmir Himalaya. *Nat. Hazards* **2015a,** *77* (1), 153–175.

Meraj, G.; Romshoo, S. A.; Yousuf, A. R.; Altaf, S.; Altaf, F. Assessing the Influence of Watershed Characteristics on the Flood Vulnerability of Jhelum Basin in Kashmir Himalaya: Reply to Comment by Shah 2015. *Nat. Hazards* **2015b,** *78* (1), 1.

Meraj, G.; Romshoo, S. A.; Altaf, S. Inferring Land Surface Processes from Watershed Characterization. In *Geostatistical and Geospatial Approaches for the Characterization of Natural Resources in the Environment*; Springer: Cham, 2016; pp 741–744.

Meraj, G.; Singh, S. K.; Kanga, S.; Islam, M. N. Modeling on Comparison of Ecosystem Services Concepts, Tools, Methods and Their Ecological-Economic Implications: A Review. *Model. Earth Syst. Environ.* **2021,** 1–20.

Nathan et al. Drought Network News; Tamil Nadu, 1998.

Nathawat, M. S. et al. Monitoring & Analysis of Wastelands and Its Dynamics Using Multiresolution and Temporal Satellite Data in Part of Indian State of Bihar. *Int. J. Geomat Geosci.* **2010,** *1* (3), 297–307.

Olajire, M. A.; Matthew, O. J.; Omotara, O. A. et al. Assessment of Indigenous Climate Change Adaptation Strategies and Its Impacts on Food Crop Yields in Osun State, South Western Nigeria. *Agric. Res.* **2020,** *9*, 222–231. https://doi.org/10.1007/s40003-01900424-8

Olayide, O. E.; Tetteh, I. K.; Popoola, L. Differential Impacts of Rainfall and Irrigation on Agricultural Production in Nigeria: Any Lessons for Climate-Smart Agriculture? *Agric. Water Manage* **2016,** *178*, 30–36.

Oyiga et al. Implication of Climate Change on Crop Yield and Food Accessibility in SubSaharan Africa. Interdisciplinary Term Paper, Center for Development Research. University of Bonn, Germany, 2011.

Pall, I. A.; Meraj, G.; Romshoo, S. A. Applying Integrated Remote Sensing and Field-Based Approach to Map Glacial Landform Features of the Machoi Glacier valley, NW Himalaya. *SN Appl. Sci.* **2019,** *1* (5), 488.

Pandey, A. C.; Singh, S. K.; Nathawat, M. S. Waterlogging and Flood Hazards Vulnerability and Risk Assessment in Indo Gangetic Plain. *Nat. Hazards* **2010,** *55* (2), 273–289.

Pandey, A. C.; Singh, S. K.; Nathawat, M. S.; Saha, D. Assessment of Surface and Subsurface Waterlogging, Water Level Fluctuations, and Lithological Variations for Evaluating Groundwater Resources in Ganga Plains. *Int. J. Digital Earth* **2013,** *6* (3), 276–296.

Pandey, R. P.; Ramasastri, K. S. Incidence of Droughts in Different Climatic Regions. *Hydrol. Sci. J.* **2002,** *47*, 31–40.

Paredes, F. J.; Barbosa, H. A.; Guevara, Y. E. Análisisespacial y temporal de lassequías en el nordeste de Brasil. *Agriscientia* **2015,** *1*, 1–14.
Quach, Q.; Bernhard, J. Immersive Visualization with Bar Graphics. *Cartogr. Geog. Inf. Sci.* **2020,** *47* (6), 471–480. DOI: 10.1080/15230406.2020.1771771
Raju, B. M. K.; Rao, K. V.; Venkateswarlu, B.; Rao, A. V. M. S.; Rama Rao, C. A.; Rao, V. U. M.; Bapuji Rao, B.; Ravi Kumar, N.; Dhakar, R.; Swapna, N.; Latha, P. Revisiting Climatic Classification in India: A District-level Analysis. *Curr. Sci.* **2013,** *104* (4), 492–495.
Ramachandran, A.; Praveen Dhanya, R. J.; D. R.; K. P. Spatiotemporal Analysis of Projected Impacts of Climate Change on the Major C3 and C4 Crop Yield under Representative Concentration Pathway 4.5: Insight from the Coasts of Tamil Nadu, South India. *PLoS One* **2018,** *12* (7), e0180706. https://doi.org/ 10.1371/journal.pone.0180706
Rao, K. N.; George, C. J.; Ramasastri, K. S. Potential Evopotranspiration (PE) over India. *Sci. Rep.* **1971,** No. 136, Drought Study Report, p 24.
Rather, M. A. et al. Remote Sensing and GIS Based Forest Fire Vulnerability Assessment in Dachigam National Park, North Western Himalaya. *Asian J. Appl. Sci.* **2018,** *11* (2), 98–114.
Romshoo, S. A. et al. Satellite-Observed Glacier Recession in the Kashmir Himalaya, India, from 1980 to 2018. *Environ. Monitor. Assess.* **2020,** *192* (9), 1–17.
Rossato, L.; Alvalá, R. Cd. S.; Marengo, J. A.; Zeri, M.; Cunha, A. P. M. d. A; Pires, L. B. M.; Barbosa, H. A.; Impact of Soil Moisture on Crop Yields over Brazilian Semiarid. *Front. Environ. Sci.* **2017,** *5*, 73. DOI: 10.3389/fenvs.2017.00073
Saeid, A. V.; Abbaspour, K. C.; Faramarzi, M.; Srinivasan, R.; Arnold, J. G. Modeling Crop Water Productivity Using a Coupled SWAT–MODSIM Model. *Water* **2017,** *9*, 157, 1–15. DOI: 10.3390/w9030157.
Sharma, A.; Kanga, S. Surface Runoff Estimation of Sind River Basin Using SCS-CN Method and GIS Technology, 2020.
Singh, S. K.; Pandey, A. C. Geomorphology and the Controls of Geohydrology on Waterlogging in Gangetic Plains, North Bihar, India. *Environ. Earth Sci.* **2014,** *71* (4), 1561–1579.
Thornthwaite, C. W.; Mather, J. R. The Water Balance. *Publ. Climatol. Lab. Climatol. Dresel Inst. Technol.* **1955,** *8* (8), 104.
UNCCD; FAO; WMO. High Level Meeting on National Drought Policy (HMNDP). Science Document: Best Practices on National Drought Management Policy; UNCCD, FAO, WMO: Geneva, 2013.
UNCCD. Drought Resilience, Adaptation and Management Policy Framework: Supporting Technical Guidelines; Editor: Tsegai, D. (UNCCD) Author: Neville D. Crossman United Nations Convention to Combat Desertification (UNCCD); Bonn, Germany, Aug 2019
Usha, R. R.; Rajarshi, K.; Gupta, D.; Chatterjee, R.; Shaw, R. Managing the Indian Coast in the Face of Disasters & Climate Change: A Review and Analysis of India's Coastal Zone Management Policies. *J. Coastal Conserv.* **2014.**
World Bank. The Costs to Developing Countries of Adapting to Climate Change. New Methods and Estimates. The Global Report of the Economics of Adaptation to Climate Change Study. 1818 H Street, NW Washington, DC 20433, USA, 2010.
World Water Assessment Programme (WWAP). *The United Nations World Water Development Report 4: Managing Water under Uncertainty and Risk*; World Water Assessment Programme: Paris, France, 2012.

CHAPTER 12

Impact Assessment of Urban Growth on Landscape Dynamics for Sustainable Management Using Geospatial Approach

MAYANK PANDEY[1*], ALKA MISHRA[1], and S. L. SWAMI[2]

[1]*Department of Rural Technology and Social Development, Guru Ghasidas University, Bilaspur, Chhattisgarh, India*

[2]*College of Agriculture and Research Station, Korba, IGKV, Chhattisgarh, India*

[*]*Corresponding author. E-mail: makepandey1995@gmail.com*

ABSTRACT

Landscapes are heterogeneous landmass characterized by diverse ecosystems representing the forest, grasslands, agriculture, aquatic, and urban settings. Understanding the complex interaction between landscape pattern and process is essential for sustainable management of land resources to ensure a continuous supply of multiple ecosystems like food, water quality, wood, climate regulation, pollination of the crop, etc. In this research work, various datasets such as LANDSAT- TM 4-5, and LANDSAT-8 have been used to generate the land use/land cover maps to monitor urban growth's impact on landscape dynamics and natural resources in Bilaspur, Chattisgarh, India. The built-up area of Bilaspur between 1990 and 2020 is increasing due to the migration of people from rural regions mainly for jobs, education, or to maintain the living standard. Changes in the built-up area are a severe problem. It is also challenging to manage the natural resources appropriately.

GIScience for the Sustainable Management of Water Resources. Gowhar Meraj, Shruti Kanga, Majid Farooq, Suraj Kumar Singh, & Sudhanshu (Eds.)
© 2023 Apple Academic Press, Inc. Co-published with CRC Press (Taylor & Francis)

12.1 INTRODUCTION

Extension of the urban area due to increase in population day by day and migration of rural people toward the metropolitan area to maintain the status of living life, for education purpose of children, to get the job and so on is nowadays grave concern. Its impact is bound to have on an urban area in terms of water, food, infrastructure, other vital resources, etc. Change monitoring, which requires comparing two images of different times, is a critical analysis technique for urban studies. The spatiotemporal analysis helps in studying urban sprawl and the growth process. The physical changes in the distribution of urban activities can be monitored directly through optical and microwave data. The size of settlements is related to the aerial extent of the built-up area, which is associated with the size of the population.

The allometric growth model and central place theory have been extensively used to estimate population on the basis of the built-up area extent (Kanga et al., 2017a, 2017b; Rather et al., 2018; Hassanin et al., 2020; Kanga et al., 2021). In many cases, urbanization indicates the transformation of an economy from a traditional one to a modern one. Most of the world's population now resides in cities for the first time in the history of humanity (Miller and Small, 2003). Natural resources are essential things and products of the earth, providing us with the required natural product and fulfilling the requirement of crucial resources to easily survive day-to-day life (Meraj 2020; Meraj and Kumar, 2021; Meraj et al., 2021). Nowadays human beings are inappropriately utilizing these natural resources. Thus, the result in their rate of reduction is swift (Rashid et al., 2013; Singh and Kanga, 2017). National Natural Resource Management System (NNRMS) was initiated by the Department of Space (DOS) for generating a proper and systematic inventory of natural resources, adopting various advanced technologies such as remote sensing, GIS (geographical information system), GPS (global positioning system), technology-database-networking framework, and various ground-based surveying approaches.

Finding out which resources and quantifying and location are necessary for effective management, as well as information on their utilization and interaction with human along with other natural phenomena (Smyth and Yong, 1998). Because of this, a thorough mapping and monitoring of natural resources are required on a regular basis. In the field of resource management, there is a common idea that assessment and inventory are regarded as necessary precursor activities for management decisions (Sebastian et al., 1985). We need detailed information and knowledge about a few items like land use and land cover of the research region to properly manage natural resources. While the terms land cover and land use are often used synonymously, they are actually distinct. However, land cover refers to the material that covers the earth's surface, whereas land use refers to whatever the land is used for. Land cover groups include water, ice, pastureland, forest, and bare land. Land use example includes wildlife management area, agriculture land, urban, recreation area, etc. and another thing like: topography, shape, and size of the watershed area, drainage parameters, geomorphology, lithological structure of the study area, etc. All the above mentioned information should be collected by the research for research work to manage the natural resources in a proper manner (Singh et al., 2017; Kanga et al., 2020a, 2020b).

Traditional and ancient time methods of resource survey such as ground-based surveying and mapping method, information through detailed field inventorying, including some demerits like lack of information, lack of infrastructure facilities, too much workforce, and the temporal distribution of the survey work, not the availability of real-time datasets (Pandey et al., 2010, 2013; Singh and Pandey, 2014; Meraj et al., 2015a, 2015b, 2016; Bhatt et al., 2017; Sharma and Kanga, 2020). Nowadays, remote sensing and GIS technique are providing us accurate and reliable information related to the earth's surface. It can be used as the most efficient and sophisticated tool for mapping natural resources and monitoring as well managed in a better way. Due to satellite sensors' ability to monitor the earth's features with high spatial resolution, high spectral resolution, coverage of the large area with a large swath of sensors as well collection of multitemporal data also. They help us to collect various seasons and time-series data of so many years. Even in rough and inaccessible terrain, geospatial technology has proven its ability to provide natural resource knowledge (Jensen, 1983; NRSA, 2000). For decades, there has been a strong interest in using earth observation data in urban areas (Haack et al., 1997; Tuyahov et al., 1973; Saxena et al., 1991, Meraj et al., 2013, 2018a, 2018b).

Early attempts to link satellite data reflectance to demographic parameters (Forster, 1983) resulted in the development of a categorization for satellite images applied to urban areas in order to generate an urban quality index. LO, 1986, 2001) used remote sensing data to forecast population and measure urban development with landscape usage (Mesev et al., 1995; Stefanov et al., 2001). A decade of aerial images and satellite images is used to track and analyze urban growth using remote sensing data to provide accurate information on urbanization and existing land-use changes. Al-Awadhi and

Azaz (2004). A variety of indicators and classification techniques, such as normalized differential vegetation index (NDVI), normalized difference water index (NDWI) and others, were used to analyze a picture. Popular physical and biological features related to the urbanized landscape pattern were identified to locate and classify the regions. The results were combined with a GIS, allowing users to verify the legal status of the classified region. Such methods are extremely useful for detecting complex irregularities in organic growth. Enhanced thematic TM and LANDSAT-7 ETM images were also used in a report on urban growth in Istanbul (Kaya, 2007). Eight key land use/land cover groups were obtained after the photos were categorized separately, and the classified production was compared to determine the degree of urban development and land-use transition (Farooq and Muslim, 2014; Nathawat et al., 2010; Kumar et al., 2018; Joy et al., 2019). Different remote sensing from 1976 to 2005, as well as Shiraz city's population census during this time span, were used in another analysis for Shiraz city in Iran, where the land use coverage for various dates was categorized and calculated (Ibrahim and Sarvestani, 2009).

In this chapter, a study has been done to monitor the changes in urban growth during (1990–2020) for this research work (LANDSAT-8 and LANDSAT TM 4-5 data with a spatial resolution of 30 m) has been used and indices have been used for impact assessment of urban growth on the landscape dynamics and natural resources such as vegetation water body.

12.2 STUDY AREA

The city of Bilaspur is about 400 years old and named after the fisherman, "Bilaspur." With time and hard work, Bilaspur has done a great deal. It is situated at 21.47 to 23.8° N and 81.15 to 83.15 E. Gaurela-Pindra in the north, and Bhatapara-Korba to the west, and the Chhattisgarh districts of Korba and Janjgir-Champa are located in the southeast of the district. This part of the area of the district is 350.48 km^2. The district population is 165,002,500 people. Bilaspur district presently consists of 5 tehsils, 4 blocks, and 708 villages. It is one of the two largest metropolitan areas in the Raipur-Durg metropolitan district. The High Court of Chhattisgarh was given the title "capital of justice" (capital of the state) because of its location in the village of Bodri. The Chhattisgarh Administrative Tribunal is the largest court in Asia. The District Headquarters of Bilaspur is where you are to be found (Figure 12.1).

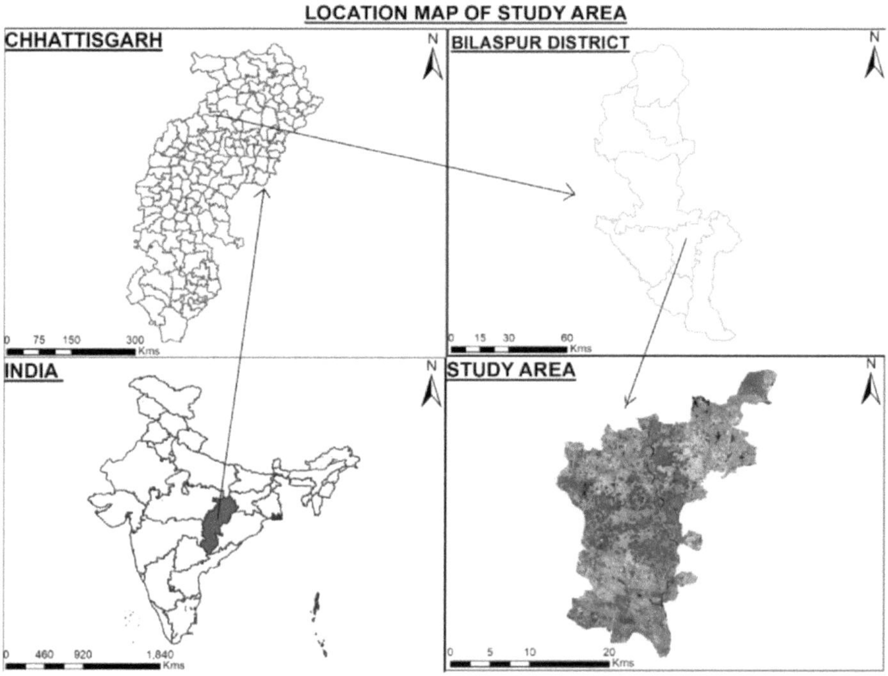

FIGURE 12.1 Study area map.

12.3 MATERIAL AND METHODS

ArcGIS 10.4 software has been used for the generation of the various maps. In this research work, data of different years have been collected to monitor the changes in urban growth sprawl between (1990–2020). Data have been downloaded from USGS (United states of Geological survey). LANDSAT-8 and LANDSAT TM-4,5 data (30 m of spatial resolution) have been used for the preparation of various maps such as follows.

12.3.1 LAND USE/LAND COVER MAPPING

Land use/land cover (LU/LC) map of the different year has been generated using SCT Maximum Likelihood method, to extract the urban area and its growth. This map has five-LU/LC classes such as—built up, water body, vegetation cover, sandy soil, cropland. All the maps have the same classes (Figure 12.2).

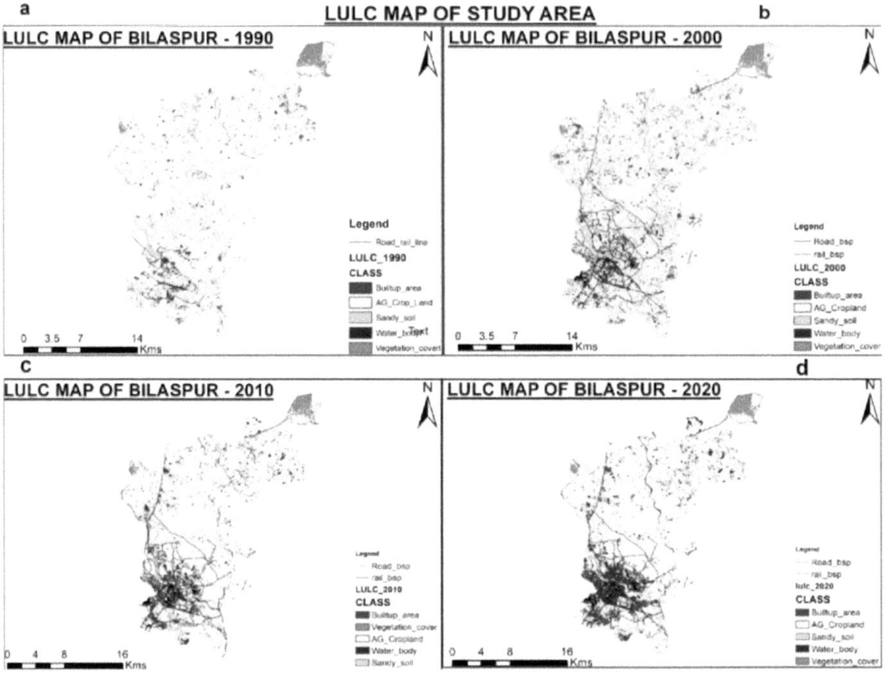

FIGURE 12.2 LULC map of study area (1990–2020).

LULC Map of 1990

When we do any kind of research or study it is very essential to prepare the LU/LC map of the study area, this will definitely help us to know about the smallest things associated with our study area and help us to find out every change that has been taken place in the study period. In this study, LULC map helps us to know about the changes that have been taken place during (1990–2020). LU/LC map of Bilaspur has been generated using supervised classification techniques, five classes have been generated such as follows—built-up area, agriculture cropland, vegetation cover, sandy soil (mainly river sand), water body, and road-rail network has been prepared using digitization method. Mainly three types of feature are used in digitization such as—point, line, and polygon. Use of this featured to digitize any things is depending up on what type of feature you are going to digitize like if you have to extract line featured such as—road, rail, drainage network, pipeline so you should use line feature for digitization, likewise if you want to extract any area so you should go for polygon and point is for location of any feature like—ATM, school, or building that you want to show as tagged featured. However, this method is very time-consuming that is why it is not very likely to do by the GIS expert than the classification method for the preparing LULC map.

LULC Map of 2000

LULC map of Bilaspur, 2000 has been generated using supervised classification techniques, five classes have been generated such as follows built-up area, agriculture crop land, vegetation cover, sandy soil (mainly river sand), water body, and road rail network has been prepared using digitization method.

LULC Map of 2010

LULC map of Bilaspur, 2010 has been generated using supervised classification techniques, five classes have been generated such as follows built-up area, agriculture crop land, vegetation cover, sandy soil (mainly river sand), water body, and road rail network has been prepared using digitization method.

LULC Map of 2020

LU/LC map of Bilaspur in the year 2020 has been generated using supervised classification techniques, five classes have been generated such as follows—built-up area, agriculture crop land, vegetation cover, sandy soil (mainly river sand), water body, and road rail network has been prepared using digitization method.

NDVI Map *(Figure 12.3a–d)*

NDVI map of two different years has been generated to monitor the changes in vegetation cover due to the urban area expansion. To generate the NDVI map, two bands have been used such as NIR (near-infrared) and ed. for *LANDSAT-8* data band number *(5—NIR and 4—Red) and for LANDSAT TM-4,5 band number (4—NIR and 3—Red)* has been used, then this map has been reclassified in two-class—vegetation and other class, then area calculation of both the classes have been done using attribute table. The formula to generate the NDVI map is

$$\text{NDVI} = \frac{(\text{NIR} - \text{Red})}{\text{NIR} + \text{Red}}$$

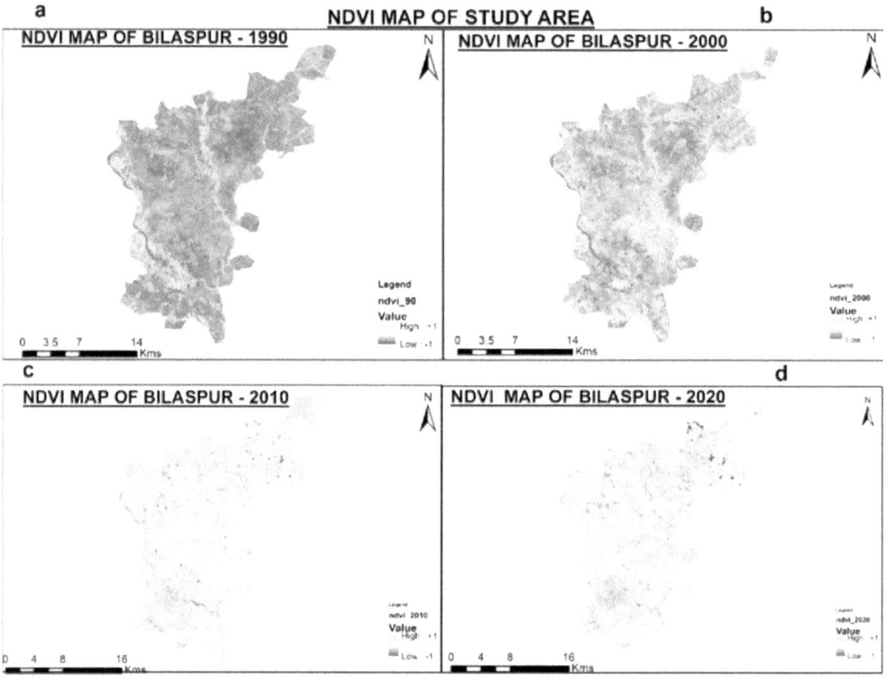

FIGURE 12.3 Study area NDVI maps.

NDVI Map of 1990

NDVI map of Bilaspur in the year of 1990 has been generated using map algebra (raster calculator), this method is very usefull and easy to calculate the area covered by the vegeatation, they reflect strongly in the portion of NIR region, so the two classes have been generated in this map such as vegetation and other classes (+1 values are showing the heighy vegetation covered area were as the 1 values are showng the lowest vegetation covered area).

NDVI Map of 2000

NDVI map of Bilaspur in the year of 2000 has been generated using map algebra (raster calculator), this method is very usefull and easy to calculate the area covered by the vegetation, they reflect strongly in the portion of NIR region, so the two classes have been generated in this map such as vegetation and other classes. (+1 values are showing the heighy vegetation covered area were as the −1 values are showng the lowest vegetation covered area).

NDVI Map of 2010

NDVI map of Bilaspur in the year of 2010 has been generated using map algebra (raster calculator), this method is very usefull and easy to calculate the area covered by the vegeatation, they reflect strongly in the portion of NIR region, so the two classes have been generated in this map such as vegetation and other classes. (+1 values are showing the heighy vegetation covered area were as the −1 values are showng the lowest vegetation covered area).

NDVI Map of 2020

NDVI map of Bilaspur in the year of 2020 has been generated using map algebra (raster calculator), this method is very usefull and easy to calculate the area covered by the vegeatation, they reflect strongly in the portion of NIR region, so the two classes have been generated in this map such as vegetation and other classes. (+1 values are showing the heighy vegetation covered area were as the −1 values are showng the lowest vegetation covered area).

Land Use and Land Cover Analysis (Figure 12.4)

LU/LC graph of the study area has been generated using Excel software, with the help of the attribute table, this method is very helpful and easy to handle. The overall analysis of LU/LC graph has shown that all classes have been changed frequently in between intervals of 10 years such as—1990, 2000, 2010, 2020. Study area majorly covered with the agriculture cropland class, two major changes have occurred in the classes of study area such as—agriculture cropland and built-up area, and some changes also occurred in the rest of the classes like- water body, vegetation cover, sandy soil. In this work, km² is used as (square km.) Built-up area class in the year (1990 covered with 20.002 km² area), in the year (2000 covered with 28.695 km² area), in the year (2010 covered with 35.865 km² area) and in the year of (2020 covered with 57.489 km² area). Whereas the agriculture cropland class in the year of (1990 covered with 410.535 km² area), in the year of (2000 covered with 375.125 km² area) in the year of (2010 covered with 389.44 km² area) and for year, 2020 covered with 367.708 km² area. This all changes have shown that the growth of the built-up area has a major impact on agriculture cropland, this class has been occupied by the concrete jungles due to day by day increasing of population rate and expansion of the built-up area.

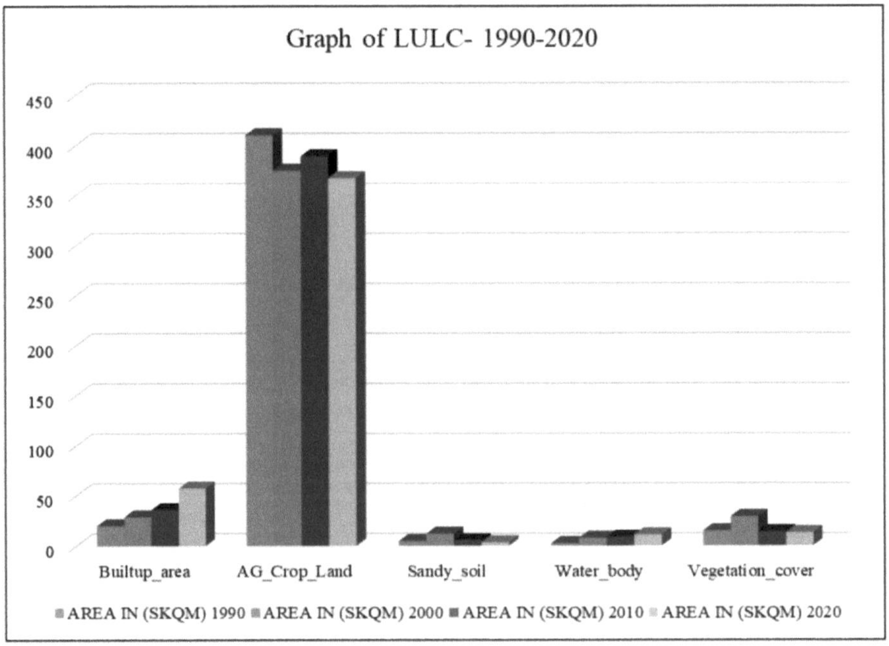

FIGURE 12.4 Graph of LULC (1990–2020).

NDVI Analysis (Figure 12.5)

NDVI graph of the study area has been generated using the Excel software, with the help of the attribute table, this method is very helpful and easy to handle. The overall analysis of the NDVI graph has shown that the vegetation classes frequently change in all the intervals of datasets such as—1990, 2000, 2010, 2020. Mostly negative changes have been seen in the class of vegetation cover. Few of the changes also has been observed in other classes. Vegetation area class in the year of (1990 covered with 96.454 km^2 area), in the year of (2000 covered with 81.033 km^2 area), in the year of (2010 covered with 106.458 km^2 area), in the year of (2020 covered with 67.88 km^2 area), whereas the other class in the year of (1990 covered with 356.126 km^2 area), in the year of (2000 covered with 371.547 km^2 area), in the year of (2010 covered with 346.122 km^2 area), in the year of (2020 covered with 384.7 km^2 area), the above-mentioned all the changes have shown that the negative changes in vegetation cover class and it has major impact on natural resources, because vegetation cover area comes under the natural resources; however, the overall graph analysis shows that the expansion of other classes which also include built-up area has greater impact on the vegetation cover area,incursion in the population and the demand of the forest product in the form of fuel wood, making of furniture, making of house or used as medicinal plant such as *Azadirachta indica*, *Phyllanthus emblica*, *Withania somnifera*, or used as any requirement lead to deforestation, to fulfill the needs, so this vegetation cover class has been converted into the built-up area and finally occupied as concrete jungle (Table 12.1).

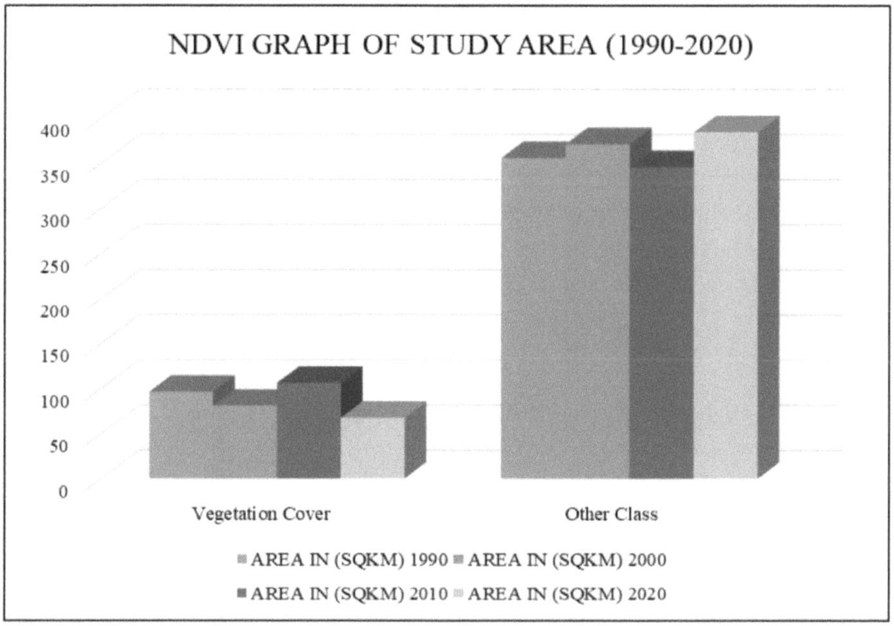

FIGURE 12.5 Graph of NDVI (1990–2020).

TABLE 12.1 Result and Discussion.

LU/LC classes	1990 area in (km^2)	2000 area in (km^2)	2010 area in (km^2)	2020 area in (km^2)
Built-up area	20.002	28.695	35.865	57.489
Agriculture crop land	410.535	375.125	389.44	367.708
Vegetation cover area	96.454	81.033	106.458	67.88
Water body	2.4	7.736	8.492	11.017
Sandy soil	4.685	11.9	5.084	3.85

12.4 RESULTS AND DISCUSSION

An attempt was made to capture as accurately as possible five major land use and land cover classes as they change through time. One of the most valuable natural resources is land. The pattern of land use and its geographic heterogeneity are the most important factors in developing an effective policy framework for efficient planning of any region. The land use map was prepared using the above-mentioned methodology. The various LULC classes delineated from the available data include such as follows:

12.4.1 BUILT-UP AREA CHANGE ANALYSIS

These structures appear as block-like structures on satellite images and show a bluish tone on the image, in all the datasets of 10-year interval built-up area has been increased frequently, in between the years 1990–2000 (8.693 km^2 area) it has been increased. The in-between years 2000–2010 it has been increased by 7.17 km^2 area. The in-between years 2010–2020 it has been increased by 21.624 km^2 area.

12.4.2 AGRICULTURE CROPLAND CHANGE ANALYSIS

Class has only one class such as agricultural cropland. Most of the study area has been covered with the agriculture class, it has been identifying using its texture, in satellite image agriculture filed seen as rough texture due to its individual boundaries, they are shown in the image as bunds like structures, in all the datasets of the 10-year interval agriculture cropland area has been decreased frequently, in between the years 1990–2000 it has been decreased by 35.41 km^2 area. In between the years 2000–2010 it has been increased by

14.315 km^2 area and in-between years 2010–2020 it has been decreased by 21.732 km^2 area.

12.4.3 VEGETATION COVER AREA

NDVI has been generated for vegetation cover monitoring, according to the NDVI map of the study area some changes have been found that the vegetation cover is decreased during all the datasets of 10-year interval. The in-between years 1990–2000 it has been decreased by (15.421 km^2 area). The in-between years 2000–2010 it has been increased by 25.425 km^2 area and in-between years 2000–2020 it has been decreased by 38.578 km^2 area. Due to some seasonal variation, value also has been increased or decreased.

12.4.4 WATER BODY

Water has been increased during all the year of datasets frequently, this indicates toward positive changes has been taken place in this particular class. The in-between years 1990–2000 (5.336 km^2 area) it has been increased. The in-between years 2000–2010 it has been increased by 0.756 km^2 area, and in-between years 2010–2020 it has been increased by 2.525 km^2 area.

12.4.5 SANDY SOIL

Mainly this class has focused on river sand, Arpa river is one of the major rivers comes under the study area, in this class both increase and decrease value has been observed in all the datasets of the study area such as follows, in between the years 1990–2000 it has been increased by 7.215 km^2 area. The in-between years 2000–2010 it has been decreased by 6.816 km^2 area and between the years 2010–2020 it has been decreased by 1.234 km^2 area).

12.5 CONCLUSION

The built-up area of Bilaspur changes drastically during the period 1990–2020, during this time built-up area has been extended in all the directions of Bilaspur city, major changes have been found in Bilaspur city area and in almost area of Bilaspur, agriculture cropland, and vegetation cover area have been converted into a built-up area. It is not a significant change, this alarming situation and expansion level of urban sprawl will have adversely affect the natural resources of the study area, because of excess utilization of natural resources by the people to survive their life in an easy way or to maintain the dignity of life, to give the better education to children they are frequently migrating toward the urban area this would definitely lead to more demand of those natural resources by the people. Both the positive and negative changes have observed between 1990 and 2020 in the study area, positive changes in a water body are a good sign of water resource management by the people for survival of life and for future planning also. Combined application of remote sensing and GIS will provide as accurate and reliable information in a better way, we can monitor urban sprawl inaccurate way by using this multi-temporal data, and we can make a proper plan to manage the urban area in a good way, and also we can manage our natural resources.

KEYWORDS

- **remote sensing**
- **geographical information system**
- **indices**
- **landscape dynamics**
- **urban growth**
- **natural resources**

REFERENCES

Bhatt, C. M. et al. Satellite-Based Assessment of the Catastrophic Jhelum Floods of September **2014,** Jammu & Kashmir, India. *Geomat. Nat. Hazards Risk* **2017,** *8* (2), 309–327.

Davies, W. K. Centrality and the Central Place Hierarchy. *Urban Studies* **1967,** *4* (1), 61–79.

Farooq, M.; Muslim, M. Dynamics and Forecasting of Population Growth and Urban Expansion in Srinagar City-a Geospatial Approach. *Int. Arch. Photogrammetry, Remote Sens. Spatial Inf. Sci.* **2014,** *40* (8), 709.

Forster, B. Some Urban Measurements from Landsat Data. *Photogrammetric Eng. Remote Sens.* **1983,** *49*, 1693–1707.

Haack, B. N.; Guptill, S. C.; Holz, R. K.; Jampoler, S. M.; Jensen, J. R.; Welch, R. A. Urban Analysis and Planning. *Manual f Photogr. Interpretation* **1997,** *2*, 517–554.

Hassanin, M.; Kanga, S.; Farooq, M.; Singh, S. K. Mapping of Trees Outside Forest (ToF) from Sentinel-2 MSI Satellite Data Using Object-Based Image Analysis. *Gujarat Agric. Univ. Res. J.* **2020,** 207.

http://wiki.gis.com/wiki/index.php/Digitizing

https://desktop.arcgis.com/en/arcmap/10.3/tools/spatial-analyst-toolbox/how-maximum-likelihood-classification-works.htm

https://www.usgs.gov/centers/eros/science/usgs-eros-archive-landsat-archives-landsat-4-5-thematic-mapper-tm-level-1-data

https://www.usgs.gov/core-science-systems/nli/landsat/landsat-8?qt-science_support_page_related_con=0#qt-science_support_page_related_con

Jensen, J. R. *Urban/Suburban Land Use Analysis, Manual of Remote Sensing*; American Society of Photogrammetry: Falls Church, Virginia, 1983; pp 1571–1666.

Joy, J.; Kanga, S.; Singh, S. K. Kerala Flood 2018: Flood Mapping by Participatory GIS Approach, Meloor Panchayat. *Int. J. Emerg. Technol.* **2019,** *10* (1), 197–205.

Kanga, S.; Rather, M. A.; Farooq, M.; Singh, S. K. GIS Based Forest Fire Vulnerability Assessment and its Validation using field and MODIS Data: A Case Study of Bhaderwah Forest Division, Jammu and Kashmir (India). *Indian Forester* **2021,** *147* (2), 120–136.

Kanga, S.; Singh S. K.; Sudhanshu. Delineation of Urban Built-Up and Change Detection Analysis Using Multi-Temporal Satellite Images. *Int. J. Recent Res. Aspects* **2017a,** *4*(3), 1–9.

Kanga, S.; Kumar, S.; Singh, S. K. Climate Induced Variation in Forest Fire Using Remote Sensing and GIS in Bilaspur District of Himachal Pradesh. *Int. J. Eng. Comput. Sci.* **2017b,** *6* (6), 21695–21702.

Kanga, S. et al. Modeling the Spatial Pattern of Sediment Flow in Lower Hugli Estuary, West Bengal, India by Quantifying Suspended Sediment Concentration (SSC) and Depth Conditions Using Geoinformatics. *Appl. Comput. Geosci.* **2020a**, 100043.

Kanga, S.; Sheikh, A. J.; Godara, U. GI Science for Groundwater Prospect Zonation. *J. Crit. Rev.* **2020b,** *7* (18), 697–709.

Kaya, S. Multitemporal Analysis of Rapid Urban Growth in Istanbul Using Remotely Sensed Data. *Environ. Eng. Sci.* **2007,** *24* (2), 228–233.

Kumar, S.; et al. Delineation of Shoreline Change along Chilika Lagoon (Odisha), East Coast of India Using Geospatial Technique. 2018.

Lo, C. P. Accuracy of Population Estimation from Medium-Scale Aerial Photography. In *American Congress on Surveying and Mapping and American Society for Photogrammetry and Remote Sensing, Annual Convention*, Washington, DC, 1986; pp 1–10.

Lo, C. P. Modeling the Population of China Using DMSP Operational Linescan System Nighttime Data. *Photogrammetric Eng. Remote Sens.* **2001,** *67* (9), 1037–1047.

Meraj, G. 2020. Ecosystem Service Provisioning—Underlying Principles and Techniques. *SGVU J. Clim. Change Water Res.* **2020,** *7*, 56–64.

Meraj, G.; Yousuf, A. R.; Romshoo, S. A. Impacts of the Geo-Environmental Setting on the Flood Vulnerability at Watershed Scale in the Jhelum Basin. MPhil Dissertation, 2013.

Meraj, G.; Romshoo, S. A.; Yousuf, A. R.; Altaf, S.; Altaf, F. Assessing the Influence of Watershed Characteristics on the Flood Vulnerability of Jhelum Basin in Kashmir Himalaya: Reply to Comment by Shah 2015. *Nat. Hazards* **2015b,** *78* (1), 1.

Meraj, G.; Singh, S. K.; Kanga, S.; Islam, M. N. Modeling on Comparison of Ecosystem Services Concepts, Tools, Methods and Their Ecological-Economic Implications: A Review. *Model. Earth Syst. Environ.* **2021,** 1–20.

Meraj, G.; Kumar, S. Economics of the Natural Capital and the Way Forward. *Preprints* **2021,** 2021010083. DOI: 10.20944/preprints202101.0083.v1.

Meraj, G. et al. Geoinformatics Based Approach for Estimating the Sediment Yield of the Mountainous Watersheds in Kashmir Himalaya, India. *Geocarto Int.* **2018a,** *33* (10), 1114–1138.

Meraj, G. et al. An Integrated Geoinformatics and Hydrological Modelling-Based Approach for Effective Flood Management in the Jhelum Basin, NW Himalaya. *Multidiscipl. Digital Publ. Inst. Proc.* **2018b,** 7 (1), 8.

Meraj, G. et al. Assessing the Influence of Watershed Characteristics on the Flood Vulnerability of Jhelum Basin in Kashmir Himalaya. *Nat. Hazards* **2015a,** *77* (1), 153–175.

Meraj, G.; Romshoo, S. A.; Altaf, S. Inferring Land Surface Processes from Watershed Characterization. In *Geostatistical and Geospatial Approaches for the Characterization of Natural Resources in the Environment*; Springer: Cham, 2016; pp 741–744.

Mesev, T. V.; Longley, P. A.; Batty, M.; Xie, Y. Morphology from Imagery: Detecting and Measuring the Density of Urban Land Use. *Environ. Plan. A* **1995,** *27* (5), pp 759–780.

Miller, R. B.; Small, C. Cities from Space: Potential Applications of Remote Sensing in Urban Environmental Research and Policy. *Environ. Sci. Policy* **2003,** *6* (2), 129–137.

Nathawat, M. S. et al. Monitoring & Analysis of Wastelands and Its Dynamics Using Multiresolution and Temporal Satellite Data in Part of Indian State of Bihar. *Int. J. Geomat. Geosci.* **2010,** *1* (3), 297–307.

Pandey, A. C.; Singh, S. K.; Nathawat, M. S. Waterlogging and Flood Hazards Vulnerability and Risk Assessment in Indo Gangetic Plain. *Nat. Hazards* **2010,** *55* (2), 273–289.

Pandey, A. C.; Singh, S. K.; Nathawat, M. S.; Saha, D. Assessment of Surface and Subsurface Waterlogging, Water Level Fluctuations, and Lithological Variations for Evaluating Groundwater Resources in Ganga Plains. *Int. J. Digital Earth* **2013,** *6* (3), 276–296.

Rashid, I. et al. Assessing the Impact of Anthropogenic Activities on Manasbal Lake in Kashmir Himalayas. *Int. J. Environ. Sci.* **2013,** *3* (6), 2036–2047.

Rather, M. A. et al. Remote Sensing and GIS Based Forest Fire Vulnerability Assessment in Dachigam National Park, North Western Himalaya. *Asian J. Appl. Sci.* **2018,** *11* (2), 98–114.

Remote Sensing Application, National Remote Sensing Center, 22/07/2019 at 2:00 pm. https://www.nrsc.gov.in/index.php/Knowledge_EBooks?language_content_entity=en

Sharma, A.; Kanga, S. Surface Runoff Estimation of Sind River Basin Using SCS-CN Method and GIS Technology, 2020.

Singh, S. K.; Pandey, A. C. Geomorphology and the Controls of Geohydrology on Waterlogging in Gangetic Plains, North Bihar, India. *Environ. Earth Sci.* **2014,** *71* (4), 1561–1579.

Singh, S. K.; Kanga, S. Role of Geoinformatics in Site Suitability Analysis of Infrastructures Using PRA Approach. *Am. Int. J. Res. Sci. Technol. Eng. Math.* **2017a,** *1* (18), 81–85.

Singh, S. K.; Mishra, S. K.; Kanga, S. Delineation of Groundwater Potential Zone Using Geospatial Techniques for Shimla City, Himachal Pradesh (India). *Int. J. Sci. Res. Dev.* **2017b,** *5* (4), 225–234.

Stefanov, W. L.; Ramsey, M. S.; Christensen, P. R. Monitoring Urban Land Cover Change: An Expert System Approach to Land Cover Classification of Semiarid to Arid Urban Centers. *Remote Sens. Environ.* **2001,** *77* (2), 173–185.

Tuyahov, A. J.; Davies, S.; Holz, R. K. Detection of Urban Blight Using Remote Sensing Techniques. *Remote Sens. Earth Resour.* **1973,** 213–226.

CHAPTER 13

Assessing the Impact of Coastal Morphodynamics on Water Chemistry Using Integrated GIScience and Modeling Approaches

G. SREENIVASULU[1*], N. JAYARAJU[1], B. UPENDRA[2], and B. LAKSHMANNA[1]

[1]*Department of Geology, Yogi Vemana University, Kadapa 516005, Andhra Pradesh, India*

[2]*ESSO-National Centre for Earth Science Studies, Thiruvananthapuram 695011, Kerala, India*

[*]*Corresponding author. E-mail: seenu9441@gmail.com*

ABSTRACT

An aerial study was conducted on a major proposed port site (Dugarajapatnam port, Tupilipalem, Nellore District of Andhra Pradesh, India). It showed the phenomenal mouth dynamics periodically. Landsat 8 OLI/TIRS multitemporal satellite images dated October 27, 2014 (monsoon) and August 27, 2015 (pre-monsoon) were chosen to demonstrate the closing and opening of the lagoon mouth during monsoon and pre-monsoon, respectively. Twelve water samples from each season (monsoon and pre-monsoon) were collected by using a shallow water sampler from both lagoon and open sea for comparison. The physico-chemical parameters viz., pH, air temperature, water temperature, electrical conductivity (EC), salinity, dissolved oxygen (DO), organic matter (OM), and silica were estimated. The results revealed

GIScience for the Sustainable Management of Water Resources. Gowhar Meraj, Shruti Kanga, Majid Farooq, Suraj Kumar Singh, & Sudhanshu (Eds.)
© 2023 Apple Academic Press, Inc. Co-published with CRC Press (Taylor & Francis)

that the pH ranged between 7.1 and 7.4 during both the seasons (monsoon and pre-monsoon). Variation in air temperature and water temperature across all the stations was observed. However, these variations were not significantly different across the stations. In addition, around the stations, these differences were not significant. Both the lower and higher values of DO were recorded from the brackish environment and it may be absorbed from the waste products of plant photosynthesis. The study area is covered with full of mangrove swamps and *Avicennia marina* is the dominant among all the mangroves. A higher value of salinity was observed at the lagoon side of the Tupilipalem coast during lagoon mouth closing period. From monsoon to pre-monsoon, the concentration of silica increased. Multivariate statistics, the most general multivariate statistical techniques, was used to minimize data and derive a limited number of latent factors to evaluate the relationship among coastal morphodynamics and the distribution of physico-chemical parameters. The significant seasonal variation among the studied physico-chemical parameters is the proxy for the influence of seasonal lagoon mouth closure. The relevant satellite datasets revealed the change detection of lagoon mouth dynamics periodically. The details are discussed in this paper.

13.1 INTRODUCTION

Different physical, chemical, and biological mechanisms mainly impact the marine ecosystem as a dynamic system (Sreenivasulu et al., 2015). The open sea is more stable compared to the near shore, where the interaction with the terrestrial zone is more effective in bringing out the variations in different physicochemical parameters (Strayer and Findlay, 2010). Thus, a detailed knowledge of hydrography is necessary to estimate the quality of the aquatic ecosystem and its effect on biological fertility (Bhadja and Kundu, 2012). Both natural as well as anthropogenic processes adversely affect the quality of water (Naithani and Pandey, 2012; Jayaraju et al., 2011; Khatri and Tyagi, 2015; Bojarczuk et al., 2018; Peters et al., 2005). Changes in physical, chemical, and biological parameters (water quality) arise from many contaminants such as chemicals, heavy metals, synthetic organics, toxic isotopes, a significant amount of organic and inorganic substances, oil and refined products, etc. (Dhir, 2013; Naja and Volesky, 2011). In the biogeochemistry of the water body, physical parameters play a major role (Bhateria and Jain, 2016; Megonigal and Neubauer, 2018). Subtle changes in physical conditions may have major implications for the water quality of the study system, which may, in turn, affect the spatial and temporal distribution of nutrients and/or biological populations (Carpenter et al., 2011). The physical and chemical properties of water have an enormous effect on the uses of a water body for the distribution and richness of biota, as well as on the horizontal migration of aquatic organisms (Balasubramanian et al., 2017). The quality of water in every environment offers significant information about the available resources that sustain life in that ecosystem (Umerfaruq et al., 2015; Meraj and Kumar, 2021; Meraj et al., 2021). Good quality of water resources depends on a large number of physicochemical parameters and biological characteristics (Singh et al., 2017b; Kanga et al. 2020a, 2020b). The monitoring of these parameters is important in order to evaluate the magnitude and source of any pollution load (Thirupathaiah et al., 2012; Rashid et al., 2013; Singh et al., 2017a).

Remote sensing (RS) satellite data capable of delivering an accurate, synoptic view of a reasonably wide region at regular intervals with a short turnaround time integrated with the geographical information system (GIS) allows it to be suitable and perfect for tracking and studying river erosion and its bankline shifting (Sreenivasulu et al., 2016; Jayappa and Vijaya Kumar, 2003; Pandey et al., 2010, 2013; Singh and Pandey, 2014; Meraj et al., 2016; Bhatt et al., 2017; Sharma and Kanga 2020). RS data are primary sources extensively used for change detection in recent decades on meso scales (Lu et al., 2004; Fu et al., 2020). GIS and RS can play an important role in the management of these coastal resources (Jiang and Hao, 2016; da Silva et al., 2012). RS data are a helpful way of understanding the dynamics of coastal processes, where the images allow a synoptic view of the landscape, detect shoreline shifts, and identify different causes at the global, regional, and local levels (Guariglia et al., 2006). Changes in the shoreline are primarily correlated with waves, tides, winds, periodic floods, changes in the sea level, geomorphological processes such as erosion and accretion, and human activities (De Swart and Calvete, 2003; Van and Binh, 2009).

Earlier Sreenivasulu et al. (2017) reported on the periodical lagoon mouth closure from the study area and its impact on the distribution of heavy metals in the water during both lagoon mouth closure and opening (Sreenivasulu et al., 2018). In this paper, for the first time, the variation among the physico-chemical parameters for the influence of seasonal lagoon mouth closure was studied. Moreover, the closing of the lagoon mouth has a socio-economic effect on human activities such as navigation, livestock, fishing, aquaculture, transport, and connectivity. Shoreline change detection and measurement are essential activities for environmental protection and coastal zone management.

13.1.1 STUDY AREA

Tupilipalem is the proposed site for constructing a major port to be named Dugarajapatnam Port, geographically located on the southeastern part of Andhra Pradesh, lying between 14°0'10"–14°02'30" N and 80°08'20"–80°19'00" E (Figure 13.1). Tupilipalem is 20 km away from Dugarajapatnam, Nellore district of Andhra Pradesh, India, and more than 120 km away from the Pulicat Lake and the satellite launching center at Sriharikota. The open bay, which forms the study area, is connected by this lagoon (mouth). With a mean annual rainfall of 1041 mm, the subtropical climate prevails over the study area. The mean temperature of the most intense catchment temperatures ranges from 30°C to 32°C and from 22.5°C to 25°C at the very least. However, the streams are often influenced by the tidal cycles, the movement of tides, the geography of the shore line, and the existence of various water volumes assuming predominantly a SW–NE direction. The study area consists of laterites and Quaternary alluvium. Triassic sandstone is also found in some parts. Alluvial deposits in the area are recent sediments of basement gneiss, amphibolites, and migmatized Pelitic schists.

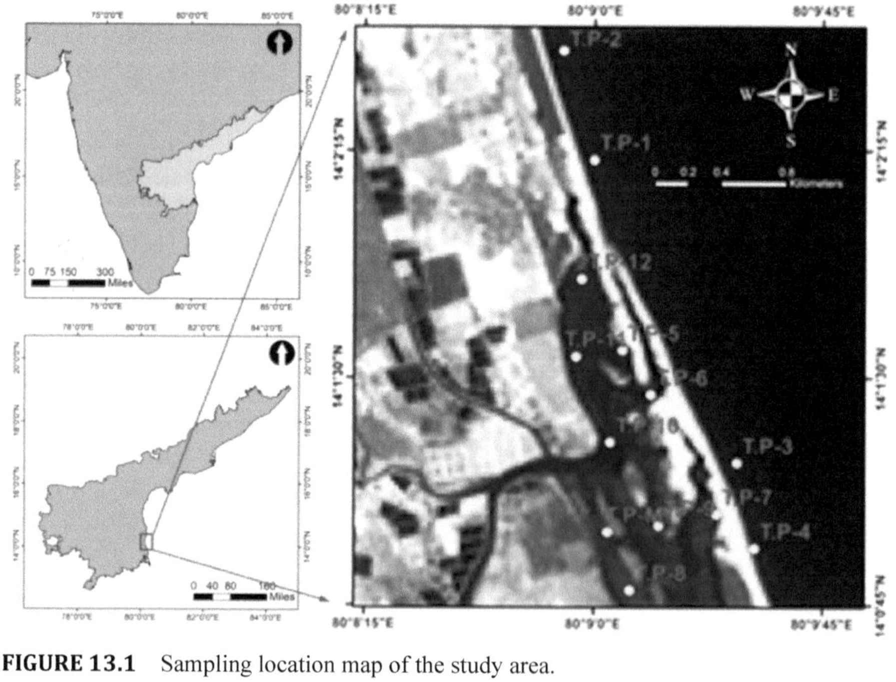

FIGURE 13.1 Sampling location map of the study area.

13.2 EXPERIMENTAL METHODS AND MATERIALS

The baseline map was prepared using Survey of India (SOI) Toposheet Nos. 66B4, 66C1 and 66C5 on 1: 50,000 scale. The Landsat 8 OLI/TIRS images were acquired for lagoon mouth observation. Then, each image was cropped and geometrically corrected by applying the UTM-WGS 84 projection and coordinate system (Alcaras et al., 2020; Chenthamilselvan et al., 2014). The lagoon mouth dynamics were identified in the images dated October 27, 2014 (monsoon) and August 27, 2015 (pre-monsoon). Twelve water samples from each season (monsoon and pre-monsoon) were collected using shallow water sampler from both lagoon and open sea from the study area. By using the hand-held global positioning system (GPS), the locations of the sampling stations were precisely located. The air temperature and the bottom water temperature were calculated and reported using the thermometer on the spot. The Atago refractometer was used to calculate salinity (Arumugam and Kumar, 2014). PyeUnicam model 292 m (after standardization with buffer solution at pH4.0, 7.0, and 9.0) was used for pH and electrical conductivity (EC) determination (Ibrahim et al., 2010). Dissolved oxygen (DO), organic matter (OM), and silica were measured using standard laboratory methods (APAH, 1995). The data, thus obtained, was statistically evaluated using XL STAT-2018 for Pearson correlation matrix and Factor loadings (Jayaraju et al., 2010).

13.3 RESULTS AND DISCUSSION

13.3.1 COASTAL MORPHOLOGY

For the analysis of the coastal morphology in the region, the multitemporal Landsat 8 OLI/TIRS images acquired from October 27, 2014 (monsoon) and August 27, 2015 (pre-monsoon) were used. During the monsoon (October 27, 2014), the lagoon mouth closing was observed and the image on August 27, 2015 has two opening points in the northern and southern side of the lagoon (Figure 13.2). Of the two, one is a natural opening (north side) and the other is an artificial opening (southern side). The artificial one is opened by the villagers after the closing of the lagoon mouth for easy transportation of their fishing vessels. Sreenivasulu et al. (2017, 2018) carried out the study of Tupilipalem coast using RS and GIS, which provides a viable source of data for monitoring and assessing the coastal changes. The study showed that the intensity of erosion and accretion represents coastal patterns and, based on wave energy and littoral currents, the loss or gain of sediment induces the development of young beaches, berms, sand dunes, and sea cliffs. Accretion during the last 5 years (2011–2015) is more than corrosion, which traps the sediments in the mouth of the lagoon, triggering its closing. It is found that an oblique wave approach to the shoreline allows the dominant northern wind to change the direction of the mouth of the lagoon.

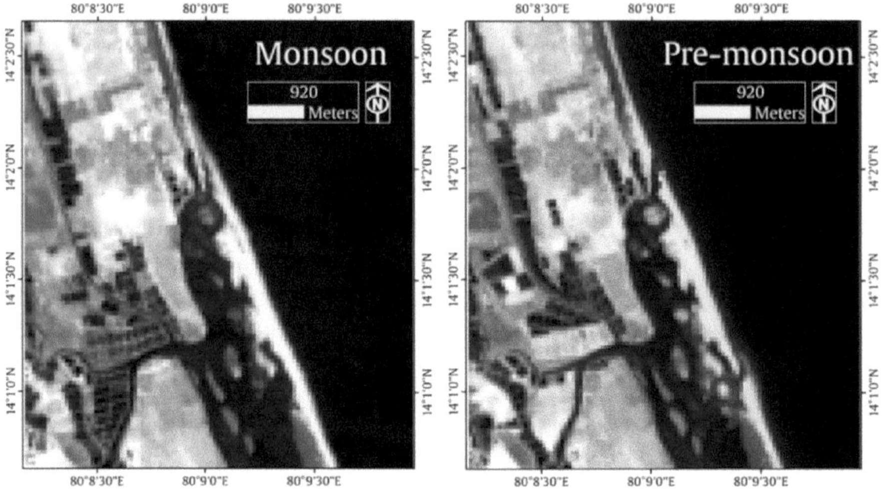

FIGURE 13.2 Morphological changes in the study area between monsoon and pre-monsoon.

13.3.2 PHYSICO-CHEMICAL PARAMETERS

Physico-chemical parameters play an essential role in the biochemistry of the water body (El Morhit and Mouhir, 2014; Umerfaruq et al., 2015). Subtle changes in physical conditions can have significant impact on the water quality of the study system, which may in turn affect the spatial and temporal distribution of nutrients and/or biological communities (Twomey et al., 2009). The common physicochemical parameters of water like pH, temperature, EC, salinity, DO, OM, and silica have been determined (Figure 13.3). The significant variation in physicochemical parameters along these transects is shown in Tables 13.1 and 13.2.

13.3.2.1 pH

The amount of free CO_2, carbonates, and bicarbonate controls the changes of pH in the water. These changes are followed by the changes in other

TABLE 13.1 Physicochemical Parameters of Tupilipalem Coast in Monsoon.

Sample no.	Air temperature (°C)	Water temperature (°C)	pH	Ec at 25°C (µmho/cm)	Salinity (ppt)	Dissolved oxygen (mg/L)	Organic matter (mg/L)	Silica (mg/L)
TP-1	34.9	34.1	7.4	79	35.7	3.6	0.78	26
TP-2	34.1	33.7	7.3	67	35.2	4.5	1.09	24
TP-3	34.5	33.7	7.1	82	35.2	4.2	1.00	21
TP-4	34.5	33.9	7.3	81	35.9	4.7	0.97	19
TP-5	34.2	34.0	7.2	69	35.4	4.1	0.72	18
TP-6	34.6	33.9	7.2	72	36.1	4.6	0.86	28
TP-7	34.5	33.9	7.2	79	35.6	4.5	0.98	35
TP-8	33.9	33.5	7.3	87	35.2	4.2	1.01	15
TP-9	34.5	33.5	7.2	84	35.2	3.9	1.01	14
TP-10	34.7	33.9	7.2	62	35.9	3.2	0.82	28
TP-11	34.5	33.8	7.4	65	36.1	4.2	1.05	29
TP-12	34.2	33.8	7.2	82	35.4	4.1	1.07	30
TP-M	34.3	34.1	7.2	79	35.7	3.9	0.74	32

TABLE 13.2 Physicochemical Parameters of Tupilipalem Coast in Pre-monsoon.

Sample no.	Air temperature (°C)	Water temperature (°C)	pH	Ec at 25°C (µmho/cm)	Salinity (ppt)	Dissolved oxygen (mg/L)	Organic matter (mg/L)	Silica (mg/L)
TP-1	34.7	33.2	7.2	66	34.9	4.1	1.20	27
TP-2	34.4	32.9	7.1	88	35.4	4.6	0.98	23
TP-3	34.2	33.9	7.1	84	35.1	4.4	1.30	23
TP-4	34.7	33.7	7.4	77	35.7	4.2	1.10	20
TP-5	34.5	33.9	6.9	60	34.9	4.8	1.00	18
TP-6	34.4	33.6	7.3	66	35.8	4.5	0.78	27
TP-7	34.2	33.8	7.3	84	35.8	4.5	0.86	32
TP-8	34.8	33.3	7.1	85	35.6	4.3	0.98	14
TP-9	34.2	33.9	7.2	80	35.1	4.1	1.00	17
TP-10	34.9	33.7	7.4	59	35.6	4.0	1.00	30
TP-11	34.3	33.4	7.4	69	35.8	4.1	1.20	27
TP-12	34.3	33.6	7.1	80	35.6	4.4	1.10	33
TP-M	34.3	33.9	7.3	74	35.2	4.0	0.85	31

physicohemical aspects that in turn influence the quality of water. In all sites, virtually neutral pH is observed during the sampling events. Boyd and Lichtkoppler (Boyd and Lichtkoppler, 1979) reported a pH range of 6.09–8.45 as being ideal for supporting aquatic life, including fish, and calcareous forms will not survive at a pH less than 7. In this study, the pH ranged between 7.1 and 7.4 with an overall mean value of 7.25 in the monsoon, and it was 6.9 and 7.4 obtained from TP-1 and TP-M, respectively, with an overall value of 7.21 in pre-monsoon.

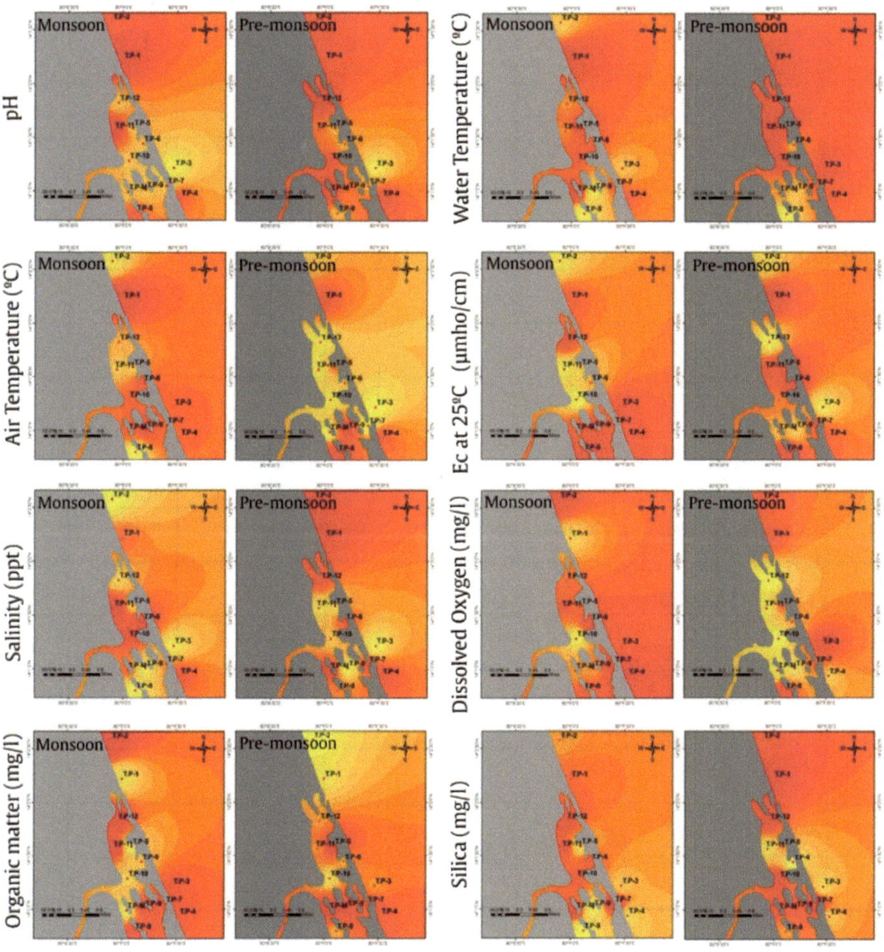

FIGURE 13.3 Contour map showing the distribution of physico-chemical parameters during lagoon mouth closure and open (monsoon and pre-monsoon).

13.3.2.2 TEMPERATURE (°C)

Bottom water temperature recorded in the monsoon period ranged between 33.5°C and 34.1°C. The lower temperature (33.5°C) was recorded in two stations (TP-8 and TP-9) and the higher (34.1°C) was recorded at two stations (TP-1 and TP-M). In pre-monsoon, bottom water temperature ranged between 32.9°C and 33.9°C. The lower temperature (32.9°C) was recorded at station TP-2 and the higher (33.9°C) was recorded in four stations (TP-3, TP-5, TP-9, and TP-M).

The air temperature in the monsoon period ranged between 33.9°C and 34.9°C at TP-8 and TP-1, respectively. In pre-monsoon period, air temperature recorded between 34.2°C and 34.9°C. The lower temperature (34.2°C) was recorded in three stations (TP-3, TP-7, and TP-9) and the higher (34.9°C) was recorded at one station only (TP-10). Among all the stations, there were differences in air and underwater temperature. However, around the stations, their differences were not substantially different. The comparatively higher bottom water temperatures in both the seasons may be inferred due to shallow depths of the water body.

13.3.2.3 ELECTRICAL CONDUCTIVITY (MMHO/CM)

EC is a measure of the capacity of a substance or solution to conduct electric current. With regard to various study locations, the EC indicates seasonal variance. It highly depends on the amount of dissolved solids in water (Rusydi, 2018). Water conductivity is impaired by the impurities suspended and often relies on the volume of ions in the water (Joshi et al., 2009). The maximum conductivity (87 µmho/cm) was observed at station TP-8 and minimum (62 µmho/cm) at station TP-10 during monsoon. In pre-monsoon, the maximum conductivity (88 µmho/cm) was recorded at station TP-2 and minimum (59 µmho/cm) at station TP-10.

13.3.2.4 SALINITY (PPT)

Salinity is a basic water quality parameter monitored by freshwater and marine ecologists because of its influence on the biota (Twomey et al., 2009; Smyth and Elliott, 2016). The majority of marine species are adaptive to only a narrow range of salinity, above which their osmotic and ionic equilibrium cannot be maintained (Timothy, 2010). Some species only withstand intermediate salinity amounts, while broadly adopted species can acclimatize to variable salinity varying from fresh to seawater. The level of salinity in brackish water is usually from 0.5 to 30 ppt (Ern et al., 2014). In this study, salinity during the monsoon ranged between 35.2 and 36.1 ppt. Lower value (35.2 ppt) was recorded at four locations (TP-2, TP-3, TP-8, and TP-9) in the study area. The higher value of (36.1 ppt) was observed at the stations TP-6 and TP-11. In pre-monsoon, salinity ranging between 34.9 and 35.8 ppt was recorded. The lower salinity value (34.9 ppt) was recorded in two stations (TP-1 and TP-5) and the higher value (35.8 ppt) was observed in three stations (TP-6, TP-7, and TP-9).

13.3.2.5 DISSOLVED OXYGEN (mg/L)

DO concentrations above 5 mg/L are commonly viewed as promoting aquatic life, while concentrations below this are possibly harmful (Abdus-Salam et al., 2010). Bottom fishes may begin to leave the region at about 3 mg/L, and the growth of sensitive species such as crab larvae is reduced (Stoklosa et al., 2018). The larvae of the less-sensitive crustacean species may start to die at 2.5 mg/L, and the growth of crab species is more severely limited (EPA, 2010). Below 2 mg/L, some juvenile fish and crustaceans that cannot leave the area may die, and below 1 mg/L, fishes totally avoid the area or begin to die in large numbers (44). In the study area, the DO ranged from 3.2 to 4.7 mg/L with overall mean value of 4.15 mg/L during monsoon. The highest value (4.7 mg/L) is observed at TP-4 and lowest (3.2 mg/L) is observed at TP-10. In the pre-monsoon, the recorded DO ranged from 4.0 to 4.8 mg/L. The lower value (4.0 mg/L) was recorded in two stations (TP-10 and TP-M) and the higher value (4.8 mg/L) was observed at TP-5.

13.3.2.6 ORGANIC MATTER (mg/L)

In river water, high levels of pollutants (mainly OM) induce a rise in the demand for biological oxygen (Wen et al., 2017; Gualtieri, 1998). Natural dissolved OM comprises of transformation process of biogenic substance and bacterial excretion products (Thornton, 2014). In this study, the OM in monsoon ranged between 0.72 and 1.09 mg/L obtained from TP-5 and TP-2, respectively, with an average mean value of 0.946 mg/L OM at pre-monsoon varied between 0.78 and 1.30 mg/L from TP-6 and TP-3, respectively.

13.3.2.7 SILICA (mg/L)

Silica is a key nutrient for the marine environment (Sospedra et al., 2018; Officer and Ryther, 1980; DeMaster, 2001). Diatoms, radiolarian, siliceous sponges, and silico-flagellates use this nutrient for growth and the formation of skeletal material (Ehrlich, 2011). It supplied to the ocean from the porous waters of coastal sediments (DeMaster, 1981), and it ranges between 14 and 35 mg/L obtained from TP-9 and TP-7, respectively, during the monsoon with an overall mean value of 23.92 mg/L. Finally, sudden increase of silica was noticed at TP-7, TP-M, TP-12, TP-11, TP-10, and TP-6 with the values of 35, 32, 30, 29, 28, and 28 mg/L, respectively. In pre-monsoon, silica ranged between 14 and 33 mg/L obtained from TP-8 and TP-12, respectively.

13.3.3 HIERARCHICAL CLUSTER ANALYSIS

Hierarchical clustering analysis for different sampling stations was performed. Four clusters were established, that is, TP-1 to TP-9; TP-7 to TP-12; TP-2 to TP-6, and TP-5 in monsoon (Figure 13.4). Cluster 1, formed by stations TP-1, TP-3, TP-4, TP-8 and TP-9, showed the highest concentration of EC in all the stations. These stations were located in the southern part of the study area. The majority of the stations (TP-1, TP-3, and TP-4) in this cluster were from the open sea and the remaining stations TP-8 and TP-9 were exactly opposite from lagoon environment. Cluster 2, formed with station TP-7, TP-M, and TP-12, showed the highest concentrations of salinity, DO, and silica in a brackish environment. In Cluster 3, TP-2, TP-10, TP-11, and TP-6 showed moderate contamination with all the studied physico-chemical parameters. All the stations in this cluster were located in the either side of lagoon mouth (TP-2 from open sea and TP-10, TP-11, and TP-6 from brackish environment). Finally, Cluster 4, generated by station TP-5, showed the lowest concentrations of EC, OM, and silica in the lagoon side of the study region.

In pre-monsoon, four clusters were observed, that is, TP-1 to TP-10; TP-5; TP-2 to TP-8; and TP-7 to TP-12 (Figure 13.4). Cluster 1, formed by stations TP-1, TP-6, TP-11, TP-M, and TP-10, showed moderate-to-high concentration of all the studied parameters in the lagoon environment. Cluster 2, formed by station TP-5 only, showed the highest concentrations of temperature and DO. Remaining all the studied physico-chemical parameters at the station were low. In Cluster 3, TP-2, TP-3, TP-4, TP-9, and TP-8 showed highest concentrations of EC, salinity, DO, and Om, and lowest values of silica and pH. Further, the stations from cluster 4 were distributed near to both inlets. Though TP-7 and TP-12 formed cluster 4, which showed the highest concentration of salinity and silica. On the seasonal scale, station TP-5, which is located in the middle of the study area, showed stable concentrations in both the seasons. The stations that showed higher and lower values of physico-chemical parameters have changed their clusters from monsoon to pre-monsoon.

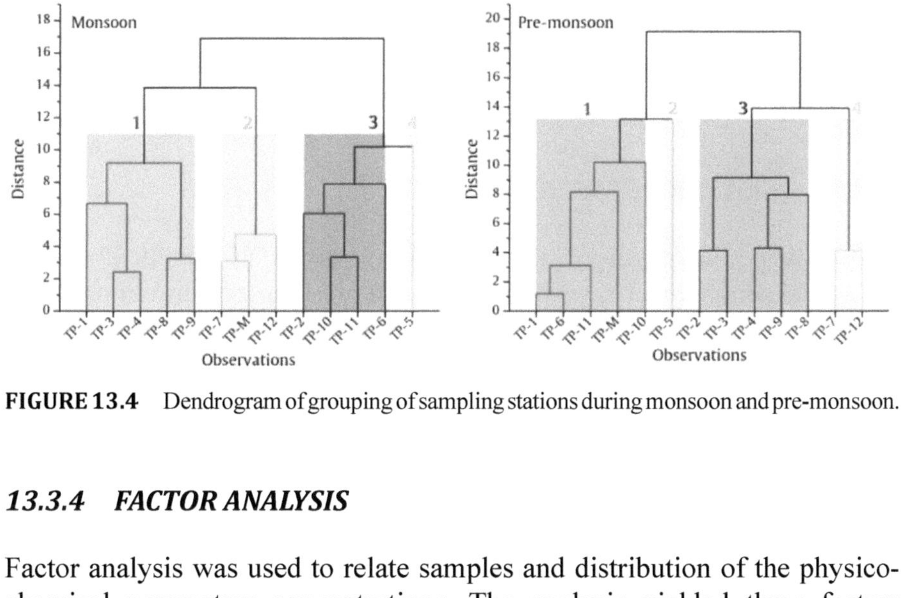

FIGURE 13.4 Dendrogram of grouping of sampling stations during monsoon and pre-monsoon.

13.3.4 FACTOR ANALYSIS

Factor analysis was used to relate samples and distribution of the physico-chemical parameters concentrations. The analysis yielded three factors during monsoon (Figure 13.5). Water temperature, silica, and salinity are the predominant parameters in Factor-1. Factor-2 was composed of DO, EC, and OM. In addition, Factor-3 was represented by pH only during monsoon. In pre-monsoon, Factor 1 enveloped three variables, namely, silica, salinity, and pH. Factor 2 was composed of two variables, OM and EC. Factor 3 included two variables, water temperature and DO. Finally, Factors analysis gave distinct information about the distribution of the physico-chemical parameters and set cluster of similar factor during both the seasons. The distribution indicates that the physico-chemical parameters have a good affinity with marine water samples of the Tupilipalem coast, the southeast coast of India.

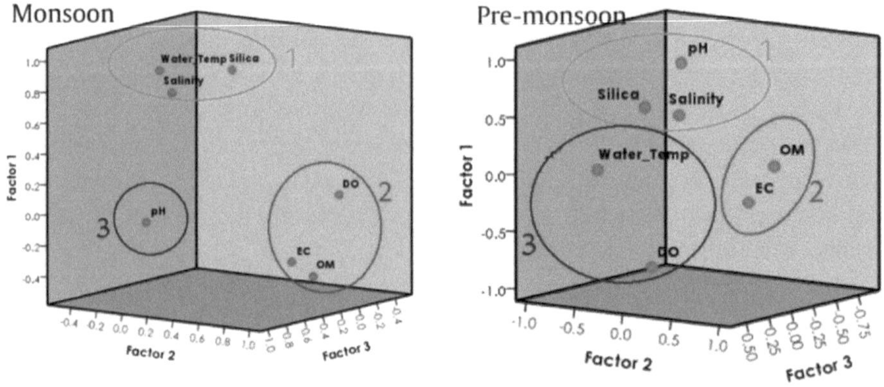

FIGURE 13.5 A plots of factor loadings during monsoon and pre-monsoon.

13.3.5 CHANGE DETECTION STUDY

The change detection of the concentrations of the studied physico-chemical parameters between monsoon and pre-monsoon is summarized in Figure 13.6. Water temperature shows significant decreasing values in 84% of the stations, whereas DO shows significant increasing concentrations in 77% of the stations. OM, silica, and pH show increasing concentrations in 62%, 54%, and 46% of the stations, respectively. EC, salinity, and pH show deceasing concentrations in 77%, 69%, and 54% of the stations, respectively. During the lagoon mouth opening period, the majority of studied physicochemical concentrations were decreased in the lagoon environment due to inflow and outflow of water. At TP-6, majority of the studied physico-chemical concentrations were decreased from monsoon to pre-monsoon except pH (even though the pH is within the permissible limit). The station TP-6 is located at northern side of the study area where the lagoon mouth joined in the open sea. Water temperature, EC, and salinity decreased in almost all the stations except stations TP-2 and TP-7. DO and OM increased in the majority of the stations after lagoon mouth opening. Finally, the substantial differences in the concentrations among the studied physicochemical parameters were found from the stations of the lagoon environment during lagoon mouth closure to opening. The highest concentrations were reported during closure from the lagoon due to the load of Buckingham Canal, waste products of plant photosynthesis released from mangrove plants, and other domestic sewage inputs. As a consequence of the opening of the lagoon mouth, the concentrations of the studied physico-chemicals were decreased due to inflow and outflow of water from the Bay of Bengal to lagoon and vice-versa.

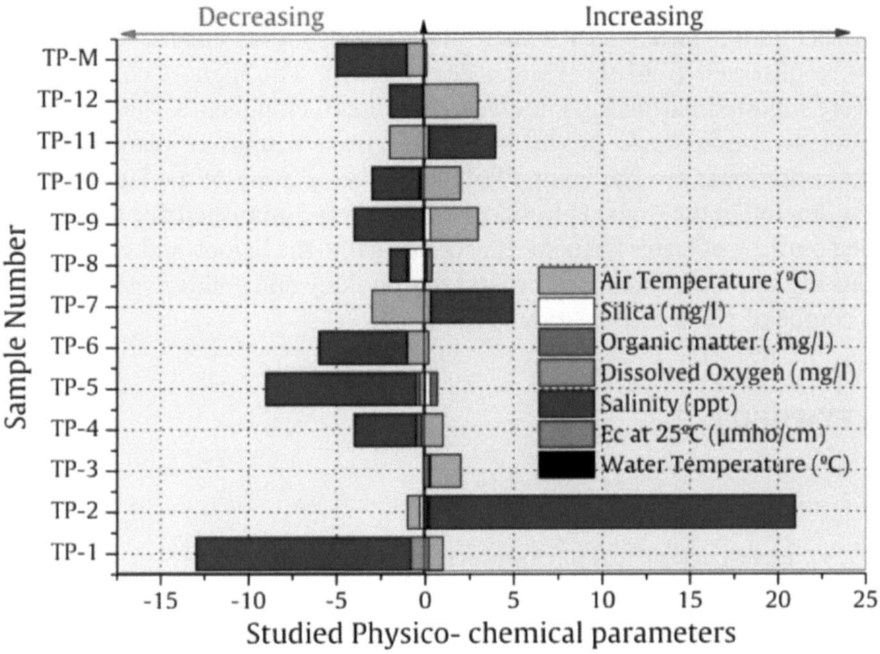

FIGURE 13.6 Variations in the physico-chemical parameters from monsoon to pre-monsoon.

13.4 CONCLUSIONS

Coastal morphology of Tupilipalem coast using Landsat 8 OLI/TIRS multi-temporal satellite imagery offers a viable source for tracking and evaluating coastal changes. The prevailing northern wind creates an oblique wave approach to the shoreline, allowing the lagoon mouth to close. Twelve water samples from each season (monsoon and pre-monsoon), were collected using shallow water sampler from both lagoon and open sea for comparison. The physicochemical parameters viz., pH, air temperature, water temperature, EC, salinity, DO, OM, and silica were analyzed. The pH ranged between 7.1 and 7.4 during both the seasons (monsoon and pre-monsoon). Variation in air temperature and bottom water temperature across all the stations was observed. However, these variations were not significantly different across the stations. The comparatively higher water temperature in both the seasons may be inferred to shallow water depth. The higher value of salinity was observed at lagoon side of the Tupilipalem coast during lagoon mouth closing period. Both the lower and higher values of DO were recorded from the brackish environment. During lagoon mouth closing to opening, major concentration differences among the studied physicochemical parameters were observed from the lagoon environment. The highest concentrations were reported during lagoon closure due to Buckingham Canal load, plant photosynthesis waste products released from mangrove plantations, and other domestic sewage inputs. Following the opening of the lagoon mouth, the amount of the studied physicochemicals was diminished due to the inflow and outflow of water from the Bay of Bengal to the lagoon and vice versa. In this region, the influence of coastal morphology and anthropogenic input as a source of water composition is alarming.

KEYWORDS

- **coastal morphodynamics**
- **physicochemical parameters**
- **Tupilipalem coast**
- **geospatial tools**

REFERENCES

Abdus-Salam, N.; Adekola, F. A.; Apata, A. O. A Physicochemical Assessment of Water Quality of Oil Producing Areas of Ilaje, Nigeria. *Adv. Nat. Appl. Sci.* **2010,** *4* (3), 333–345.

Alcaras, E.; Parente, C.; Vallario, A. The Importance of the Coordinate Transformation Process in Using Heterogeneous Data in Coastal and Marine Geographic Information System. *J. Mar. Sci. Eng.* **2020,** *8* (9).

APAH. *Standard Methods for the Examination of Water and Wastewater*; 19th ed.; American Public Health Association Inc., References–Scientific Research Publishing: New York, 1995.

Arumugam, A.; Kumar, S. P. Evaluation of Physico-Chemical Parameters and Nutrients in the Mangrove Ecosystem of Manakudy Estuary, Southwest Coast of India. *Int. J. Latest Res. Sci. Technol.* **2014,** *3* (6), 205–209.

Balasubramanian, R.; Babu, R.; Ravi, V.; Kavitha, V.; Anandha, R. Influence of Seasonal Changes on Physico-Chemical Characteristics of Uppanar Estuary, Southeast Coast of India. **2017,** *6* (March), 1–8.

Bhadja, P.; Kundu, R. Status of the Seawater Quality at Few Industrially Important Coasts of Gujarat (India) off Arabian Sea. *Indian J. Geo-Marine Sci.* **2012,** *41* (1), 90–97.

Bhateria, R.; Jain, D. Water Quality Assessment of Lake Water: A Review. *Sustain. Water Resour. Manage. [Internet].* **2016,** *2* (2), 161–173. https://link.springer.com/article/10.1007/s40899-015-0014-7

Bhatt, C. M. et al. Satellite-Based Assessment of the Catastrophic Jhelum Floods of September 2014, Jammu & Kashmir, India. *Geomat. Nat. Hazards Risk* **2017,** *8* (2), 309–327.

Bojarczuk, A.; Jelonkiewicz, Ł.; Lenart-Boroń, A. The Effect of Anthropogenic and Natural Factors on the Prevalence of Physicochemical Parameters of Water and Bacterial Water Quality Indicators along the River Białka, Southern Poland. *Environ. Sci. Pollut. Res. [Internet].* Apr 1, **2018** [cited Oct 27, 2020]**,** *25* (10), 10102–10114. /pmc/articles/PMC5891572/?report=abstract

Boyd, C. E.; Lichtkoppler, F. Water Quality Management in Pond Fish Culture Research and Development Series. *Int. Cent. Aquac. Agric. Exp. Station. Auburn. Univ*. **1979,** *22* (22), 1–30.

Carpenter, S. R.; Stanley, E. H.; Vander Zanden, M. J. State of the World's Freshwater Ecosystems: Physical, Chemical, and Biological Changes. *Annu. Rev. Environ. Resour*. **2011,** *36*, 75–99.

Chenthamilselvan, S.; Kankara, R. S.; Rajan, B.; Assessment of Shoreline Changes along Karnataka Coast, India Using GIS & Remote Sensing Techniques [Internet]. *Indian J. Geo-Marine Sci.* **2014** [cited Oct 27, 2020]**,** *43*. http://nopr.niscair.res.in/handle/123456789/34441

da Silva, T. S.; Luiza, M.; Fari, F. GIS Applied to Integrated Coastal Zone and Ocean Management: Mapping, Change Detection and Spatial Modeling for Coastal Management in Southern Brazil. In: Application of Geographic Information Systems [Internet]. *InTech*, 2012 [cited Oct 27, 2020]. http://dx.doi.org/10.5772/47840

De Swart, H. E.; Calvete, D. Non-Linear Response of Shoreface-Connected Sand Ridges to Interventions. *Ocean Dyn.* **2003,** *53* (3), 270–277.

DeMaster, D. J.; Marine Silica Cycle. In *Encyclopedia of Ocean Sciences*, 2nd ed.; Elsevier Inc. 2001; pp 678–685.

DeMaster, D. J. The Supply and Accumulation of Silica in the Marine Environment. *Geochim. Cosmochim. Acta* **1981,** *45* (10), 1715–1732.

Dhir, B. Introduction. In: *Phytoremediation: Role of Aquatic Plants in Environmental Clean-Up* [Internet]. Springer: India, 2013 [cited Oct 27, 2020]; pp 1–20. https://link.springer.com/chapter/10.1007/978-81-322-1307-9_1

Ehrlich, H. Silica Biomineralization, Sponges. In: *Encyclopedia of Earth Sciences Series* [Internet]. Springer: Netherlands 2011; pp 796–808. https://link.springer.com/referencework entry/10.1007/978-1-4020-9212-1_31

El Morhit, M.; Mouhir, L. Study of Physico-Chemical Parameters of Water in the Loukkos River Estuary (Larache, Morocco). *Environ. Syst. Res*. **2014,** *3* (1), 1–9.

EPA. Hypoxia in Gulf of Mexico and Long Island Sound, *Report on the Environment, United States Environmental Protection Agency* [Internet]. 2010. https://www.epa.gov/roe/ Hypoxia

Ern, R.; Huong, D. T. T.; Cong, N. V.; Bayley, M.; Wang, T. Effect of Salinity on Oxygen Consumption in Fishes: A Review. *J. Fish Biol.* **2014,** *84* (4), 1210–1220.

Fu, W.; Ma, J.; Chen, P.; Chen, F. Remote Sensing Satellites for Digital Earth. In: *Manual of Digital Earth* [Internet]; Springer: Singapore, 2020 [cited Oct 27, 2020]; pp 55–123. https://link.springer.com/chapter/10.1007/978-981-32-9915-3_3

Gualtieri, C. Sediment Oxygen Demand Modeling in Dissolved Oxygen Balance. In *Environmental Engineering and Renewable Energy*; Elsevier, 1998; pp 313–322.

Guariglia, A.; Buonamassa, A.; Losurdo, A.; Saladino, R.; Trivigno, M. L.; Zaccagnino, A. A Multisource Approach for Coastline Mapping and Identification of Shoreline Changes. *Ann. Geophys*. **2006,** *49* (1), 295–304.

Ibrahim, B.; Auta, J.; Balogun, J. An Assessment of the Physico-Chemical Parameters of Kontagora Reservoir, Niger State, Nigeria. *Bayero J. Pure Appl. Sci.* 2010, *2* (1), 64–69.

Jayappa, K. S.; Vijaya Kumar, G. T.; Subrahmanya, K. R. Influence of Coastal Structures on the Beaches of Southern Karnataka, India. *J. Coast Res.* **2003,** *19* (2), 389–408.

Jayaraju, N.; Raja Reddy, B. C. S.; Reddy, K. R. Anthropogenic Impact on Andaman Coast Monitoring with Benthic Foraminifera, Andaman Sea, India. *Environ. Earth Sci.* **2011,** *62* (4), 821–829.

Jayaraju, N.; Sundara Raja Reddy, B. C.; Reddy, K. R. Impact of Dredging on Sediment of Krishnapatnam Port, East Coast of India: Implications for Marine Biodiversity. *J. Environ. Sci. Eng.* **2010,** *4*.

Jiang, D.; Hao, M.; Fu, J. Monitoring the Coastal Environment Using Remote Sensing and GIS Techniques. In *Applied Studies of Coastal and Marine Environments* [Internet]; *InTech*, 2016 [cited Oct 27, 2020]. http://dx.doi.org/10.5772/62242

Joshi, D. M.; Kumar, A.; Agrawal, N. Studies on Physicochemical Parameters to Assess the Water Quality of River Ganga for Drinking Purpose in Haridwar District. *Rasayan J. Chem.* **2009,** *2* (1), 195–203.

Kanga, S.; Sheikh, A. J.; Godara, U. GIscience for Groundwater Prospect Zonation. *J. Crit. Rev.* **2020b,** *7* (18), 697–709.

Kanga, S. et al. Modeling the Spatial Pattern of Sediment Flow in Lower Hugli Estuary, West Bengal, India by Quantifying Suspended Sediment Concentration (SSC) and Depth Conditions Using Geoinformatics. *Appl. Comput. Geosci.* **2020a,** 100043.

Khatri, N.; Tyagi, S. Influences of Natural and Anthropogenic Factors on Surface and Groundwater Quality in Rural and Urban Areas. *Front. Life Sci. [Internet].* Jan 2, **2015** [cited Oct 27, 2020]**,** *8* (1), 23–39. https://www.tandfonline.com/doi/abs/10.1080/21553769.2014.933716

Lu, D.; Mausel, P.; Brondízio, E.; Moran, E. Change Detection Techniques. *Int. J. Remote Sens. [Internet].* Jun 20, **2004** [cited Oct 27, 2020]**,** *25* (12), 2365–2401. https://www.tandfonline.com/doi/abs/10.1080/0143116031000139863

Megonigal, J. P.; Neubauer, S. C. Biogeochemistry of Tidal Freshwater Wetlands. In *Coastal Wetlands: An Integrated Ecosystem Approach*; Elsevier, 2018; pp 641–683.

Meraj, G.; Singh, S. K.; Kanga, S.; Islam, M. N. Modeling on Comparison of Ecosystem Services Concepts, Tools, Methods and Their Ecological-Economic Implications: A Review. *Model. Earth Syst. Environ.* **2021,** 1–20.

Meraj, G.; Kumar, S. Economics of the Natural Capital and The Way Forward. Preprints 2021, 2021010083. DOI: 10.20944/preprints202101.0083.v1.

Meraj, G.; Romshoo, S. A.; Altaf, S. Inferring Land Surface Processes from Watershed Characterization. In *Geostatistical and Geospatial Approaches for the Characterization of Natural Resources in the Environment*; Springer: Cham, 2016b; pp 741–744.

Naithani, R.; Pandey, I. Physicochemical Assessment of Water Quality with Respect to Remote Sensing and GIS Techniques. *Adv. Pure Appl. Chem.* **2012,** *1*: 1–6.

Naja, G. M.; Volesky, B. Constructed Wetlands for Water Treatment. In Comprehensive Biotechnology, 2nd ed.; Elsevier Inc.**, 2011**; pp 353–369.

Officer, C. B.; Ryther, J. H. The Possible Importance of Silicon in Marine Eutrophication. *Marine Ecol. Progress Ser.* **1980,** *3*.

Pandey, A. C.; Singh, S. K.; Nathawat, M. S. Waterlogging and Flood Hazards Vulnerability and Risk Assessment in Indo Gangetic Plain. *Nat. Hazards* 2010, *55* (2), 273–289.

Pandey, A. C.; Singh, S. K.; Nathawat, M. S.; Saha, D. Assessment of Surface and Subsurface Waterlogging, Water Level Fluctuations, and Lithological Variations for Evaluating Groundwater Resources in Ganga Plains. *Int. J. Digit. Earth* 2013, *6* (3), 276–296.

Peters, N. E.; Meybeck, M.; Chapman, D. V. Effects of Human Activities on Water Quality. In *Encyclopedia of Hydrological Sciences*; John Wiley & Sons, Ltd, 2005.

Rashid, I. et al. Assessing the Impact of Anthropogenic Activities on Manasbal Lake in Kashmir Himalayas. *Int. J. of Environ. Sci.* 2013, *3* (6), 2036–2047.

Rusydi, A. F. Correlation between Conductivity and Total Dissolved Solid in Various Type of Water: A Review. In *IOP Conference Series: Earth and Environmental Science* [Internet]; Institute of Physics Publishing, 2018 [cited Oct 27, 2020]; p 012019. https://iopscience.iop.org/article/10.1088/1755-1315/118/1/012019

Sharma, A.; Kanga, S. Surface Runoff Estimation of Sind River Basin Using SCS-CN Method and GIS Technology, 2020.

Singh, S. K.; Pandey, A. C. Geomorphology and the Controls of Geohydrology on Waterlogging in Gangetic Plains, North Bihar, India. *Environ. Earth Sci. 2014*, *71* (4), 1561–1579.

Singh, S. K.; Kanga, S. Role of Geoinformatics in Site Suitability Analysis of Infrastructures Using PRA Approach. *Am. Int. J. Res. Sci. Technol. Eng. Math.* **2017a,** *1* (18), 81–85.

Singh, S. K.; Mishra, S. K.; Kanga, S. Delineation of Groundwater Potential Zone Using Geospatial Techniques for Shimla City, Himachal Pradesh (India). *Int. J. Sci. Res. Dev.* 2017, *5* (4), 225–234.

Smyth, K.; Elliott, M. Effects of Changing Salinity on the Ecology of the Marine Environment. In *Stressors in the Marine Environment*; Oxford University Press: Oxford, 2016; pp 161–174.

Sospedra, J.; Niencheski, L. F. H.; Falco, S.; Andrade, C. F. F.; Attisano, K. K.; Rodilla, M. Identifying the Main Sources of Silicate in Coastal Waters of the Southern Gulf of Valencia (Western Mediterranean Sea). *Oceanologia* **2018,** *60* (1), 52–64.

Sreenivasulu, G.; Jayaraju, N.; Sundara Raja Reddy, B. C.; Lakshmanna, B.; Prasad, T. L. Influence of Coastal Morphology on the Distribution of Heavy Metals in the Coastal Waters of Tupilipalem Coast, Southeast Coast of India. *Remote Sens. Appl. Soc. Environ.* **2018,** *10*, 190–197.

Sreenivasulu, G.; Jayaraju, N.; Sundara Raja Reddy, B. C.; Lakshmi Prasad, T. Physico-Chemical Parameters of Coastal Water from Tupilipalem Coast, Southeast Coast of India. *J. Coast Scencces* **2015,** *2* (2), 34–39.

Sreenivasulu, G.; Jayaraju, N.; Sundara Raja Reddy, B. C.; Lakshmi Prasad, T.; Lakshmanna, B.; Nagalakshmi, K. River Mouth Dynamics of Swarnamukhi Estuary, Nellore Coast, Southeast Coast of India. *Geod Geodyn.* **2016,** *7* (6), 387–395.

Sreenivasulu, G.; Jayaraju, N.; Sundara Raja Reddy, B. C.; Lakshmi Prasad, T.; Lakshmanna, B.; Nagalakshmi, K. Coastal Morphodynamics of Tupilipalem Coast, Andhra Pradesh, Southeast Coast of India. *Curr. Sci.* **2017,** *112* (4), 823–829.

Stoklosa, A. M.; Keller, D. H.; Marano, R.; Horwitz, R. J.; A Review of Dissolved Oxygen Requirements for Key Sensitive Species in the Delaware Estuary. Final Report, 2018.

Strayer, D. L.; Findlay, S. E. G. Ecology of Freshwater Shore Zones. *Aquat. Sci.* **2010,** *72* (2), 127–163.

Thirupathaiah, M.; Samatha, C.; Sammaiah, C. Analysis of Water Quality Using Physico-Chemical Parameters in Lower Manair Reservoir of Karimnagar District, Andhra Pradesh. *Int. J. Environ. Sci.* **2012,** *3* (1), 172–180.

Thornton, D. C. O. Dissolved Organic Matter (DOM) Release by Phytoplankton in the Contemporary and Future Ocean. *Eur. J. Phycol.* [Internet]. **2014,** *49* (1), 20–46.

Timothy, J. B. Transport of Ions and Water in Epithelia: Molecular Insights Transport of Ions and Water in Epithelia: Molecular Insights. In *Animal Osmoregulation*; Oxford University Press, 2010; pp 146–155.

Twomey, J.; Piehler, M.F,; Paerl, H. W.; Twomey, L. J. *Priority Parameters for Monitoring of Freshwater and Marine Systems, and their Measurement, Environmental Monitoring*, Vol. I; The University of North Carolina: Chapel Hill, 2009.

Umerfaruq, Q.; Rupesh, M.; Sandip, G.; Hitesh, S. Studies on the Physico-Chemical Parameters and Correlation Coefficient of Sarkhej Roza Lake. *J. Environ. Anal. Toxicol.* **2015,** *5*, 284.

Van, T. T.; Binh, T. T.; Application of Remote Sensing for Shoreline Change Detection in Cuu Long Estuary. *Earth Sci.* **2009,** *25*, 217–222.

Wen, Y.; Schoups, G.; Van De Giesen, N. Organic Pollution of Rivers: Combined Threats of Urbanization, Livestock Farming and Global Climate Change. *Sci. Rep.* [Internet]. Feb 23, **2017** [cited Oct 27, 2020]**,** *7*. /pmc/articles/PMC5322379/?report=abstract

CHAPTER 14

Soil Erosion Estimation Using Revised Universal Soil Loss Equation (RUSLE) Model and GIS

PAWAN THAPA*

Department of Geomatics Engineering, Kathmandu University, Dhulikhel, Nepal

**Corresponding author. E-mail: pawan.thapa@ku.edu.np*

ABSTRACT

Soil loss is a major causal factor for the critical watershed of a mountainous region. It is performed to estimate the soil erosion rate of Charnawati Watershed, Nepal using the revised universal soil loss equation (RUSLE), integrating remote sensing data and geographic information system (GIS) techniques. Its erosion rate is 58.55% in the low zone area and approximately 18% area on the verge of high-to-very-severe intensity classes. The watershed had scattered high severity of soil loss. This study identifies vulnerable areas and soil erosion. That information will be useful for the conservation of catchment areas supporting sustainable watershed and ecological management planning.

14.1 INTRODUCTION

Over the last few decades, soil erosion causes watershed degradation. Also, poor watershed management causes soil loss (Teng et al., 2019; Issaka and Ashraf, 2017; Chen et al., 2019). The primary concern is soil fertile eroded nearer the catchment areas that affect crop production. That leads to low output

GIScience for the Sustainable Management of Water Resources. Gowhar Meraj, Shruti Kanga, Majid Farooq, Suraj Kumar Singh, & Sudhanshu (Eds.)
© 2023 Apple Academic Press, Inc. Co-published with CRC Press (Taylor & Francis)

and low-quality surface runoff (Singh and Kanga, 2017). It is global issues that raise the mismanagement of land and watershed. Proper management of these hazards can estimate by soil erosion modeling using revised universal soil loss equation (RUSLE) and geographic information system (GIS) techniques.

A study over 20 years shows a 10% degraded Nepal's watershed area (Dahal and Guragain, 2013; Poudel, 2005; Pravongviengkham et al., 2003). Due to soil loss, floods, landslides, and farming practices (Farhan and Nawaiseh, 2015; S. Lee, 2004; Pradhan et al., 2012; Singh et al., 2017; Kanga et al. 2020a, 2020b). High mountains people's primary water resource is watershed for their livelihoods. Natural disasters, such as soil erosion, landslides cause losses of life and property that are habitat near watershed areas. Much of occurs in the past few years due to a significant increase in monsoon rainfall. About 45.5% of Nepal's land suffers from water erosion, mostly through sheet and rill erosion (Addisu, 2009; Koirala et al., 2019). Several studies conducted in Nepal use remote sensing and GIS techniques to estimate soil erosion and distribution (Thapa, 2020). This integrated method provides an accurate estimation of a cell by cell in larger areas with reasonable cost (Lehner et al., 2006; Lu et al., 2004). This study uses the RUSLE model with GIS to quantify and understand the soil erosion impact on watershed using the RUSLE model (Mallick et al., 2014; Terranova et al., 2009; Thapa, 2020; Tirkey et al., 2013). This research produces soil erosion that provides a framework for decision makers for planning activities to control erosion and fill a gap of soil loss information of a particular district.

14.2 MATERIALS AND METHODS

14.2.1 STUDY AREA

Dolakha is located in the Bagmati Province and extends from the altitude of 2088 m in latitude 27°46′42.24″ North and longitude 86°10′30.72″ East in mountain physiographic area, in the central-east part of Nepal (Khezri, 2019; Pawan, 2020; Thapa, 2020). In recent years, this district is prone to soil erosion, deforestation, and landslides result in watershed degradation (Chalise et al., 2019; Prasannakumar et al., 2011; Woldemariam et al., 2018; Pandey et al., 2010; Pandey et al., 2013; Singh and Pandey, 2014; Bhatt et al., 2017; Sharma and Kanga 2020; Kanga et al., 2017a, 2017b; Rather et al., 2018; Hassanin et al., 2020; Kanga et al., 2021). Changes in land use and occupation caused by human actions affect land (Foley et al., 2005; Lambin et al., 2001; Meyer and Turner, 1992) (Figure 14.1).

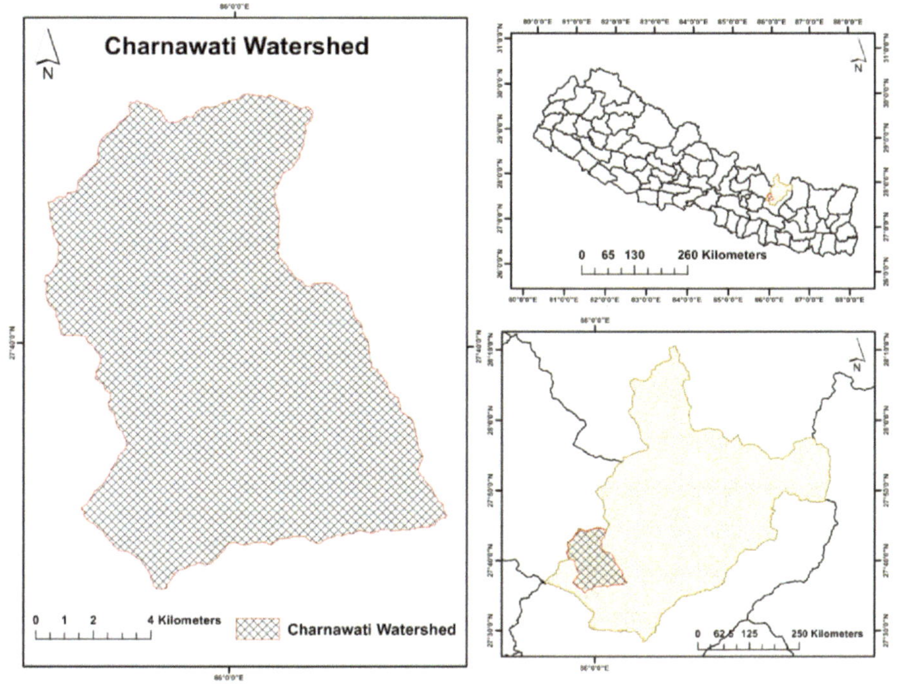

FIGURE 14.1 Study area used for estimating soil erosion.

14.2.2 DATA COLLECTION

The study used various spatial datasets that are shown in Table 14.1 (Thapa, 2020).

14.2.3 METHODS

The RUSLE model computes the mean annual soil loss of the Charnawati watershed using combined Remote sensing and Geographic Information System (GIS) technology ((Tang et al., 2015; Thapa, 2020) (Figure 14.2). An equation expresses the RUSLE

$$A = [R] \times [K] \times [LS] \times [C] \times [P], \quad (14.1)$$

where A = soil loss (mg/ha/yr), R = rainfall erosivity factor (mm/h/yr), K = soil erodibility factor (mg/ha/yr), LS = slope-length and slope steepness factor (dimensionless), C = land management factor (dimensionless), and P = conservation practice factor (dimensionless).

TABLE 14.1 The Datasets Used for the RUSLE Modeling.

Datasets	Data source
	SRTM DEM⁺
DEM×	https://earthexplorer.usgs.gov/
	A product of METI[$] and NASA[#]
	Digital Soil Map of the World (2019)
Soil map	www.fao.org/geonetwork/srv/en/metadata.show
	Produced by FAO-UNESCO¥ (1:5,000,000 scale)
Land cover map	Land cover Map of Nepal (2010)
	Produced by ICIMOD^
	TRMM Data[@]
Rainfall map	https://gpm.nasa.gov/?q=TRMM
	Produced by NASA

+DEM—Digital elevation model; †SRTM DEM—Shuttle Radar Topography Mission, Digital Elevation Model; $METI—Ministry of Economy, Trade and Industry, Japan; #NASA—National Aeronautics and Space Administration, US; ¥FAO–UNESCO-Food and Agriculture Organization, the United Nations Educational, Scientific and Cultural Organization; ^ICIMOD—International Center for Integrated Mountain Development; @TRMM—Tropical Rainfall Measuring Mission.

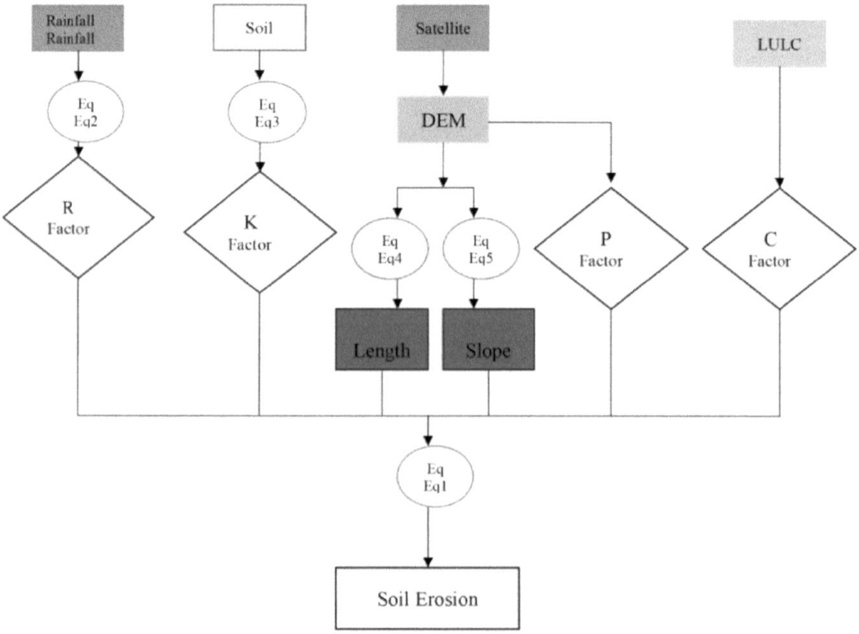

FIGURE 14.2 The methodology used to model RUSLE for the estimation of soil loss.

14.2.3.1 RUSLE PARAMETERS COMPUTATION

(a) Rainfall Erosivity Factor (R)

The rainfall erosivity factor (R) is a rainfall amount that affects a particular location of soil (Diodato, 2004; J.-H. Lee and Heo, 2011; Mannaerts and Gabriels, 2000; Wischmeier and Smith, 1978). It quantified the raindrop amount and runoff rate associated with rainfall and expressed in mm/ha/h/yr-1. During this study, the rainfall map produced by the National Aeronautics and Space Administration (NASA) was used for generating a rainfall erosivity factor (Daly et al., 2000; Gafforov et al., 2020; Mondal et al., 2015). The rainfall map represents mean annual precipitation over the district; the equation integrated to create the R-factor is given by reference (Irvem et al., 2007; Parveen and Kumar, 2012; Sheikh et al., 2011; Thapa, 2020).

$$R = 38.5 + 0.35P, \tag{14.2}$$

where R = Rainfall Erosivity Factor, P = Mean Annual Rainfall in mm.

(b) Soil Erodibility Factor (K)

The soil erodibility factor (K) estimates suspect on soil occurs through rainfall and runoff (Hellin, 2006; Maliehe, 2012; Thapa, 2020). One of the primary soil properties that impact K-factor are its organic matter, texture, soil structure, and permeability of the soil profile. The equation provided is stated below.

$$K = \text{Fcsand} \times \text{Fsi-cl} \times \text{Forgc} \times \text{Fhisand} \times 0.1317, \tag{14.3}$$

where

$$\text{Fcsand} = [0.2 + 0.3 \exp(-0.0256 \text{ SAN } (1 - \frac{SIL}{100}))], \tag{14.4}$$

$$\text{Fsi-cl} = [\frac{\text{SIL}}{\text{CLA} + \text{SIL}}]^{0.3}, \tag{14.5}$$

$$\text{Forgc} = [1.0 - \frac{0.25 \text{ C}}{C + \exp(3.72 - 2.95C)}], \tag{14.6}$$

$$\text{Fhisand} = =[1.0 - \frac{0.70 \text{ SN!}}{\text{SN1} + \exp(-5.51 + 22.9 \text{ SN1})}], \tag{14.7}$$

where SAN, SIL, and CLA are % sand, silt, and clay, respectively. C is the organic carbon content; and

SN1 is sand content subtracted from 1 and divided by 100.

Fcsand = it gives a low soil erodibility factor for soil with coarse sand and a high value for soil with little sand content.

Fsi-cl = it gives a low soil erodibility factor with high clay to silt ration.

Forgc = it is the factor that reduces soil erodibility for soil with high organic content.

Fhisand = it is the factor that reduces soil erodibility for soil with extremely high sand content.

(c) Topographic Factor (LS)

Digital elevation model (DEM) is used to create a slope gradient factor (S) and a slope-length factor (L). It is one of the significant characteristics of soil loss modeling. Both represent slope length and steepness, respectively, on erosion, which increases soil loss per unit area as slope increases and vice versa. The slopes gradient and slope length factors combined result computed using the following relation.

$$\mathrm{L} = \left(\frac{\lambda}{22.13}\right)^{\mathrm{m}}, \tag{14.8}$$

where L = slope length factor, λ = slope length (m), m = slope-length exponent

$$\mathrm{m} = \frac{F}{1+F'}, \tag{14.9}$$

$$\mathrm{F} = \frac{\sin / 0.0896}{3\left(\sin\right)_{0.8} + 0.56}, \tag{14.10}$$

where F = ratio of rill erosion to inter rill erosion, β = slope angle (°).
In ArcGIS, L was calculated as

$$L = \frac{\mathrm{flow}_{\mathrm{acc}} + 625)^{(m+1)} - \mathrm{flow}_{\mathrm{acc}}^{(m+1)}}{25^{(m+2)} \times 22.13^{m}}, \tag{14.11}$$

For slope gradient factor

$$\mathrm{S} = \mathrm{Con}\ ((\mathrm{Tan(slope} \times 0.01745) < 0.09), (10.8 \times \mathrm{Sin}\ (\mathrm{slope} \times 0.01745) + 0.03), (16.8 \times \mathrm{Sin(slope} \times 0.01745) - 0.5), \tag{14.12}$$

$$\mathrm{Final,\ LS} = L \times S. \tag{14.13}$$

(d) Cover Management Factor (C)

The ICIMOD data of land use/land cover (LULC) is used for preparing a C-factor map (Gemechu, 2014; Pawan, 2020; Uddin et al., 2018). For each land use-type, C values are assigned ranging from 0 to 1 through references (Stocking, 1984). The higher values of C mean the strong chances of soil erosion and vice versa (Wischmeier and Smith, 1978).

TABLE 14.2 Cover Management Factor.

Land use	C-Factor
Forest	0.03
Grassland	0.01
Cultivated Area	0.21
Barren Land	0.45
Water Body	0.00
Snow Glacier	0.00

(e) Support Practice Factor (P)

Agriculture practice provides support practice for soil damage. It can reduce by contours, cropping, and terrace (Cestti et al., 2003; Clay, 2013; Sklenicka et al., 2015; Tarolli et al., 2014), in this P ranging from 0 to 1, where 1 indicates non-anthropic properties and vice versa (Table 14.3, (Thapa, 2020; Wischmeier and Smith, 1978)).

TABLE 14.3 P-Factor Values for Slope as Per Agricultural Practice (Angima et al., 2003).

Slope %	Contouring
0–7	0.55
7–11.3	0.60
11.3–17.6	0.80
17.6–26.8	0.95
>26.8	1.00

14.2.3.2 POTENTIAL EROSION MAP

The five-factor maps were prepared through an integrated ArcGIS environment using the RUSLE relation for a composite map, including soil loss of the estimated erosion loss within the study area (Figure 14.3, (Aiello et al., 2015; Saptari et al., 2015)). Potential erosion rates are calculated from an area-weighted mean for slopes and LULC (Kayet et al., 2018; Olabode, 2019).

14.3 RESULTS

14.3.1 FACTOR MAPS

The estimated result shows topographic factor (LS) value between 0 and 340.659 (Figure 14.3a), cover management factor (C) ranged 0–0.45 (Figure 14.3b), Factor P value ranged from 0.55 to 1 (Figure 14.3c); Soil erodibility factor (K) value ranged from 0.2 to 0.32 (Figure 14.3d). Rainfall erosivity factor (R) value ranges between 764.8 and 1126.3 mm/ha/yr (Figure 14.3e).

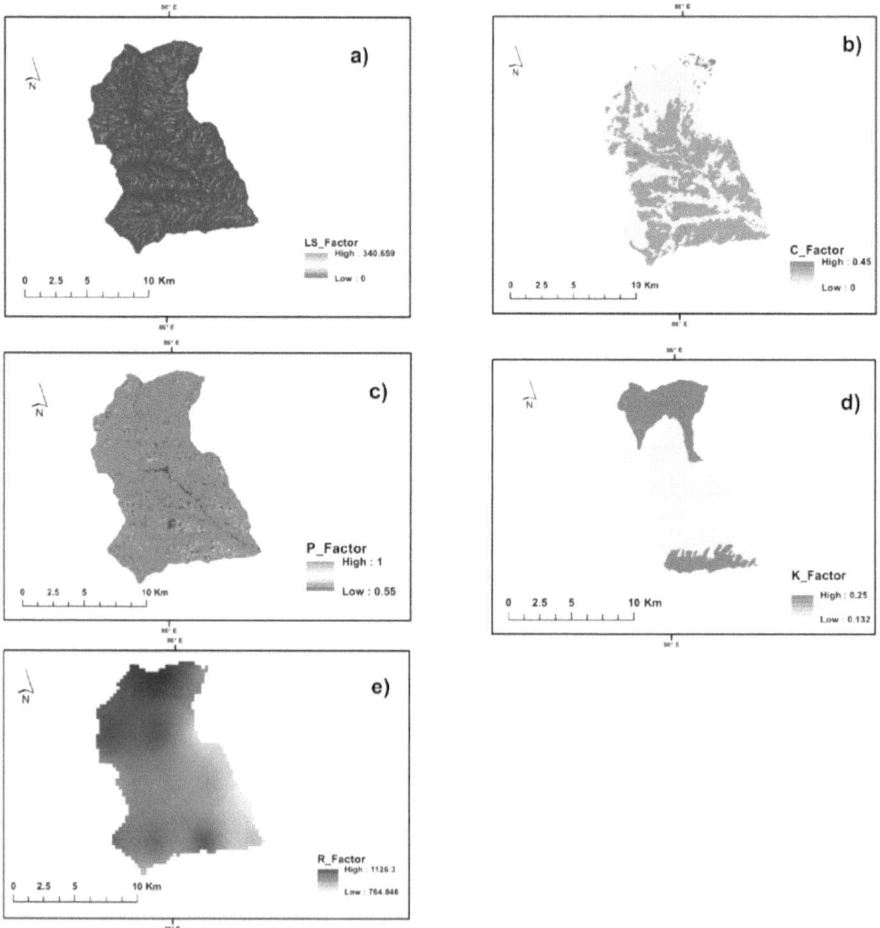

FIGURE 14.3 Factor maps of potential soil loss. (a) Topographic factor map. (b) Cover management factor map. (c) Support practice factor. (d) Soil erodibility factor map. (e) Rainfall erosivity factor map.

14.3.2 POTENTIAL SOIL EROSION RATES OF NEPAL

The potential soil loss map of Charnawati watershed is shown in Figure 14.4 and Table 14.4. The rate of loss classifies into six classes. More than 80% of soil erosion is 5.82% for a particular area (Table 14.4). Furthermore, it shows that more than 10% of the area at high severity of erosion. The steeper slopes and barren lands observe a higher risk of erosion, followed by agriculture, grasslands, and forests.

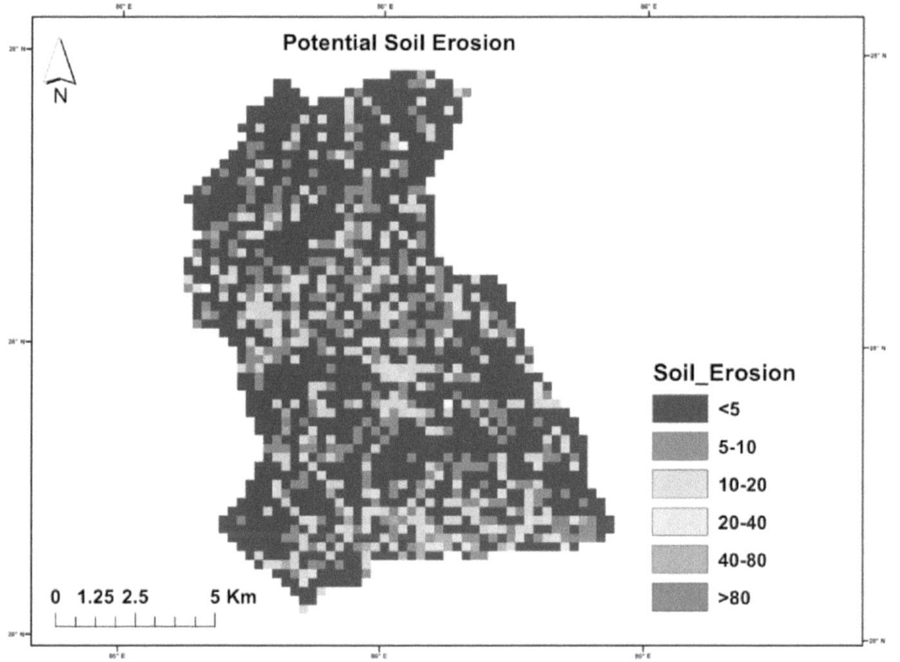

FIGURE 14.4 Potential map of soil loss of Charnawati watershed.

14.4 DISCUSSION

An empirical-based RUSLE model predicts soil loss's mean annual rate for a longer time (Borrelli et al., 2017; Brooks et al., 2014; Renschler and Harbor, 2002; Thapa, 2020). This research used the RUSLE model and its five factors to estimate Charnawati watershed erosion using integrated remote sensing and ArcGIS techniques. Such a study is an initial for a particular area with this approach and techniques. Overall, this provides information about erosion risk areas, which helps to formulate control plans and policies. Other similar studies use the same method to estimate soil loss (Jain and Das, 2010; Koirala et al., 2019; Tamene and Le, 2015; Thapa, 2020; Uddin et al., 2016). However, the methodology contains certain limitations, such as few weather stations, overestimation by model. For better estimation results, other or combine approaches, such as Rich Mesic Forest (RMF), and Mesic Forest (MF) models predict a loss in steeper slope (Collins and Jenkins, 1996; Gardner and Gerrard, 2003; Shrestha, 1997).

TABLE 14.4 Estimated Soil Loss Rate of Charnawati Watershed.

Class	Rate of erosion (Mg/ha/yr)	Area (km²)	Percentage of area	Severity
1	0–5	80.701	58.55	Low
2	5–10	17.746	12.87	Moderate
3	10–20	14.286	10.36	High
4	20–40	10.357	7.51	Very high
5	40–80	6.714	4.87	Severe
6	>80	8.024	5.82	Very severe

14.5 CONCLUSIONS

Soil loss significantly impacts the watersheds of mountainous regions. This study's main objective is to estimate the soil loss of the Charnawati watershed. Information can be a guideline for land-use planners and decision makers to formulate control erosion plans. The outcome represents a region of around 25% in high-to-very-severe prone risk of soil erosion. That might be increasing in upcoming years, taking it as the basis for further research and analysis needed for the area to reduce its vulnerability from soil loss.

KEYWORDS

- **soil erosion**
- **GIS**
- **Charnawati watershed**
- **mountainous region**

REFERENCES

Addisu, S. Assessment of the Impact of Road Construction on Physical Land Degradation in Central Highlands of Ethiopia: The Case of Two Selected Projects; Unpublished Master's Thesis, Addis Ababa University, Addis Ababa, 2009.

Aiello, A.; Adamo, M.; Canora, F. Remote Sensing and GIS to Assess Soil Erosion with RUSLE3D and USPED at River Basin Scale in Southern Italy. *Catena* **2015,** *131*, 174–185.

Angima, S. D.; Stott, D. E.; O'Neill, M. K.; Ong, C. K.; Weesies, G. A. Soil Erosion Prediction Using RUSLE for Central Kenyan Highland Conditions. *Agric. Ecosyst. Environ.* **2003,** *97* (1–3), 295–308.

Bhatt, C. M. et al. Satellite-Based Assessment of the Catastrophic Jhelum Floods of September **2014,** Jammu & Kashmir, India. *Geomat. Nat. Hazards Risk* **2017,** *8* (2), 309–327.

Borrelli, P.; Robinson, D. A.; Fleischer, L. R.; Lugato, E.; Ballabio, C.; Alewell, C.; Meusburger, K.; Modugno, S.; Schütt, B.; Ferro, V. An Assessment of the Global Impact of 21st Century Land Use Change on Soil Erosion. *Nat. Commun.* **2017,** *8* (1), 1–13.

Brooks, A.; Spencer, J.; Borombovits, D.; Pietsch, T.; Olley, J. Measured Hillslope Erosion Rates in the Wet-Dry Tropics of Cape York, Northern Australia: Part 2, RUSLE-Based Modeling Significantly Over-Predicts Hillslope Sediment Production. *Catena* **2014,** *122*, 1–17.

Cestti, R.; Srivastava, J. P.; Jung, S. *Agriculture Non-Point Source Pollution Control: Good Management Practices—The Chesapeake Bay Experience*; The World Bank, 2003.

Chalise, D.; Kumar, L.; Kristiansen, P. Land Degradation by Soil Erosion in Nepal: A Review. *Soil Syst.* **2019,** *3* (1), 12.

Chen, J.; Xiao, H.; Li, Z.; Liu, C.; Wang, D.; Wang, L.; Tang, C. Threshold Effects of Vegetation Coverage on Soil Erosion Control in Small Watersheds of the Red Soil Hilly Region in China. *Ecol. Eng.* **2019,** *132*, 109–114.

Clay, J. *World Agriculture and the Environment: A Commodity-by-Commodity Guide to Impacts and Practices*; Island Press, 2013.

Collins, R.; Jenkins, A. The Impact of Agricultural Land Use on Stream Chemistry in the Middle Hills of the Himalayas, Nepal. *J. Hydrol.* **1996,** *185* (1–4), 71–86.

Dahal, K. R.; Guragain, H. P. Local Response to Conservation Practices in Use for the Protection of Tinau River, Nepal. *Hydro Nepal: J. Water Energy Environ.* **2013,** *12*, 32–38.

Daly, C.; Taylor, G. H.; Gibson, W. P.; Parzybok, T. W.; Johnson, G. L.; Pasteris, P. A. High-Quality Spatial Climate Data Sets for the United States and Beyond. *Trans. ASAE* **2000,** *43* (6), 1957.

Diodato, N. Estimating RUSLE's Rainfall Factor in the Part of Italy with a Mediterranean Rainfall Regime. 2004.

Farhan, Y.; Nawaiseh, S. Spatial Assessment of Soil Erosion Risk Using RUSLE and GIS Techniques. *Environ. Earth Sci.* **2015,** *74* (6), 4649–4669.

Foley, J. A.; DeFries, R.; Asner, G. P.; Barford, C.; Bonan, G.; Carpenter, S. R.; Chapin, F. S.; Coe, M. T.; Daily, G. C.; Gibbs, H. K. Global Consequences of Land Use. *Science* **2005,** *309* (5734), 570–574.

Gafforov, K. S.; Bao, A.; Rakhimov, S.; Liu, T.; Abdullaev, F.; Jiang, L.; Durdiev, K.; Duulatov, E.; Rakhimova, M.; Mukanov, Y. The Assessment of Climate Change on Rainfall-Runoff Erosivity in the Chirchik–Akhangaran Basin, Uzbekistan. *Sustainability* **2020,** *12* (8), 3369.

Gardner, R. A. M.; Gerrard, A. J. Runoff and Soil Erosion on Cultivated Rainfed Terraces in the Middle Hills of Nepal. *Appl. Geog.* **2003,** *23* (1), 23–45.

Gemechu, T. *Estimating the Impacts of Land Use and Land Cover Change on Runoff and Sediment Yield Using SWAT Model: Case Study of Keleta Watershed*; Upper Awash Sub-Basin, 2014.

Hassanin, M.; Kanga, S.; Farooq, M.; Singh, S. K. Mapping of Trees Outside Forest (ToF) from Sentinel-2 MSI Satellite Data using Object-Based Image Analysis. *Gujarat Agric. Univ. Res. J.* **2020,** *207*.

Hellin, J. *Better Land Husbandry: From Soil Conservation to Holistic Land Management*; Science Publishers, 2006.

Irvem, A.; Topaloğlu, F.; Uygur, V. Estimating Spatial Distribution of Soil Loss over Seyhan River Basin in Turkey. *J. Hydrol.* **2007,** *336* (1–2), 30–37.

Issaka, S.; Ashraf, M. A. Impact of Soil Erosion and Degradation on Water Quality: A Review. *Geol. Ecol. Landscapes* **2017,** *1* (1), 1–11.

Jain, M. K.; Das, D. Estimation of Sediment Yield and Areas of Soil Erosion and Deposition for Watershed Prioritization Using GIS and Remote Sensing. *Water Resour. Manage.* **2010,** *24* (10), 2091–2112.

Kanga, S.; Rather, M. A.; Farooq, M.; Singh, S. K. GIS Based Forest Fire Vulnerability Assessment and Its Validation using field and MODIS Data: A Case Study of Bhaderwah Forest Division, Jammu and Kashmir (India). *Indian Forester* **2021,** *147* (2), 120–136.

Kanga, S. et al. Modeling the Spatial Pattern of Sediment Flow in Lower Hugli Estuary, West Bengal, India by Quantifying Suspended Sediment Concentration (SSC) and Depth Conditions using Geoinformatics. *Appl. Comput. Geosci.* **2020a,** 100043.

Kanga, S.; Sheikh, A. J.; Godara, U. GIscience for Groundwater Prospect Zonation. *J. Crit. Rev.* **2020b,** *7* (18), 697–709.

Kanga, S.; Singh S. K.; Sudhanshu. Delineation of Urban Built-Up and Change Detection Analysis using Multi-Temporal Satellite Images. *Int. J. Recent Res. Aspects* **2017a,** *4* (3), 1–9.

Kanga, S.; Kumar, S.; Singh, S. K. Climate Induced Variation in Forest Fire Using Remote Sensing and GIS in Bilaspur District of Himachal Pradesh. *Int. J. Eng. Comput. Sci.* **2017b,** *6* (6), 21695–21702.

Kayet, N.; Pathak, K.; Chakrabarty, A.; Sahoo, S. Evaluation of Soil Loss Estimation Using the RUSLE Model and SCS-CN Method in Hillslope Mining Areas. *Int. Soil Water Conserv. Res.* **2018,** *6* (1), 31–42.

Khezri, A. Climate Adaptation Services: Integrating Land Information to Support Mountain Communities; PhD Thesis, University of Twente, 2019.

Koirala, P.; Thakuri, S.; Joshi, S.; Chauhan, R. Estimation of Soil Erosion in Nepal Using a RUSLE Modeling and Geospatial Tool. *Geosciences* **2019,** *9* (4), 147.

Lambin, E. F.; Turner, B. L.; Geist, H. J.; Agbola, S. B.; Angelsen, A.; Bruce, J. W.; Coomes, O. T.; Dirzo, R.; Fischer, G.; Folke, C. The Causes of Land-Use and Land-Cover Change: Moving beyond the Myths. *Global Environ. Change* **2001,** *11* (4), 261–269.

Lee, J.-H.; Heo, J.-H. Evaluation of Estimation Methods for Rainfall Erosivity Based on Annual Precipitation in Korea. *J. Hydrol.* **2011,** *409* (1–2), 30–48.

Lee, S. Soil Erosion Assessment and Its Verification Using the Universal Soil Loss Equation and Geographic Information System: A Case Study at Boun, Korea. *Environ. Geol.* **2004,** *45* (4), 457–465.

Lehner, B.; Döll, P.; Alcamo, J.; Henrichs, T.; Kaspar, F. Estimating the Impact of Global Change on Flood and Drought Risks in Europe: A Continental, Integrated Analysis. *Clim. Change* **2006,** *75* (3), 273–299.

Lu, D.; Li, G.; Valladares, G. S.; Batistella, M. Mapping Soil Erosion Risk in Rondonia, Brazilian Amazonia: Using RUSLE, Remote Sensing and GIS. *Land Degrad. Dev.* **2004,** *15* (5), 499–512.

Maliehe, M. Assessment of Water Availability for Competing Uses in South Phuthiatsana Catchment in Lesotho by 2035 Using SWAT and WEAP, 2012.
Mallick, J.; Alashker, Y.; Mohammad, S. A.-D.; Ahmed, M.; Hasan, M. A. Risk Assessment of Soil Erosion in Semi-Arid Mountainous Watershed in Saudi Arabia by RUSLE Model Coupled with Remote Sensing and GIS. *Geocarto Int.* **2014,** *29* (8), 915–940.
Mannaerts, C. M.; Gabriels, D. Rainfall Erosivity in Cape Verde. *Soil Till Res.* **2000,** *55* (3–4), 207–212.
Meyer, W. B.; Turner, B. L. Human Population Growth and Global Land-Use/Cover Change. *Annu. Rev. Ecol. Systemat.* **1992,** *23* (1), 39–61.
Mondal, A.; Khare, D.; Kundu, S.; Meena, P. K.; Mishra, P. K.; Shukla, R. Impact of Climate Change on Future Soil Erosion in Different Slope, Land Use, and Soil-Type Conditions in a Part of the Narmada River Basin, India. *J. Hydrol. Eng.* **2015,** *20* (6), C5014003.
Olabode, O. F. Potential Groundwater Recharge Sites Mapping in a Typical Basement Terrain: A GIS Methodology Approach. *J. Geovisual. Spatial Analy.* **2019,** *3* (1), 5.
Pandey, A. C.; Singh, S. K.; Nathawat, M. S. Waterlogging and Flood Hazards Vulnerability and Risk Assessment in Indo Gangetic Plain. *Nat. Hazards* **2010,** *55* (2), 273–289.
Pandey, A. C.; Singh, S. K.; Nathawat, M. S.; Saha, D. Assessment of Surface and Subsurface Waterlogging, Water Level Fluctuations, and Lithological Variations for Evaluating Groundwater Resources in Ganga Plains. *Int. J. Digital Earth* **2013,** *6* (3), 276–296.
Parveen, R.; Kumar, U. Integrated Approach of Universal Soil Loss Equation (USLE) and Geographical Information System (GIS) for Soil Loss Risk Assessment in Upper South Koel Basin, Jharkhand, 2012.
Pawan, T. Evaluating a Fit-For-Purpose Integrated Service Platform for Climate Change and Land Information of Rural Mountain Community. *J. Remote Sens. GIS* **2020,** *9*, 278.
Poudel, K. Watershed Management in Nepal: Challenges and Constraints. Preparing for the Next Generation of Watershed Management Programmes and Projects **2005,** 119.
Pradhan, B.; Chaudhari, A.; Adinarayana, J.; Buchroithner, M. F. Soil Erosion Assessment and Its Correlation with Landslide Events Using Remote Sensing Data and GIS: A Case Study at Penang Island, Malaysia. *Environ. Monitor. Assess.* **2012,** *184* (2), 715–727.
Prasannakumar, V.; Shiny, R.; Geetha, N.; Vijith, H. Spatial Prediction of Soil Erosion Risk by Remote Sensing, GIS and RUSLE Approach: A Case Study of Siruvani River Watershed in Attapady Valley, Kerala, India. *Environ. Earth Sci.* **2011,** *64* (4), 965–972.
Pravongviengkham, P.; Khamhung, A.; Sysanhouth, K.; Qwist-Hoffmann, P. Integrated Watershed Management for Sustainable Upland Development and Poverty Alleviation in Lao People's Democratic Republic. Preparing for the next Generation of Watershed Management Programmes and Projects, Asia. Proceedings of the Asian Regional Workshop Kathmandu, Nepal **2003**; pp 11–13.
Rather, M. A. et al. Remote Sensing and GIS Based Forest Fire Vulnerability Assessment in Dachigam National Park, North Western Himalaya. *Asian J. Appl. Sci.* **2018,** *11* (2), 98–114.
Renschler, C. S.; Harbor, J. Soil Erosion Assessment Tools from Point to Regional Scales—The Role of Geomorphologists in Land Management Research and Implementation. *Geomorphology* **2002,** *47* (2–4), 189–209.
Saptari, A. Y.; Supriadi, A.; Wikantika, K.; Darmawan, S. Remote Sensing Analysis in RUSLE Erosion Estimation. *Indonesian J. Geospatial* **2015,** *4* (1), 34–45.
Sharma, A.; Kanga, S. Surface Runoff Estimation of Sind River Basin Using SCS-CN Method and GIS Technology, 2005.

Sheikh, A, A. H.; Palria, S.; Alam, A. Integration of GIS and Universal Soil Loss Equation (USLE) for Soil Loss Estimation in a Himalayan Watershed. *Recent Research in Science and Technology*, 2011.

Shrestha, D. P. Assessment of Soil Erosion in the Nepalese Himalaya: A Case Study in Likhu Khola Valley, Middle Mountain Region. *Land Husbandry* **1997,** *2* (1), 59–80.

Singh, S. K.; Pandey, A. C. Geomorphology and the Controls of Geohydrology on Waterlogging in Gangetic Plains, North Bihar, India. *Environ. Earth Sci.* **2014,** *71* (4), 1561–1579.

Singh, S. K.; Kanga, S. Role of Geoinformatics in Site Suitability Analysis of Infrastructures using PRA Approach. *Am. Int. J. Res. Sci. Technol. Eng. Math.* **2017a,** *1* (18), 81–85.

Singh, S. K.; Mishra, S. K.; Kanga, S. Delineation of Groundwater Potential Zone Using Geospatial Techniques for Shimla City, Himachal Pradesh (India). *Int. J. Sci. Res. Dev.* **2017b,** *5* (4), 225–234.

Sklenicka, P.; Molnarova, K. J.; Salek, M.; Simova, P.; Vlasak, J.; Sekac, P.; Janovska, V. Owner or Tenant: Who Adopts Better Soil Conservation Practices? *Land Use Policy* **2015,** *47*, 253–261.

Stocking, M. Rates of Erosion and Sediment Yield in the African Environment. *Challenges Afr. Hydrol. Water Resour.* **1984,** 285–295.

Tamene, L.; Le, Q. B. Estimating Soil Erosion in Sub-Saharan Africa Based on Landscape Similarity Mapping and Using the Revised Universal Soil Loss Equation (RUSLE). *Nutr. Cycl. Agroecosyst.* **2015,** *102* (1), 17–31.

Tang, Q.; Xu, Y.; Bennett, S. J.; Li, Y. Assessment of Soil Erosion Using RUSLE and GIS: A Case Study of the Yangou Watershed in the Loess Plateau, China. *Environ. Earth Sci.* **2015,** *73* (4), 1715–1724.

Tarolli, P.; Preti, F.; Romano, N. Terraced Landscapes: From an Old Best Practice to a Potential Hazard for Soil Degradation Due to Land Abandonment. *Anthropocene* **2014,** *6*, 10–25.

Teng, M.; Huang, C.; Wang, P.; Zeng, L.; Zhou, Z.; Xiao, W.; Huang, Z.; Liu, C. Impacts of Forest Restoration on Soil Erosion in the Three Gorges Reservoir Area, China. *Science Total Environ.* **2019,** *697*, 134164.

Terranova, O.; Antronico, L.; Coscarelli, R.; Iaquinta, P. Soil Erosion Risk Scenarios in the Mediterranean Environment Using RUSLE and GIS: An Application Model for Calabria (Southern Italy). *Geomorphology* **2009,** *112* (3–4), 228–245.

Thapa, P. Spatial Estimation of Soil Erosion Using RUSLE Modeling: A Case Study of Dolakha District, 2020.

Tirkey, A. S.; Pandey, A. C.; Nathawat, M. S. Use of Satellite Data, GIS and RUSLE for Estimation of Average Annual Soil Loss in Daltonganj Watershed of Jharkhand (India). *J. Remote Sens. Technol.* **2013,** *1* (1), 20–30.

Uddin, K.; Abdul Matin, M.; Maharjan, S. Assessment of Land Cover Change and Its Impact on Changes in Soil Erosion Risk in Nepal. *Sustainability* **2018,** *10* (12), 4715.

Uddin, K.; Murthy, M. S. R.; Wahid, S. M.; Matin, M. A. Estimation of Soil Erosion Dynamics in the Koshi Basin Using GIS and Remote Sensing to Assess Priority Areas for Conservation. *PloS One* **2016,** *11* (3), e0150494.

Wischmeier, W. H.; Smith, D. D. Predicting Rainfall Erosion Losses: A Guide to Conservation Planning. Department of Agriculture, Science and Education Administration, 1978.

Woldemariam, G. W.; Iguala, A. D.; Tekalign, S.; Reddy, R. U. Spatial Modeling of Soil Erosion Risk and Its Implication for Conservation Planning: The Case of the Gobele Watershed, East Hararghe zone, Ethiopia. *Land* **2018,** *7* (1), 25.

CHAPTER 15

Analyzing the Degradation of Rivers Using GI Science

SHER SINGH BAGHEL[1], and EKTA BARANWAL[2*]

[1]*Gateway College of Architecture and Design, Sonipat 131001, Haryana, India*

[2]*Department of Civil Engineering, Jamia Millia Islamia, New Delhi 110025, India*

[*]*Corresponding author. E-mail: ektabaranwal01@gmail.com*

ABSTRACT

Water is the most essential and necessary need for every life on the earth, getting distributed irregularly and degrading for several reasons, considering it is less natural and more human-made. Because of the significance of element water in the entire ecosystem and the deuteriation of water resources over time, its management is ultimately our responsibility. The earth's water cycle explains the different water forms available on the planet and their interaction with the environment. From various authentic documents, the global distribution of water can be found as 97.5% of the total water is ocean water. The rest, 2.5% water, comes from freshwater, including glacier water, groundwater, permafrost water, and surface and atmospheric water. This surface and atmospheric water is distributed in six forms, freshwater lakes, wetlands, soil moisture, rivers, atmosphere, and plants and animals. Among all these water resources, groundwater, freshwater lakes, and rivers are the primary water sources for all human activities. The world is presently facing a fall in the overall number of the water bodies' presence on the earth and continuously waning of the water resources' dimension and quality due

GIScience for the Sustainable Management of Water Resources. Gowhar Meraj, Shruti Kanga, Majid Farooq, Suraj Kumar Singh, & Sudhanshu (Eds.)
© 2023 Apple Academic Press, Inc. Co-published with CRC Press (Taylor & Francis)

to the inadequate understanding of the management of natural resources. India is also in not much better condition while facing these consequences; however, ancient India was so rich in its hydraulic structure that it comprises natural water resources like rivers, lakes, and human-made water bodies like ponds. But over time, these water resources get decayed like either narrowed or completely lost, because of the lack of proper management and insufficient knowledge about managing it. This chapter focuses on one of the water resources, that is, the river, specifically the Yamuna River in India for the sacred Mathura district, which is waning with time. This chapter also covers the overview of water resources, causes of their degradation, and their management. Since remote sensing (RS) and Geographical Information System (GIS) make the process easy nowadays for assessing the spatiotemporal distribution of water resources, change in their dynamics, and quality of the ground and surface water. Therefore, this chapter's main objective is to assess the Yamuna River's degradation on a spatiotemporal scale through satellite imagery and an area survey that ultimately covers the causes behind the degradation of this river. Through satellite images, the assessment is done for four decades to show the changes in the river Yamuna's dimension and flow direction within this stretch.

15.1 INTRODUCTION

The chapter is about the deterioration and waning of water resources in India, focusing on the Yamuna River with its spatiotemporal analysis for the Mathura district. This chapter is starts with some facts about water resources on earth and the problems humans are facing at present. Next is an introduction about water resource management (WRM), why it is needed, how it was done in the ancient times, and how it is done nowadays. Then the chapter discusses sustainable WRM and the role of geoinformatics for WRM continued by water resource deprivation, its causes, Yamuna River for India's case study, and some of the initiatives taken for the WRM in India. This chapter has some latest ground images and the 40-year change analysis through satellite images to show the waned Yamuna River.

15.2 WATER RESOURCES ON EARTH: SOME FACTS

As an essential part of all living beings, water is the most significant element globally in various forms and available percentage. Broadly, water availability on the earth is classified as oceans, which are approximately 97.5%, and 2.5% as freshwater (WWAP, 2019; 2020). This freshwater is further subdivided into various forms and their availability is shown below in Figure 15.1:

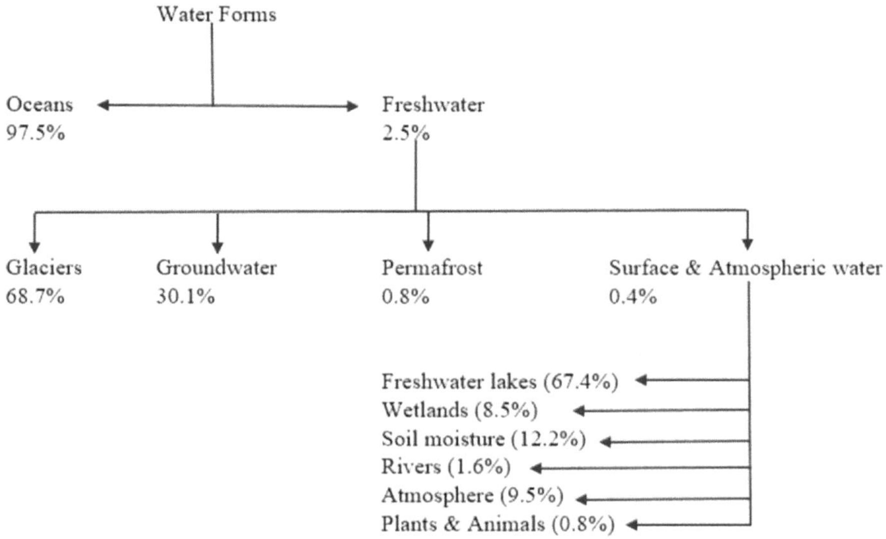

FIGURE 15.1 Water forms distribution on the earth.
Source: Author.

Much of the freshwater available is in the frozen form as glaciers and also as ice caps. The remaining are mostly below the surface as groundwater and a very few fractions above the ground as surface water r or in the atmosphere. Apart from groundwater, rivers and lakes (WWAP, 2019; 2020) are the primary surface water resources on the earth that have been used to fulfill the needs of all living beings.

15.3 WATER RESOURCE PROBLEMS AND WATER RESOURCE MANAGEMENT

The availability, flow, numbers, dimension, and quality of all types of water resources in the ecosystem change in time and space mainly because of human activities (WWAP, 2019; 2020; Loucks and Beek, 2005). That is how the problems came to the water resources mentioned as either too much water or too little apart from the low quality of water (Loucks and Beek, 2005). The water resources' problems impact the entire ecosystem, including humans, plants, animals, soils, air, and everything surrounding us (Tundisi, 2008). The water resources are:

- Decreasing in their numbers, dimensions, and availability.
- The flow of surface water is getting changed, either narrowed or overflowed.
- The level of groundwater is lowered.
- Pollution makes the water quality deteriorated.
- Water scarcity and water stress is increased.

These problems of water resources affect our surroundings in several profound ways, including humans and the environment as (Loucks and Beek, 2005; Tundisi, 2008):

- Unsafe drinking water.
- Decreasing water networks in the urban areas.
- Pressure on social and economic developments to fulfill the required demand of the water for everyone.
- Affected human health, soil health, and environmental health.
- Compromised food security due to excessive rains and droughts.
- Aquatic ecosystem health problems due to polluted water.
- Dramatic biodiversity, flora, and fauna.
- Planning and management related issues like navigation due to flood or narrowed the river and erosion of river banks.

According to the World Resource Institute (WRI), human needs for freshwater have been globally increased since the 1960s as the populations grew, resulting in water stress at highly extreme intensity. WRI has given some figures about the usages of water by every sector, and the domestic industry has already crossed the limit by reaching beyond the 600% during the last five decades (Figure 15.2a). Water withdrawals in this way is not the correct way because it leads to a shortage of quantity and waning in the water quality around the entire globe. Figure 15.2 is a graphical representation of a typical way of using water by humans and how poorly it is managed, ultimately affecting everything surrounding us.

From the above scenario, we have the data of the problems that water resources are facing and the other consequences which make it clear why we need WRM, a system to manage and conserve water resources. Since everything is related to the water, we need some strategies to decrease the intensity of water resources. WRM (Loucks and Beek, 2005) is the complete system to plan and manage all the issues related to water resources by identifying the exact problem in the area, determining the possible solutions, and finally applying them. Managing water resources includes water resources, how much they are damaged, and the causes of their deterioration, then only, the appropriate solution can be found.

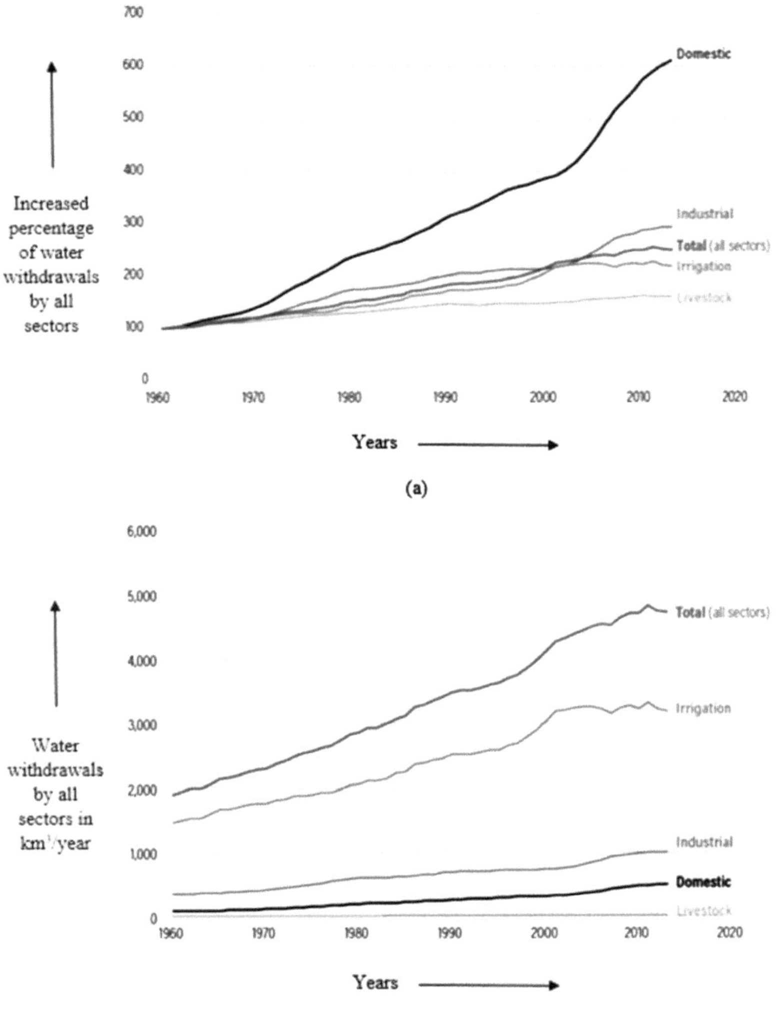

FIGURE 15.2 Water withdrawals by all sectors during 1960–2014, (a) in percentage, (b) in km^3/year.

Source: Reprinted from Otto and Schleifer (2020).

15.4 SUSTAINABILITY IN WATER RESOURCE MANAGEMENT

Sustainable management is the method of using natural resources to balance their availability with their usability. Sustainability means to achieve the even distribution of resources among all in the entire ecosystem by filling the gap between the current availability of the resources like water and vegetation and all living beings. In terms of water, it is essential for all entities in the ecosystem, that is, for humans, trees, animals, birds, and even for the atmosphere. Therefore, water must be evenly distributed to be useful for everyone and satisfy the high quantity and quality requirements. Unfortunately, the world is already facing several issues that deteriorate the water quality and create the environment for its scarcity (NITI Aayog). It can be easily seen that freshwater has been excessively used and has been recorded for the past five decades. Hence, the water resources must be managed sustainably, and initiatives (Bharat and Dkhar, 2018; UNEP, 2012) should be taken and implemented by everyone toward:

- Efficient water availability for every farm in the agriculture sector, that is, for sustainable irrigation.
- Water treatment and water recycling encouragement for industries to ensure zero untreated water discharge.
- Enhancing the potential of surface water by creating additional capacity of water resources.
- Ensuring the groundwater resources sustainability in the long-term by controlling the excessive groundwater exploitation.
- Sewage treatment.
- Afforestation to prevent flood and drought.
- Implementing the strategies for preventing water pollution and water conservation.

Several policies and programs are planned for sustainable water management in the world by the government of each country and non-government organizations. Still, the goal cannot be achieved until every individual takes responsibility for the efficient use of nature's gifts.

15.5 REMOTE SENSING AND GEOGRAPHICAL INFORMATION SYSTEM FOR WATER RESOURCES

Remote sensing (RS) and geographical information system (GIS) are the components of the whole geoinformatics system, which involves variously diversified subjects or fields to monitor and analyze the features, objects, and phenomena of the earth's surface, mainly from the space to the ground collect data. Remote sensing does not only involve the earth observation from space but also aerial and terrestrial platforms. GIS helps store geographical data and develop geographical databases and digital maps through the various data collected via remote sensing. The geoinformatics system is being used nowadays in every field of science and technology. How RS and GIS are used in the hydrology and water resources lot can be widely categorized as:

- Water resources assessment, development, management, and mapping.
- Watershed management.
- Information and decision support systems for water resources.
- Environmental impact assessment (EIA) and management.
- Flood disaster support.
- Rainfall measurements and runoff prediction.

This chapter deals with WRM, and RS and GIS can be used in combination for numerous studies and analyses to support a good WRM. Below are the detailed applications of RS-GIS in WRM, which have been developed and used for decades (Welch and Remillard, 1991; Dwivedi et al., 2020):

- Snow cover monitoring and mapping, including forecasting of snow-melt runoff.
- Glacier monitoring and mapping, including glacier mass balance and glacial lake.
- Studying and mapping surface water resources: all water bodies and wetlands.
- Irrigation planning and irrigation water management.
- Reservoir analysis and management.
- EIA studies for hydropower generation.
- Monitoring and mapping of interlinking of rivers and groundwater analysis.
- Flood and drought detection, monitoring, modeling, assessment, and management.
- Soil moisture and erosion assessment.
- Rainfall measurements, runoff simulation, modeling, and prediction.
- Water balance studies.
- Urban water supply, distribution, mapping, and performance assessment.
- Waterlogged and catchment area monitoring and mapping.
- Evapotranspiration studies.
- Water harvesting.
- Watershed management.

15.6 IN INDIA: WATER RESOURCES, ITS MANAGEMENT, AND POLICIES FOR WATER RESOURCE MANAGEMENT

In India, water resources are meant to fulfill human life's basic needs and have an extreme religious significance, especially for rivers and kunds (ponds). Water resources (natural and human-made) of India are about 4% of the globe's water resources, which include surface water resources—rivers, ponds, lakes, and tanks; groundwater resources—lagoons; and backwaters (NCERT, 2019a, 2019b). India has approximately 10,360 rivers and their tributaries, which can be categorized into four classes according to the drainage type (P. IAS, 2016; KnowIndia, 2020):

- Himalayan rivers, which include the Ganga–Brahmaputra–Meghna system and the Indus.
- Deccan Rivers mostly cover the rivers, such as the Godavari, Cauvery, Krishna, Mahanadi, Narmada, Tapti, etc.
- Coastal rivers are smaller in size comparatively but are numerous in number which comes around 600 rivers.
- Rivers of the inland drainage like the streams in Rajasthan and Gujarat named as the Sambhar and the Luni, respectively.

Ganga, also known as Bhagirathi, is the most extensive river system and has many tributaries in India. The Yamuna, Ghaghra, Kosi, Ramganga, Gandak, Mahananda, Gomti, and Hugli are some tributaries (KnowIndia, 2020). All these rivers have been the life source in India from the ancient civilization till now.

15.6.1 WRM IN ANCIENT INDIA

Ancient India was very rich in water resources and hydraulic structures with lots of natural and human-made water bodies. "From the Indian history, it can be easily seen that all the ancient civilizations were full of hydrological knowledge whose proofs can be found in one of the oldest documents, the Vedas like Yajurveda, Rigveda, and Atharvaveda followed by Harappan civilization and Mauryan empire. During Harappan civilization, that is, approximately 3000–1500 BCE, proofs have been found for water science development, construction of the hydraulic structure, and disposal and treatment systems for wastewater. Mauryan empire, that is, 322-185 BCE had the understanding of the water balance, hydrological processes, measuring rainfall, constructing reservoirs–dams–channels with spillways, and even water pricing systems were developed during this period" (Singh et al., 2020). Water had religious values starting in India. In each civilization, water was treated as either God/Goddess or a sacred item that people had worshiped. Archaeologists found architectural designs about various cultures and show the well-constructed drainage systems, wells, well-harvesting techniques, systems to manage wastewater, canals, several water networks, water tanks, stream flows, and erosion process, hydraulic structures for agriculture and irrigation, and other structures (Singh et al., 2020). So, it can be said that ancient Indian civilizations over 3000 years were always enriched with the knowledge of water resources, their management, and adequate and magnificent hydraulic structures, which are also described in the ancient scriptures.

15.6.2 INDIAN WATER POLICIES

Water is introduced as a subject of State if we consider the Constitution of India, which includes water supplies, water storage, water power, drainage, dams, irrigation, and canals as the States' jurisdiction according to Schedule VII, List II, Entry 17 from the Constitution of India. For saving the water resources in India, including surface and groundwater, there are many initiatives taken by the Indian government which is broadly a framework as the Water Policy is influenced by the time to provide a large vision, plan and objectives, and economic and political institutions. The Government of India has affirmed the first policy for water-related issues, its usability, and management in 1987 as the National Water Policy (NWP) of India whose principal priorities were about drinking water, irrigation, hydropower, navigation, and industrial and other users(Kumar and Bharat, 2014). This water policy was reformed two times after that in 2002 and 2012 including all of the prioritizations of 1987 and major inclusion of "ecological and environmental aspects" and "water treatment for safe sanitation and drinking water, efficient use and conservation of water, and lastly efficiently using of water for economic goods was given priority toward supporting poor people's basic livelihood and national food security" in NWP 2002 and NWP 2012, respectively (MWR, 2002; Seth, 2012). NWP 2012 covered the major and ongoing problems that water resources are facing and the areas affected by these water resources problems. Major issues that NWP 2012 includes are (Kumar and Bharat, 2014; Seth, 2012):

- Adaptation of climate change.
- Enhancement of water availability.

- Efficiently using water and managing water demand.
- Water pricing.
- Conservation of water bodies, river corridors, and infrastructure.
- Planning and implementation of the project.
- Floods and droughts management.
- Sanitation and water supply.
- Institutional arrangements.
- Trans-boundary rivers.
- Creating and maintaining the database information system.
- Training and research needs.
- Preparing a plan of action on NWP through the National Water Board.

Apart from the NWP 2012, the Indian policy has three other frameworks for water which include (1) National Water Mission (NWM) of India, 2008; (2) Draft National Water Framework Law (NWFL) of India, 2016; (3) Model Ground Water Bill (MGWB), 2016 (Bharat and Dkhar, 2018)." NWM is part of eight missions set up in the National Action Plan on Climate Change (NAPCC) by the Ministry of Environment, Forest and Climate Change (MoEFCC) with main objectives: minimizing the wastage of water, ensuring equitable distribution of water through better development and management of water resources, and conservation of water. NWFL includes the principles to govern the exercise of the legal framework and action for the principles of NWP 2012 to protect, conserve, regulate, and manage the water resources. MGWB was specifically introduced for groundwater-related issues, such as its protection, conservation, regulation, and management to make ground-water availability in a balanced way for everyone" (Bharat and Dkhar, 2018). The Government of India has taken several other initiatives as missions and programs under this entire Water Policy Framework, such as National Rural Drinking Water Programme (NRDWP), National River Conservation Programme (NRCP), National Mission for Clean Ganga (Namami Gange), Interlinking of Rivers, Guidelines for Improving Water Use Efficiency in Irrigation, Domestic and Industrial Sectors, and Water (Control and Prevention of Pollution) Act 1974, etc. (MWR, 2020).

15.7 YAMUNA RIVER

Yamuna river is one of the significant tributaries or subbasin of river Ganga, known as India's largest and most significant river. "Stream Yamuna starts from Yamunotri glacier in the Himalayas and navigates through Himachal

Pradesh and Uttaranchal within the upper stretch of 200 km, drawing water from a few major streams. It enters the plains at Dak Pathar in Uttranchal, where the waterway water is controlled through the weir and diverted into the canal for the control era. Here, it comes to Hathnikund/Tajewala in Yamuna Nagar locale of Haryana state, where the waterway water is occupied into the Western Yamuna Canal and Eastern Yamuna canal for the water system. Amid dry season, no water is allowed to stream within the waterway downstream to Tajewala barrage, and the waterway remains dry in a few extends between Tajewala and Delhi. The river regains water as (1) groundwater gathering and (2) commitments of bolstering the canal through Som Nadi (regular stream) upstream of Kalanaur. The catchment of the Yamuna waterway framework covers parts of Uttar Pradesh, Uttranchal, Haryana, Himachal Pradesh, Madhya Pradesh, Rajasthan, and Delhi" (Upadhya and Rai, 2013; Nallathiga, 2018).

"Concurring to legend, the Yamuna is considered as the daughter of Surya (the Sun God) and his spouse Saranya, and the twin sister of Yama (the God of Death). A plunge within the Yamuna is accepted to free an individual from all past sins and contaminations of the fabric world. As a portion of immortal civilization, the Stream Yamuna, other than a traditional water source, is also a cultural symbol and revered as a Goddess in Indian culture. Many sanctuary towns are situated along her banks, and their characters are based on the river. The Mahabharata too mentions the Yamuna as being one of the seven tributaries of the Ganges. Since time immemorial, bathing in and drinking Yamuna's waters are believed to evacuate sin. Afterward, Yamuna is related to the God Krishna. Krishna's association with the stream started right from his birth when he let her surging waters touch his feet" (DUAC, 2015).

15.7.1 YAMUNA RIVER IN MATHURA DISTRICT

Mathura, the name has a unique place in the world as the mystical and religious land of love spread by Lord Krishna. Geographically, Mathura is one of the city in Uttar Pradesh State, India, with coordinates 77.41° east and 27.28° north bounded by Haryana and Rajasthan. According to the epics in Hinduism in India, Mathura is where Lord Krishna was born and his maternal grandfather's home. Mathura as a district involves the famous Lila's (playful acts) places Gokul, Goverdhan, and Vrindavan of Lord Krishna. The whole Braj Mandal or region was Lord Krishna, which currently includes nine major centers Mathura, Goverdhan, Barsana, Vrindavan, Deeg, Kaman, Gokul, Nandgaon, and Hodal from three states, such as Uttar Pradesh, Rajasthan, and Haryana.

The Braj Mandal had several forests, subforests, and water bodies (small and big), which have deteriorated with time due to growing urbanization, lack of knowledge for their proper management, and poor management of these natural resources. The whole Braj Mandal was extremely rich in vegetation and water bodies. Among hundreds of traditional water bodies, the Yamuna River has a special significance in the Braj Mandal and Lord Krishna's entire stories. Yamuna River was the primary water resource of the time and the center of all Krishna Lilas in the mythological tales. Fables and stories about his childhood and youth are told and retold nowadays with an immaculate and the beautiful Yamuna as the scenery. This assists strengthened through most conventional and society verse and dance shapes. Agreeing to legend, Krishna hopped into the stream to tame the noxious Kaliya snake, which was contaminating the stream with his noxious fumes. Mythology could be a complex science. Century-old stories and stories can be translated by generation after era with their ever-complicated settings and changing ethos. Kaliya's advanced day equivalents are the mechanical units that discharge effluents into the waterway. Also, the Vishram Ghat in the Mathura is considered the bathing platform for going toward salvation.

15.7.2 YAMUNA RIVER DEGRADATION IN INDIA

The stream is not given a livelihood for communities living within the basin and offers a life bolster to agricultural, mechanical, and urban sectors and is the most important source of drinking water for most of the towns along its course. In view of the rapid development of the urban populace, industrialization, and lacking foundation, the waterway water quality across the nation is breaking down alarmingly. The industrial towns all along with the river discharge considerable sums of squanders into the stream. The Yamuna gets to be a drain to a lower extent, receiving primarily agribusiness, mechanical, and household effluents. Yamuna River is waning, and the issues it is facing can be grouped as the quality of the water and its surface dimension. Yamuna River is already under the overexploited groundwater development category, which continuously leads to the termination of linkage between the river runoff and the groundwater (Upadhya and Rai, 2013).The water quality, wastewater, and solid waste release into the river are the leading causes of the river's polluted water. The source of these discharges is industrial discharges, municipal waste, improper religious practice, agricultural runoff, and inadequate social practices (Nallathiga, 2018). The pollution in the Yamuna River has made its water the issue of life and death for many people who reside in the cities along the river, specifically the low-class population (Singh, 2001). Figure 15.3–155 shows an idea about the intensity of the pollution in the Yamuna River.

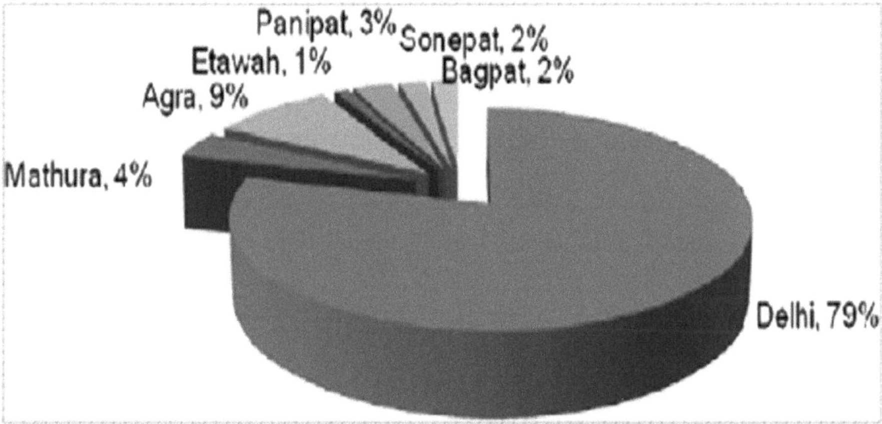

FIGURE 15.3 Percentage of pollution in the Yamuna River contributed by various cities.
Source: Reprinted from Sharma and Chaudhry (2015).

Yamuna River has several huge and industrial cities along its banks like Delhi, Sonipat, Yamuna Nagar, Panipat, Agra, and Mathura. There are 42 industrial (mechanical) units in Delhi, 22 in Haryana, and 17 in Uttar Pradesh, straightforwardly releasing and contaminating the Yamuna Stream. These businesses incorporate paper, chemicals, sugar, leather, refinery, power, pharmaceuticals, etc. Due to overall anthropogenic weight, the stream Yamuna is quickly falling apart into a sewage depletion. Delhi alone releases its untreated and treated residential and industrial squanders through small or expansive channels. 95% of the Yamuna contamination at Delhi is brought about due to the release of wastewaters from seven major channels, which are Sen Nursing Domestic, Najafgarh, Yamunapur, Barathpula, Kalkaji, Maharanibag, and Tuglakabad. In Delhi, the Yamuna stream is so contaminated that past Okhla, it barely underpins any frame of life. The natural matter and supplements but various pesticides and overwhelming metals have been found disturbing within the stream. On the premise of the distinctive geographical and environmental characteristics, the waterway has been partitioned into five major fragments, viz., the Himalayas extend (172 km), upper extend (224 km), Delhi extends (22 km), blended extend (490 km), and weakened extend (468 km) of which Delhi is the foremost contaminated extend of the waterway, it is alone mindful for 79% of the entire contamination stack within the waterway Yamuna. The Himalayan extend of the Waterway Yamuna is void of any kind of contamination. The water quality here in Tajewala is, for the most part, inside the required levels for washing quality. The mechanical belt is primarily arranged along the northeastern portion of the State along the Yamuna. The squanders from these businesses, rural runoff, and the channels carrying civil sewage of the cities enter the waterway and influence its water quality. In order to comply with environmental laws, these businesses must treat the emanating to realize endorsed benchmarks, sometimes recently releasing profluent into the environment. The biological oxygen demand (BOD) level in the Yamuna also changes at distinctive places but was moreover found over the allowable limits in Mathura, Agra, Etawah, and Juhika"(Sharma and Chaudhry, 2015). Pollution in the Yamuna River at Agra near the Taj Mahal and Mathura near Vishram Ghat can be seen in Figure 15.4.

Taking the case of the Yamuna in Mathura district, which is very significant religiously, still it is not escaped from being polluted and deteriorated as got odor, ugly, and black water with following sources which are caused by humans (Bhargava, 2006):

- Domestic wastewater.
- Industrial wastes.
- Pollution via agricultural activities.
- Solid wastes.
- Religious wastes.
- Mass bathing.
- Disposal of unburnt/half-burnt/burnt dead bodies (human and animals).
- Overexploitation of ground and surface water.
- Scattering of all kind of solid waste all over the river and sides by animals and birds.
- Siltation problems.
- Bank erosion.

15.7.3 SOME INITIATIVES THAT ARE TAKEN FOR THE YAMUNA RIVER IN INDIA

Most of the issues related to the river are covered and the NWP in India, including every river, but since the chapter is about the Yamuna River, its preservation is protected under the programs:

(a)

(b)

FIGURE 15.4 (a) Polluted Yamuna River in Agra near Taj Mahal; (b) Extremely polluted area near Vishram Ghat in Mathura.

Source: a) Reprinted from Dubey and Maurya (2018); b) Reprinted with permission from Bhargava (2006). © Springer Nature.

- Namami Gange
- Ganga rejuvenation
- Interlinking of rivers
- Research and Development Program in the water sector
- National Hydrology Project
- Implementation of the National Water Mission
- River Basin Management
- Sutlej–Yamuna Canal
- River management activities

These programs listed above are according to the Ministry of Water Resources (Jal Shakti), Government of India. Apart from these, "there is the Yamuna Action Plan (YAP) initiated in 1993 by the National River Conservation Directorate (NRCD) under the Ministry of Environment and Forests to focus on lessening the domestic and other wastewater discharge. The YAP was extended and reformed as YAP-II in 2002, where several sewerage lines were built along the Yamuna River banks to restore the water quality. YAP was the project of a one-time mechanism that involved the actions regarding direct and indirect waste disposal treatment and reducing their release into the Yamuna River (Nallathiga, 2018). Presently, several initiatives are running to protect, conserve, and restore various water bodies by government and non-government organizations.

15.8 SPATIOTEMPORAL ASSESSMENT OF YAMUNA RIVER IN MATHURA

Here, we did the Mathura district's spatial change analysis to grow urban and narrow the Yamuna river. Over time, urban is increased from the Yamuna River bank and spread toward the river's outer side. For these analyses, seven images from the Landsat series were taken each from different years but the same season of months, February and March. For the study, all the imageries were resampled to a similar resolution. Normalized difference water index (NDWI, spectral index extract water content using green and near-infrared bands) was applied to highlight the river pattern (Figure 15.5). Finally, the river was extracted as in Figurel 15.6. To find the evolution of built-up in Mathura district, classification was done, and an evolution map was developed, which was separately detailed already in another of our work but has been shown here in Figure 15.8 to establish a relation between growing built-ups and decreasing the flow and width of the water in the Yamuna

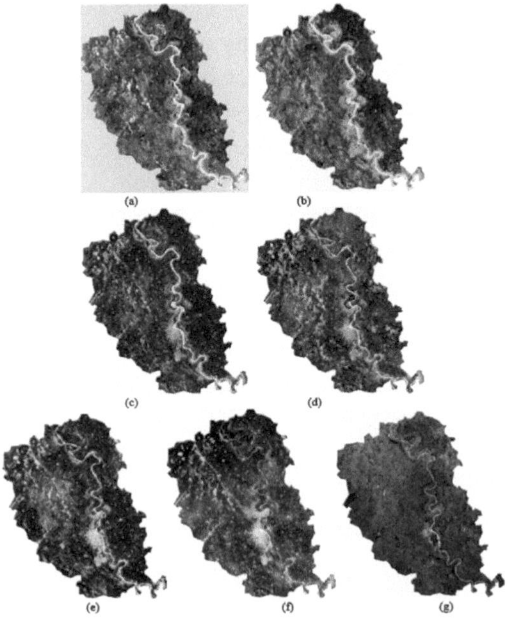

FIGURE 15.5 NDWI for Landsat series images of Mathura district: (a) Landsat-1, 02/03/1973, (b) Landsat-3, 12/02/1980, (c) Landsat-5, 20/02/1989, (d) Landsat-5, 01/03/1998, (e) Landsat-5, 05/02/20001, (f) Landsat-5, 14/02/2010, (g) Landsat-8, 24/03/2018.

Source: Author.

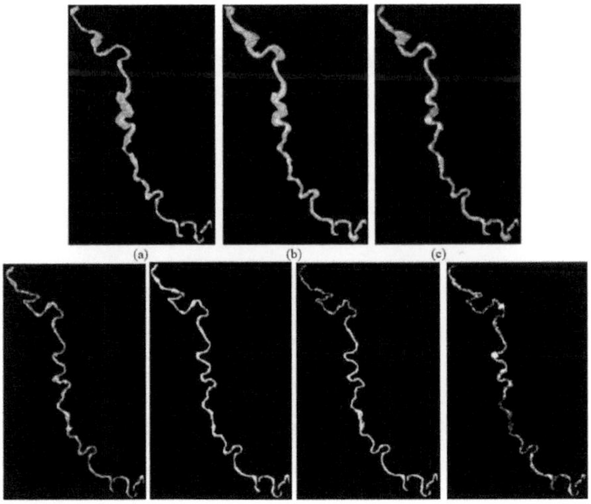

FIGURE 15.6 Yamuna River for Landsat series images of Mathura district: (a) Landsat-1, 02/03/1973, (b) Landsat-3, 12/02/1980, (c) Landsat-5, 20/02/1989, (d) Landsat-5, 01/03/1998, (e) Landsat-5, 05/02/20001, (f) Landsat-5, 14/02/2010, (g) Landsat-8, 24/03/2018.

Source: Author.

River. It can be seen from Figures 15.5 and 16.6 that the pattern of the river is changing continuously as the quantity of water has been decreased, leaving more sand residing on the bank. Apart from the water quality, the river has been narrowed at a different stretch (specifically the upper part and middle part of the river in Figure 15.6) within the Mathura district itself. This change took place gradually as the built-ups grew from Yamuna bank since 1973, as the study shows here. Figure 15.7 is about changes in the built-ups for more than four decades and approximately 8 years intervals. The number of built-ups started to grow near the river every year interval because of the

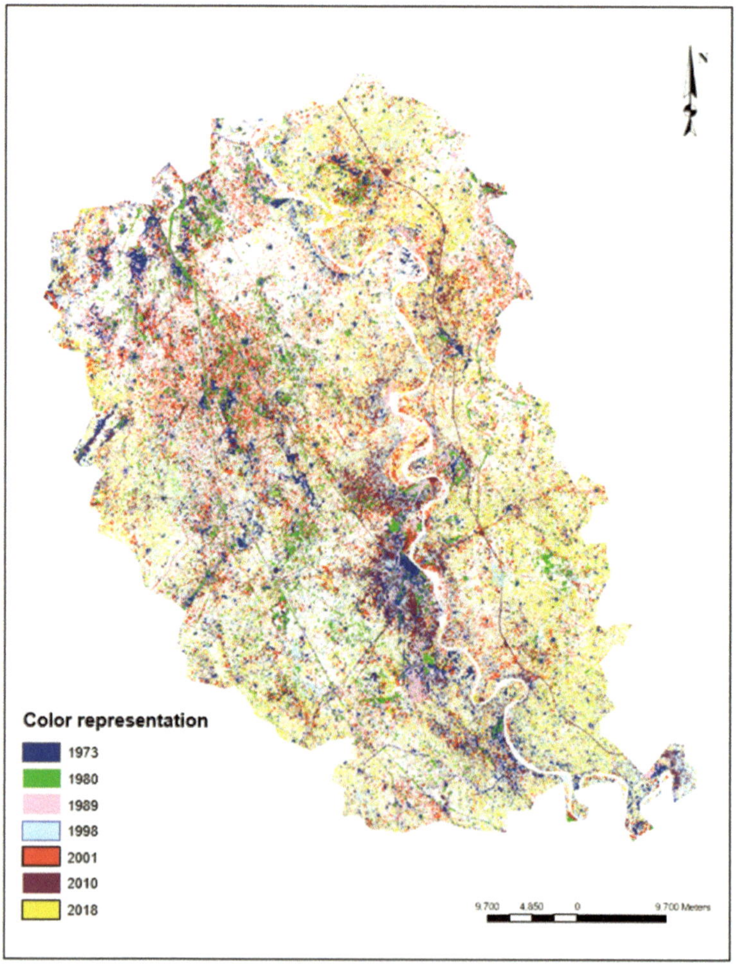

FIGURE 15.7 Spatiotemporal built-up change analysis in Mathura district since 1973.
Source: Reprinted from Baranwal et al. (2019).

easy availability of raw water needs for all human activities. The built-ups increased all over the district resulting in a degradation in all water resources and vegetation in the area. The change rate of water for this study is shown in Table 15.1, and as per our previous work (Baranwal et al., 2019), the growth rate of the built-ups since 1973 is shown in Table 15.2. This change rate of the river is in terms of narrowing its width, that is, the Yamuna River's dimension is reduced, and sand or barren area exposed too much as can be seen in ground pictures of the figure. The river's water level is also lowered and shifted away from the Ghats leaving the Ghats sandy and barren (Figure 15.8).

TABLE 15.1 Water Change Rate in the Yamuna River within Mathura District.

Year interval	Percentage reduced
1973–1980	31.9
1980–1989	41.8
1989–1998	58.9
1998–2001	27.0
2001–2010	20.7
2010–2018	13.6

TABLE 15.2 The Growth Rate of the Built-Ups in Mathura District.

Year interval	Built-up growth rate (%)
1973–1980	8.5
1980–1989	17.7
1989–1998	29.4
1998–2001	24.1
2001–2010	5
2010–2018	55.7

Source: Baranwal et al. (2019).

15.9 CONCLUSION

The current scenario of the Yamuna River is not excellent or satisfactory at every stretch, specifically in Delhi and near regions. Even it can be said that the Yamuna River is in a terrible condition that needs immediate and thorough attention to conserve its water quantity and quality because it is the primary water provider for towns and people. The waste release in the river should be controlled by the government and also by individuals. Since

Mathura is religiously significant and famous, pilgrims and tourists increase in numbers in the area every year. Also, Mathura is becoming the second home place of these tourists and pilgrims. This enhanced the commercial and residential built-ups in the area and gave the rapid growth in the urbanization and usability of land and water, which led to the degradation of the spatial arrangement in the area for everything and improper or inadequate management of the resources. Lack of knowledge and carelessness leads to the deterioration of resources across the limit, for example, people use these sandy areas at the bank and Ghats for their personal use of daily activities and not changing their opinion for "no religious waste disposal" in the river. Firm and strict actions must be implemented to revive the State of the Yamuna river. Otherwise, it will not take long to vanishing the river as per its degradation rate. The spatiotemporal analysis of the Yamuna River helps us uncover its deterioration status and explore the spots where it has been taking place in its extremity.

FIGURE 15.8 Sample pictures from Ghats of Vrindavan in Mathura district.
Source: Author.

KEYWORDS

- **ground water quality**
- **urban water crises**
- **ecorestoration**
- **resource degradation**
- **ecosystem services**
- **remote sensing**
- **GIS**

REFERENCES

Bandukwala, A. N.; Firdaus, S. Optimal Location of Smart Water Meters & Hotspot Analysis of Excessive Water Consumption: A Case Study of IIT, Bombay. In *National Seminar on Recent Advances in Geospatial Technology & Applications*; Dehradun, India, 2020.

Baranwal, E.; Ahmad, S.; Baghel, S. S. Spatiotemporal Analysis for Urban Pattern Evolution in Sacred District Mathura of India through K-means Classification. *Int. J. Town Plan. Manage.* **2019,** *5* (1), 26–35.

Basavaraj, H.; Nijagunappa, R. Applications of Geoinformatics in Water Resources Management of Semi-Arid Region, North Karnataka, India. *Int. J. Geomat. Geosci.* **2011,** *2* (2), 373–382.

Bastiaanssen, W. G. M. *Remote Sensing in Water Resources Management: The State of the Art*; International Water Management Institute (IWMI): Colombo, Sri Lanka, 1998.

Bharat, G. K.; Dkhar, N. B. Discussion Paper: Aligning India's Water Resource Policies With the SDGs, Nov 2018. www.teriin.org

Bhargava, D. S. Revival of Mathura's Ailing Yamuna River. *Environmentalist* **2006,** *26*, 111–122.

Cheema, M. J. M.; Bastiaanssen, W. G. M. Chapter 16: Remote Sensing and GIS Applications in Water Resources Management, 2017. https://www.researchgate.net/publication/321966058_Remote_Sensing_and_GIS_Applications_in_Water_Resources_Management (accessed Oct 20, 2020).

Czajkowski, K.; Lawrence, P. L. Chapter 8: GIS and Remote Sensing Applications for Watershed Planning in the Maumee River Basin, Ohio. In *Geospatial Tools for Urban Water Resources, Geotechnologies and the Environment 7*; Springer: The Netherlands, 2013; pp 131–144.

Dhamijal, S.; Joshi, H. Vulnerability Assessment of Arsenic in Groundwater and Its Occurrence in Districts of Bihar, India. In *National Seminar on Recent Advances in Geospatial Technology & Applications*; Dehradun, India, 2020.

DUAC. City Level Projects: Yamuna Riverfront Development-Indraprastha Zone, 2015. http://e.duac.org/images/pdf/13.%20ITO%20river%20front.pdf (accessed Nov 2, 2020).

Dubey, S. K.; Maurya, R. Water Quality Status for Yamuna river. *Int. J. Pure Appl. Res. Eng. Technol.* **2018,** *6* (8), 277–284.

Dwivedi, S. K.; Pandey, A.; Joshi, G.; Pandey, S.; Sharma, N. Inventory of Glacial Lakes & Detection of Potentially Dangerous Lakes for GLOF Using Remote Sensing & GIS in Pithoragarh District of Uttarakhand, India. In *National Seminar on Recent Advances in Geospatial Technology & Applications*; Dehradun, India, 2020.

Estigade, A. P.; Astuti, A. P.; Wicaksono, A.; Maitela, T.; Widyatmanti, W. Remote Sensing and GIS Application for Water Environment Suitability Evaluation in Lampung and Hurun Bay. In *International Conference on Environmental Resources Management in Global Region, IOP Conference Series: Earth and Environmental Science*, 2019.

Gebeyehu, M. N. Remote Sensing and GIS Application in Agriculture and Natural Resource Management. *Int. J. Environ. Sci. Nat. Resour.* **2019,** *19* (2), 45–49.

Harshavardhan, M.; Suganya, K. Water Harvesting Systems of the Past-A Case Study of Badami, India. In *National Seminar on Recent Advances in Geospatial Technology & Applications*; Dehradun, India, 2020.

Jain, S. K. Fresh Water Resources: A Perspective- Water Resources Management In India. http://nihroorkee.gov.in/sites/default/files/uploadfiles/WaterRes-Mgmt-India.pdf (accessed Nov 6, 2020).

Jenness, J.; Dooley, J.; Aguilar-Manjarrez, J.; Riva, C. *African Water Resource Database: GIS-Based Tools for Inland Aquatic Resource Management, 1.Concepts and Application Case Studies*; Food and Agriculture Organization of the United States: Rome, Italy, 2007.

KnowIndia. Rivers. https://knowindia.gov.in/profile/rivers.php (accessed Nov 6, 2020).

Kumar, D. N. Remote Sensing Applications to Water Resources. In *Research Perspectives in Hydromechanics and Water Resources Engineering*; World Scientific: Singapore, 2002; pp 287–316.

Kumar, D. N.; Reshmidevi, T. V. Remote Sensing Applications in Water Resources. *J. Indian Inst. Sci.-A Multidiscipl. Rev. J.* **2013,** *93* (2), 163–187.

Kumar, H.; Singh, K. Water Resources: Role of GIS and Remote Sensing. *Int. J. Eng. Res. Technol.* **2018,** *6* (11).

Kumar, N.; Yamac, S. S.; Velmurugan, A. Applications of Remote Sensing and GIS in Natural Resource Management. *J. Andaman Sci. Assoc.* **2015,** *20* (1), 1–6.

Kumar, S. V.; Bharat, G. K. Discussion Paper: Perspectives on a Water Resource Policy in India, Oct 2014 [Online].

Loucks, D. P.; Beek, E. V. *Water Resources Systems Planning and Management-An Introduction to Methods, Models and Applications*; UNESCO: Paris, 2005.

M. o. W. R. MWR. National Water Policy (2002), Apr 2002 [Online].

M. o. W. R. MWR. National Water Policy (2012), 2012 [Online].

Maniyar, C. B.; Kumar, A.; Mishra, D. R. Cloud-Based Approach for Continuous Monitoring and Assessment of Inland and Estuarine Water Environments Using Sentinel-3 OLCI Data. In *National Seminar on Recent Advances in Geospatial Technology & Applications*; Dehradun, India, 2020.

Meetei, N. P.; Ahluwalia, R. S.; Khobragade, S. D.; Sarangi, S.; Rai, S. P.; Jain, S. K. Streamflow Modelling of Pindar River Basin, Central Himalaya Using Remote Sensing Data and Temperature Index Model. In *National Seminar on Recent Advances in Geospatial Technology & Applications*; Dehradun, India, 2020.

MWR. Policy/Schemes, 2020. http://mowr.gov.in/ (accessed Oct 20, 2020).

Nallathiga, R. Institutional Intervention in River Water Management: The Study of Yamuna River Sub-Basin in India. *J. Govern. Public Policy* **2018,** *8* (2), 28–45.

Nanda, A.; Nikam, B. R.; Aggarwal, S. P. Impact Assessment of Meteorological Drought on Agricultural Parameters Using Google Earth Engine: A Case Study on Penna River Basin. In *National Seminar on Recent Advances in Geospatial Technology & Applications*; Dehradun, India, 2020.

NCERT. Chapter 3: Water Resources. In *Social Science: Contemporary India II*; National Council of Educational Research and Training (NCERT): New Delhi, India, 2019a; pp 23–33.

NCERT. Chapter 6: Water Resources. In *India People and Economy*; National Council of Educational Research and Training (NCERT): New Delhi, India, 2019b; pp 60–71.

NITI Aayog. Action Plan. https://niti.gov.in/writereaddata/files/coop/24.pdf (accessed Nov 5, 2020).

Otto, B.; Schleifer, L. Domestic Water Use Grew 600% Over the Past 50 Years. Feb 10, 2020. https://www.wri.org/blog/2020/02/growth-domestic-water-use (accessed Oct 20, 2020).

P. IAS. Classification of Drainage Systems of India, Jan 19, 2016. https://www.pmfias.com/classification-of-drainage-systems-of-india/

Rao, V. V.; Raju, P. V. Water Resources Management. In *Remote Sensing Applications*; National Remote Sensing Centre (NRSC)/ISRO: Hyderabad, India, 2010; pp 133–164.

Sanath Kumar, K. R.; Sharanya Ram, S. R. Delineation of Groundwater Potential Zone Using RS and GIS in Devanahalli Taluk, Karnataka. In *National Seminar on Recent Advances in Geospatial Technology & Applications*; Dehradun, India, 2020.

Saxena, P. R.; Prasad, N. S. The International Archives of the Photogrammetry. *Remote Sens. Spatial Inf. Sci.* **2008,** *XXXVII* (B8), 729–732.

Schultz, G. A. Use of Remote Sensing Data in a GIS Environment for Water Resources Management. In *Remote Sensing and Geographic Information Systems for Design and Operation of Water Resources Systems (Proceedings of Rabat Symposium S3)*; Rabat, Morocco, 1997.

Seth, B. L. National Water Policy, 2012 Silent on Priorities, July 4, 2015. https://www.downtoearth.org.in/news/national-water-policy-2012-silent-on-priorities--35952

Sharma, M.; Chaudhry, S. Impact of Industrial Pollution on Yamuna River: A Review, 2015. https://www.researchgate.net/publication/283443245_Impact_of_Industrial_Pollution_on_Yamuna_River_A_Review

Sharma, N.; Chaudhry, S. Impact of Industrial Pollution on Yamuna River: A Review, 2015. https://www.researchgate.net/publication/283443245_Impact_of_Industrial_Pollution_on_Yamuna_River_A_Review

Sharma, V.; Nikam, B. R.; Thakur, P. K.; Aggarwal, S. P.; Garg, V.; Shrivastava, V.; Singh, S. Generation of Historic, Present and Future Snow Cover Datasets Using Modelling Approach. In *National Seminar on Recent Advances in Geospatial Technology & Applications*; Dehradun, India, 2020.

Singh, P. K.; Dey, P.; Jain, S. K.; Mujumdar, P. P. Hydrology and Water Resources Management in Ancient India. *Hydrol. Earth Syst. Sci.* **2020,** *24*, 4691–4707.

Singh, R. B. Sustainable Regional Water Management in the Yamuna River Basin: Case Study of the Delhi Region (India). In *Regional Management of Water Resources (Proceedings of a Symposium Held during the Sixth IAHS Scientific Assembly at Maastricht, The Netherlands)*; Maastricht, Netherlands, 2001.

Tiwari, G.; Shukla, J. P. A Review on Remote Sensing and GIS Techniques in Water Resource Development and Management with Special Reference to Groundwater. *Int. J. Remote Sens. Geosci. (IJRSG)* **2015,** *4* 1, 10–16, 2015.

Tundisi, J. G. Water Resources in the Future: Problems and Solutions. *Estudos Avancados* **2008,** *22* (63), 7–16.

UNEP. *The UN-Water Status Report on the Application of Integrated Approaches to Water Resources Mangement*; United Nations Environment Programme (UNEP), 2012.

Upadhya, A.; Rai, R. K. Brief Overview of the Yamuna River Basin and Issues. In *Water Management and Public Participation*; Springer: Dordrecht, 2013; pp 13–24.

W WAP. *The United Nations World Water Development Report 2020-Water and Climate Change*; UNESCO: Paris, 2020.

WWAP. *The United Nations World Water Development Report 2019-Leaving No One Behind*; UNESCO: Paris, 2019.

Welch, R.; and Remillard, M. Remote Sensing/GIS for Water Resource Management Applications in the Southeast. In *Georgia Water Resources Conference*; Athens, GA, 1991.

Xianwei, W.; Hongjie, X. A Review on Applications of Remote Sensing and Geographic Information Systems (GIS) in Water Resources and Flood Risk Management. *Water, MDPI* 2018.

CHAPTER 16

Evaluating the Glacier Change Dynamics Using GI Science

AVINASH KANDEKAR[1], SWAPNIL VYAS[1*], VIKRANT BHADE[1], and BINAY KUMAR[2]

[1]*Department of Geography, Savitribai Phule Pune University, Pune, India*

[2]*Centre for Development of Advanced Computing, Pune, India*

[*]*Corresponding author. E-mail: svyas.unipune@gmail.com*

ABSTRACT

Mt. Kangchenjunga is the world's third highest peak and India's highest mountain. Perennial snow and ice melt from Kangchenjunga is used for irrigation, hydropower generation, bioresource production, and domestic water demand in the Brahmaputra's catchments and alluvial plains. The melting of glaciers around the world becomes one of the most common signs of global climate change. As a result, changes in the size of these glaciers are thought to be a sensitive indicator of climatic changes in the earth system. To put it another way, collecting spatial data on fresh water resources is critical for understanding the impact of climatic variations. Using remote sensing datasets, the current study was carried out to determine the temporal change and deglaciation rate in and around the Kangchenjunga region (Khangchendzonga National Park of Sikkim, India, and Kangchenjunga Conservation Area, Nepal) over a 30-year period. Accumulation areas, ice-exposed ablation areas, debris-covered ablation areas, deglaciated valleys, glacial lakes, and scarp faces were all mapped on the glacier. Over the course of 30 years, the glacial area shrank by 43.5 km^2 (at a rate of 1.50 km^2 per year,

GIScience for the Sustainable Management of Water Resources. Gowhar Meraj, Shruti Kanga, Majid Farooq, Suraj Kumar Singh, & Sudhanshu (Eds.)
© 2023 Apple Academic Press, Inc. Co-published with CRC Press (Taylor & Francis)

or 3.45% per year). The gradual impact of climate change and its impact on global warming are thus reported in the current study.

16.1 INTRODUCTION

Glaciers are vital components of climate change and global warming, as they can both be indicators of change. The longest and most important rivers in India originate in the Himalayas and are created by glaciers from that range. The Himalayan mountain system located to the north of the Indian land mass with an arcuate strike of NW–SE for approximately 2400 km is referred to as the "third pole" (Bahuguna et al., 2014, Kaushik et al., 2019, Bolch et al., 2012). The Himalayan glacier blanket is approximately 33,000 km^2 in size (SAC 2011). The existence of these frozen reservoirs in India significantly impacts the country's climate (Singh et al., 2015). At least 30–50% of the annual flow in these rivers (viz. Indus, Ganga, and Brahmaputra) is derived from meltwater from snow and glaciers (SAC 2011, Kulkarni, 2003). The changes in glaciers have been observed through centuries, where retreating and melting have been observed as the result of climate change (Kaushik et al., 2019). In the context of the global warming and climate change, the fears are that the melting of glaciers will drastically affect the landscape, slope stability, the water cycle, the sediment load in rivers, the sea level and natural hazards far beyond the historical and Holocene variability. In order to learn about the way climate change has influenced resources around the world, glaciers are monitored across the globe (Zemp, et al., 2009; Gärtner-Roer, et al., 2019).

Temperature rises in the Himalayas have ranged from 0.15°C to 0.60°C per decade over the last three decades (Shrestha et al., 2010). The recent rise in temperature and fluctuation has had significant impact on the glacial cycle in the Himalayas. Multiple studies suggest that glaciers in the Indian Himalayas are receding (Kulkarni et al., 2005, 2007, 2011). Due to their youth and fragility, the Himalayas are especially vulnerable to even minor fluctuations in the climate (Lama and Devkota, 2009; Gujree et al., 2017). In the North Indian rivers, there is an insatiable demand to constantly monitor changes in glacial areas, which are the source of water supply.

Because the Himalayas are a biodiversity hotspot and Kangchenjunga National Park is a UNESCO World Heritage site, monitoring the current state and predicting future glacier areas in this area would be extremely beneficial. The glacial extent and distribution information is needed in studies like water resource management, disaster, and hazard management, sea level change,

The glacial extent maps form the most important and common input for most of the glaciological and hydrological modeling studies (Shukla, 2009).

Global coverage of optical remote sensing data became affordable in the latter half of the 20th century, allowing for the mapping of glaciers at smaller spatial scales. Advances in satellite technology and satellite data have greatly aided estimates of region-wide glacier mass balance estimation of glacier surface elevation change, glacier lake changes, glacier area changes, mapping of debris cover estimation of surface ice velocity, ice volume change, high-altitude glacier lakes in the Eastern Himalaya, but at a glance evaluation of all of these variables is difficult (Bajracharya et al., 2007; Miller et al., 2011; Kumar et al., 2011; Rankl and Braun 2016; Kääb et al., 2012; Robson et al., 2015; Satyabala 2016; Shukla et al., 2018).

In this study, efforts were made to use a better technique for glacier delineation and detection of glacier change over time, as well as an assessment of the areas affected by this change. Some regional glacier inventories have been built in the past for the western Himalaya (Kamp et al., 2011), but only a few are available for the Himalaya's eastern extremity (Krishna 2005). Cloud cover and sensor saturation frequently obstruct satellite image acquisition in this region, owing to incorrect gain settings and the persistence of seasonal snow, which obstructs image quality. Furthermore, there is a scarcity of topographic data that can be used to detect glacier changes, making it difficult to detect subtle changes (Bhambri and Bolch 2009). The earliest Indian glacier maps were produced by topographic studies conducted by explorations in the mid-19th century, but they only cover a few glaciers. Due to its rugged topography, high altitude, and extreme weather conditions, It is difficult to study the Himalayas using traditional field-based methods. Satellite data provide accurate and reliable observations due to its synoptic view, distinct spectral properties of snow and glaciers, high temporal frequency, and advanced digital image processing and analysis techniques (SAC 2011). As a result, remote sensing (RS) methods combined with Geographical Information System (GIS) methods have proven to be extremely useful in carrying out accurate, quick inventory, mapping, and tracking of the inaccessible Himalayan terrain. The country has been developing remote sensing methods to monitor Himalayan snow and glaciers for the past two and a half decades. The extent of glaciers is usually mapped using satellite imagery from the end of the ablation period. At the end of the ablation season, when the snowline reaches its maximum altitude, snow-line images are used to define accumulation and ablation zones. The study's main goal is to determine how much the Kangchenjunga glacier has subsided over time in the Kangchenjunga—Sikkim and Nepal regions.

16.2 MATERIALS AND METHODS

16.2.1 STUDY AREA

The study area stretches over the parts of Nepal and Sikkim state of India. The study area stretches over the parts of west Sikkim, state of India and parts of the far-eastern Taplejung district of Nepal and encompasses region over 27° 22′ 48″N–27° 57′ 36″N latitudes and 87° 38′ 24″E–88° 39′ 50″E longitudes. Figure 16.1 shows the location map of study area. It has an area of 4021.78 km².

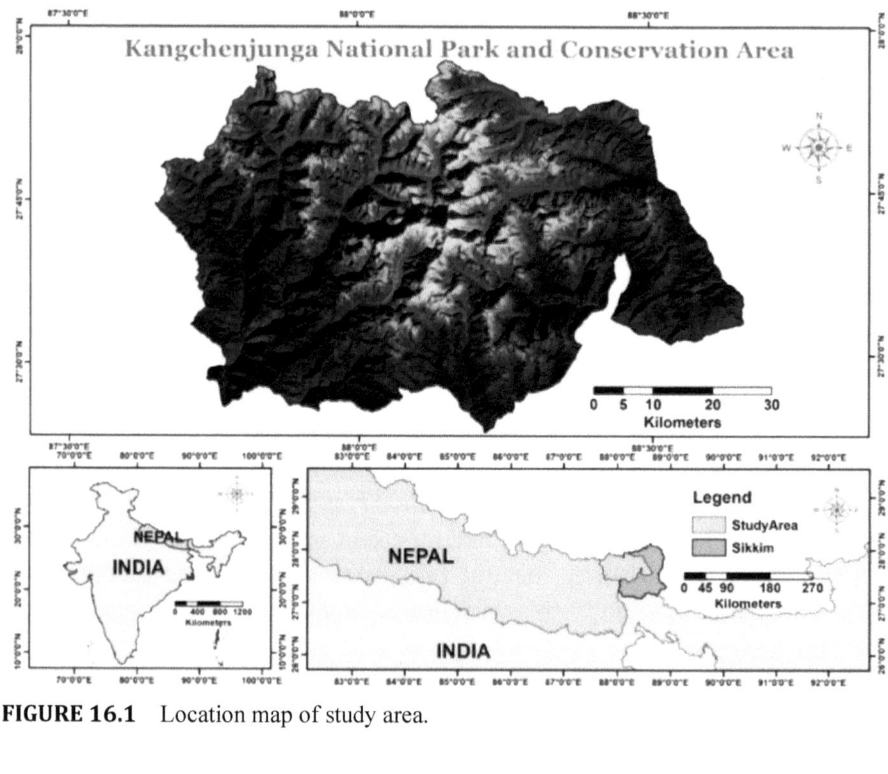

FIGURE 16.1 Location map of study area.

The Indian part is known as the Khangchendzonga National Park (KNP), which is a UNESCO "mixed" World Heritage site. The Nepalese part of the study area is known as Kangchenjunga Conservation Area (KCA).

The research area is located in the Eastern Himalayas. Relief in this area ranges from 753 m at the bottom of the valleys to 8586 m, according to ASTER data (Mt. Kangchenjunga). There are two types of glaciers: valley and mountain glaciers, with cirque and apron glaciers making up the rest (Mool et al., 2002). The South Asian summer monsoon circulation system (Bhatt and Nakamura 2005) is dominated by the inflow of moist air from the Bay of Bengal branch to the Indian subcontinent during the summer (Bhatt and Nakamura 2005, Benn and Owen 1998). The Himalayas are at their narrowest point here, resulting in highly steep terrain that emphasizes the differences between the various eco-zones that make up the region. The glaciers provide water to important rivers, creeks, and wetlands in the study area's major watersheds. Many Indian rivers, such as the Teesta, Rangit, Lachen, Poke chu, Lonak chu, and others, as well as Nepal's Tamur, Kabeli, Simbua Khola, and Ghunsa, originate in this region.

16.2.2 DATA USED

Multispectral images of Landsat-5 TM, Landsat-8 OLI, and US military topographic map data were used in the present study to delineate the glacial area and change the detection over a period of 30 years. The ASTER GDEM V2 (30 m spatial resolution) is used as reference digital elevation model (DEM) to map the drainage basin and derive glacier topographic parameters. All the images and DEM are acquired from USGS. Most of the satellite images are acquired at the end of the ablation season with minimum seasonal snow and cloud cover. The ablation period end varies across the Himalaya from west to east. End of ablation for western Himalaya corresponds mainly from September to mid-October period, whereas the corresponding period for the eastern Himalayan region (Teesta region in Sikkim) is from December to early January (Bahuguna et al., 2014). Thus, the data used were from the month of December.

16.3 METHODOLOGY

The U.S. Army Topographical sheets of 1962 were georeferenced in UTM projection (spheroid and datum being WGS 84 N, Zone 45) with nearest neighbor resampling technique. The satellite images were processed for radiometric correction or geometric distortion and were georeferenced with respect to toposheet keeping RMS error less than 0.50 or half of pixel size.

A glacier inventory was created to study the changes in the area. End of the ablation period satellite data were used for mapping the glacier extent. The end of ablation period varies across the Himalaya from west to east.

End of ablation for western Himalaya corresponds mainly from September to mid-October period, whereas the corresponding period for the eastern Himalayan region (Teesta region in Sikkim) is from December to early January (Bahuguna et al., 2014). Thus, the data used were from the month of December. Glacier outlines and extents for the years 1988 and 2017 were generated using the visual interpretation techniques. The multi-temporal lake changes for the 1988 image were assessed after a baseline data set was created for the year 2017. Digital false color composites (FCCs) of Landsat images were interpreted on-screen in different combinations of green, red, and NIR, or green, blue, and SWIR bands.

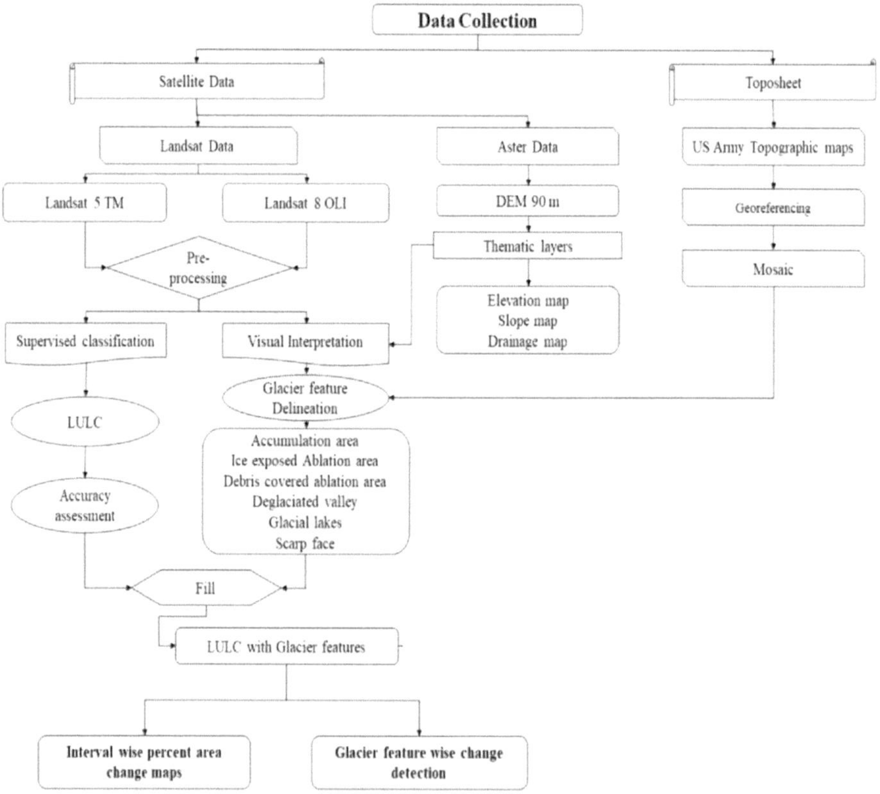

FIGURE 16.2 Methodology for temporal change over 30 years of glaciated area.

The first pairing was used to differentiate vegetated areas round the glacier snouts, whereas the SWIR band was used to distinguish snow and clouds, as well as the glaciated area from the encircling rocky areas. Mapping of glaciers with clear ice surface was relatively simpler as ice has a distinct signature than the other surrounding features. Many Himalayan glaciers, on the other hand, were covered in varying amounts of moraine, which included dust, silt, sand, gravel, cobbles, and large rocks (Pall et al., 2019; Romshoo et al., 2020). Though snout position and glacier boundary delineation were challenging for debris-covered glaciers, specific interpretation techniques have been used to accurately identify snout position as well as glacial extents in the above circumstances (Brahmbhatt et al., 2012). Because of their distinct morphological shapes and inactive lateral moraines, the valley glaciers could be identified. Extent of the glaciers were digitized using their geometrical shape. The snout positions were confirmed by the emergence point of streams from the glacier, and by the additional use of the DEM. Glacier features were divided into snout positions, accumulation area, ice-exposed ablation area, debris-covered ablation area, de-glaciated valley, glacial lakes, and scarp face. All these vector layers were generated for both the years.

16.4 RESULTS AND DISCUSSIONS

The images were classified into four broad categories or land cover classes. These include Rocky and Barren areas, Scrub area, Dense Forests, and Snow cover. The study area lies in the Himalayas and also is a National Park, so it had negligible settlements and agriculture.

16.4.1 CREATION OF AREA OF INTERESTS

Unique areas of interests for the four classes mentioned above were taken from all parts of the image as samples for classification. Google Earth images of that period of time were used as a support for more precise location of these interest areas, bearing in mind that it must be sufficiently representative and homogeneous samples of the class defined. Once a training site was demarcated, others were selected for the same category from different parts of the image. Multiple fields by categories must be selected in order to reflect the variability of the classes in the study area.

After completion of the selection of classes for the classification, the next phase was allocation. Because it tries to allocate each image pixel to one of the previously selected classes. This assignment is performed naturally according to the DN, to each of the bands that are involved in the process. The criteria used to establish the statistical boundaries between classes was:

Maximum likelihood, where the pixel is assigned to that class which has higher probability of membership. The algorithm is selected owning to its robustness and the fact that it fits with more rigor to the original layout of the data.

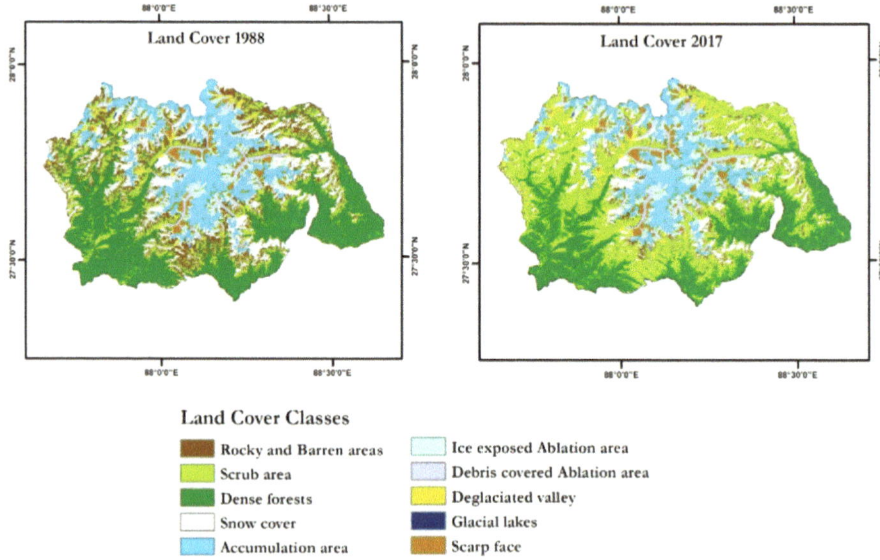

FIGURE 16.3 Land cover maps with final 10 classes.

It was not possible to demarcate exact boundaries of glacial features using supervised classification, therefore, first the glacier boundaries were digitized manually for all the years. Those were in vector format. It included six classes, viz., accumulation area, ice-exposed ablation area, debris covered ablation area, deglaciated valley, glacial lakes and sScarp face. It was necessary to demarcate exact outlines of glacial features in the classification image. Hence, the vector layers were added into the supervised classification images. "Fill" option in ERDAS IMAGINE was used to add these six layers in the supervised classification images of all the years. Now the land cover classes increased from four to ten, including glacial features.

16.4.2 CALCULATION OF TOTAL GLACIER AREA

Between 1988 and 2017, six glacier features were digitized. These features include accumulation area, ice-exposed ablation area, debris covered ablation area, deglaciated valley, glacial lakes and scarp face. Accumulation and ablation areas were counted as part of the total glacier area calculation. In many previous studies, the method for detecting changes was based on determining the change in the total area of the glacier over a period of time that included zones of accumulation as well as ablation (Bolch et al., 2010).

The area was calculated glacier-wise where each glacier was given an ID (Figure 16.4). The areas of deglaciated valleys were also calculated, as it directly tells about glacier retreat. The areas of glacial lakes and scarp face were omitted from glacier area calculation. Glacier change area, percentage change for the whole 30-year span from 1988 to 2017 was also calculated.

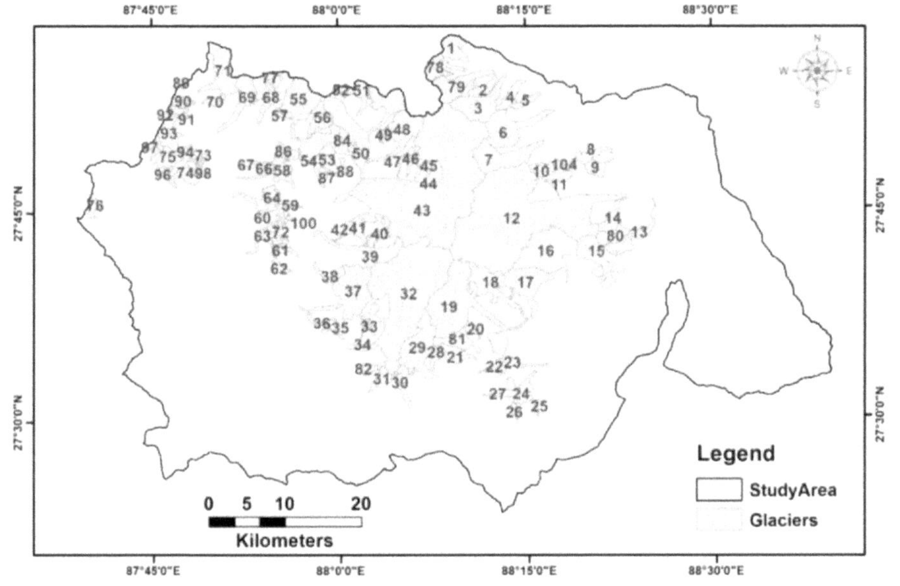

FIGURE 16.4 Map showing IDs of glaciers.

16.4.3 CALCULATION OF STREAM LENGTH

Increase in the length of streams within a fix area symbolizes glacier retreat. This is because the snout gets shifted backwards as a glacier melts over time, and streams cover a larger distance than earlier. Hence, streams were generated using "Hydrology" tool in ArcMap. Streams generated within the glaciers were omitted and length beyond the glacier boundary, that is, from snout position, was taken into consideration.

16.4.4 GLACIER CHANGE INTERPRETATION 1988–2017

Studying the glacier change for the span of 30 years (1988–2017) was one of the important aspects of the research. Analysis of such big year interval will be helpful for understanding the glacier change in the Kangchenjunga region. In 2017, the total glacier area decreased when compared with 1988. Each glacier was assigned a unique ID, and area change was calculated separately for each area. The mean glacier changes for all the glaciers in the interval 1988–2017 was −0.4619 km^2. The change in the glacial area was negative, hence showing retreat.

From Table 16.1 a sharp decrease in the areas of glacier features can be seen. Accumulation area retreated by a massive 49.29 km^2 (9.39%) in the 30-year span. Ice-exposed ablation area also show a decrease in area by 10.41 km^2 (4.22%). Deglaciated valley area increased by 14.89 km^2 (20.95%) which symbolizes retreat of glaciers. An increase of 23.56 km^2 (11.65%) in scarp face also signifies ice melting and retreat.

TABLE 16.1 Comparison of Class Areas of 1988 and 2017 Land Cover.

Classes	Area(km^2)		Change (km^2)	Percent change
	1988	2017		
Rocky and Barren areas	713.60	307.07	−406.52	−56.97
Scrub area	360.44	1165.21	804.77	223.28
Dense forests	1240.32	1001.37	−238.95	−19.26
Snow cover	503.77	349.00	−154.77	−30.72
Accumulation area	524.74	475.45	−49.29	−9.39
Ice-exposed ablation area	246.77	236.36	−10.41	−4.22
Debris-covered ablation area	156.57	172.77	16.20	10.35
Deglaciated valley	71.09	85.99	14.90	20.95
Glacial lakes	2.21	2.71	0.51	23.07
Scarp face	202.29	225.85	23.56	11.65
Total	4021.78	4021.78		

Increase in the areas of debris-covered ablation area and glacial lakes by 16.20 km^2 (10.35%) and 0.50 km^2. (23.07%) respectively also indicates retreat.

The changes in other land cover classes, such as dDense forests, snow cover, rocky and barren areas are negative. Dense forests have depleted by

238.94 km^2 (19.26%) throughout 30 years. However, there is increase in the area of scrub area.

Figure 16.5 gives a clear idea of glacier change in the study area over the span of around three decades. A majority of glaciers shows retreat. Most of the glaciers show a retreat ranging between 0% and 10%. The number of glaciers having retreat range 10% to 20% and 20 % to 30% is also significant. Some glaciers also show increase in the area. Maximum increase in the glacier area is also seen upto 10%.

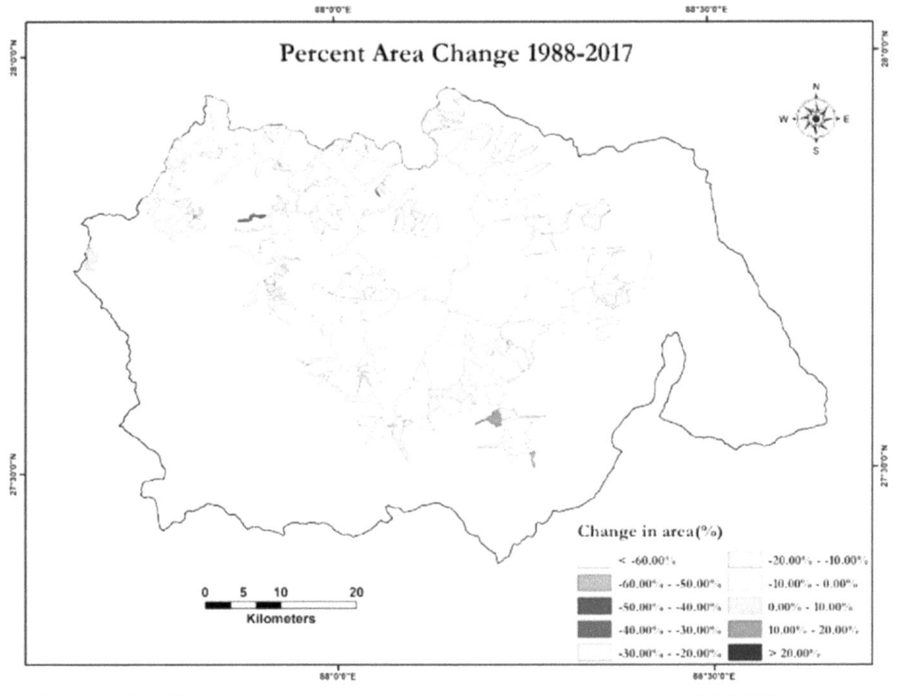

FIGURE 16.5 Glacier wise percent area change map for the interval 1988–2017.

In the interval 1988–2017, 100 out of 104 glaciers have shown retreat. Glacier no. 1, 10, 12, and 43 show a retreat of more than 2.5 km^2. Glacier no. 12 shows a maximum retreat of more than 5 km^2 as shown in Figure 16.6. Glacier no. 25, 26, and 99 show a retreat of more than 50% (Figure 16.7). One of the glaciers out of the four has break down from another glacier, hence, it is showing 100% positive area change. Remaining three glaciers have a negligible increase in the area. Most of the retreating glaciers range between 1 and 3 km^2 (upto 10%).

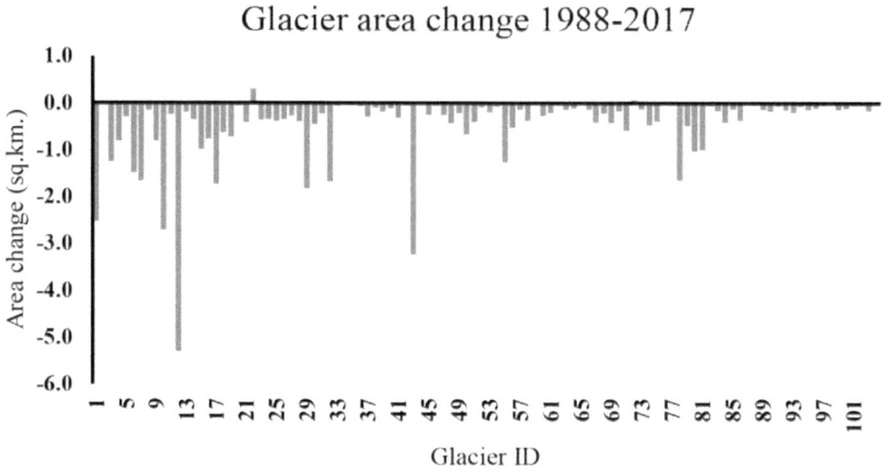

FIGURE 16.6 Glacier wise area change (km^2) in 1988–2017.

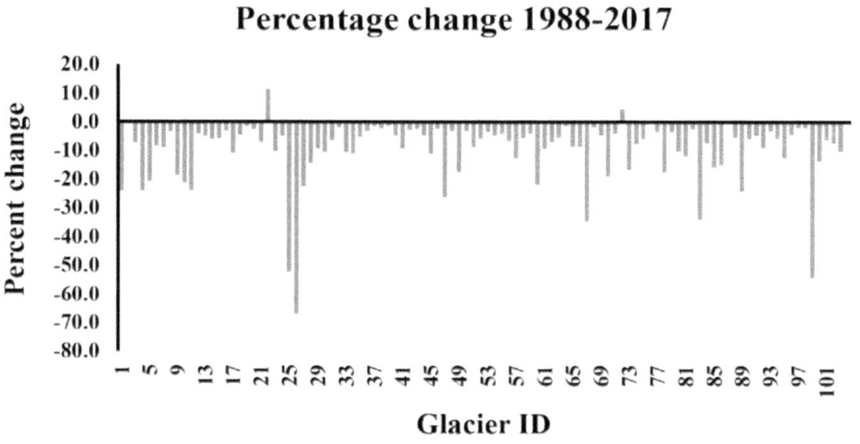

FIGURE 16.7 Glacier wise percent area change in 1988–2017.

16.4.5 TOTAL GLACIER CHANGE ANALYSIS

A decreasing trend in the glacier area has been observed. The total glacier area has drastically changed for the year 2017 as compared with the steady change in the first 4 years. In 1988, the total area was 928.08 km^2 which decreased to 923.52 km^2 in 1996. The next interval showed a slight decrease upto 920.62 km^2. In 2008, the area again showed a decreasing trend with 917.23 km^2. The last year interval of 2008–2017 had the most decrease in area, dropping upto 884.58 km^2. It means that Sikkim between 1988 and 2017, the total glacial area in the Himalayas decreased by 43.5 km^2, indicating a negative area change. It also means that the rate of deglaciation noted in this study in these 29 years was ±1.50 km^2/year or 3.4483%/year.

16.4.6 CHANGE IN GLACIATED AREAS

The changes of accumulation area and ablation areas are compared in Figure 16.8. It can be clearly seen that accumulation and ice-exposed ablation area decrease over time, whereas debris-covered ablation area shows increase in area. This proves that as the ice melts, the accumulation area and ice-exposed ablation areas get converted into debris-covered ablation area. The decreasing trend of both accumulation area and ice-exposed ablation area implies that the glaciers have been retreating. This confirms the retardation of the total glacier area. Increase in the debris-covered ablation area is a sign of ice melting, which further results in the retreat of a glacier.

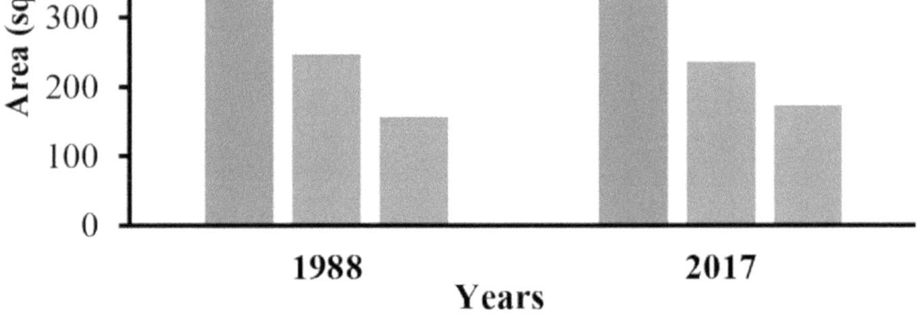

FIGURE 16.8 Bar diagram representing accumulation and ablation areas for all years.

Accumulation area was 524.74 km^2 in 1988, which further decreased to 475.45 km^2 in 2017. Figure 16.9 shows the retreat in accumulation area over the years. Retreat in the glacier no. 12 has been demonstrated. This gives a clear idea of change in the area over a period of time.

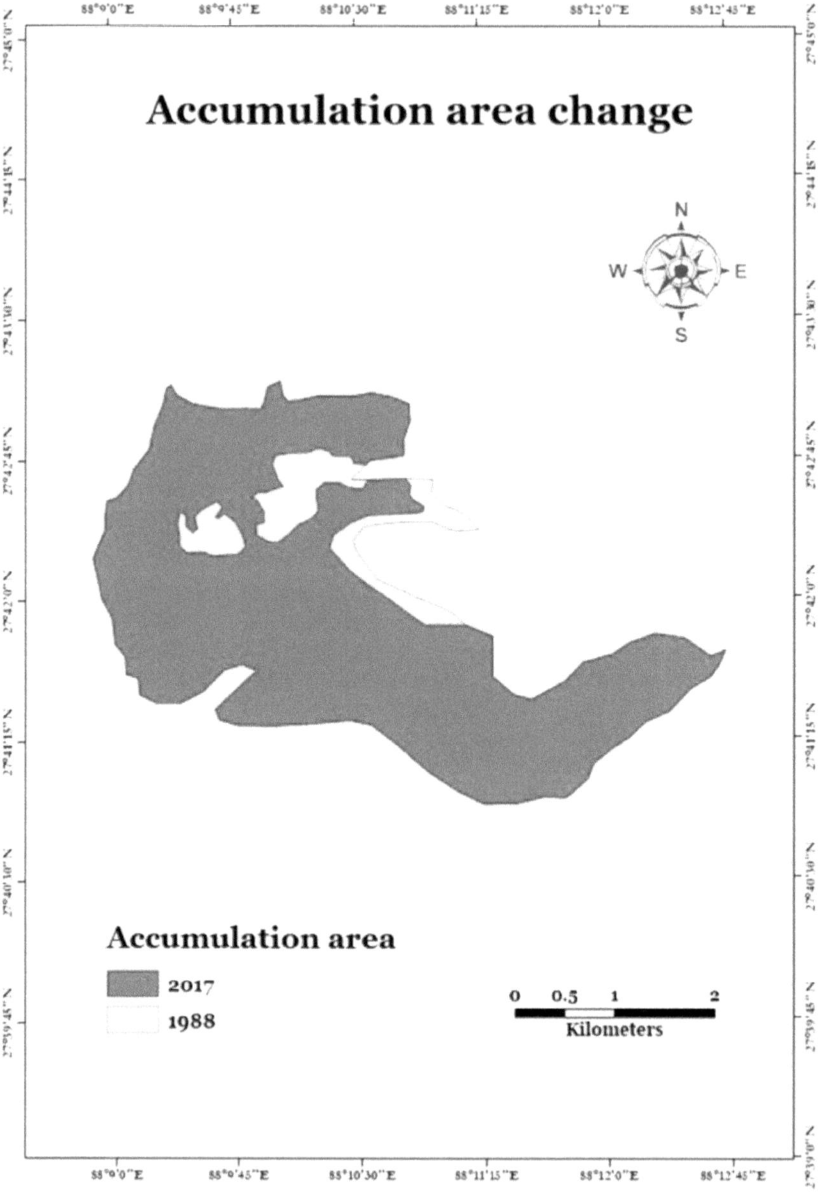

FIGURE 16.9 Map showing accumulation area retreat in glacier no. 12.

Ice-exposed ablation area also followed a decreasing trend with 246.77 km^2 in 1988 to 236.36 km^2 in 2017. This retreat in ice-exposed ablation area is highlighted in Figure 16.10 where glacier no. 23 is shown.

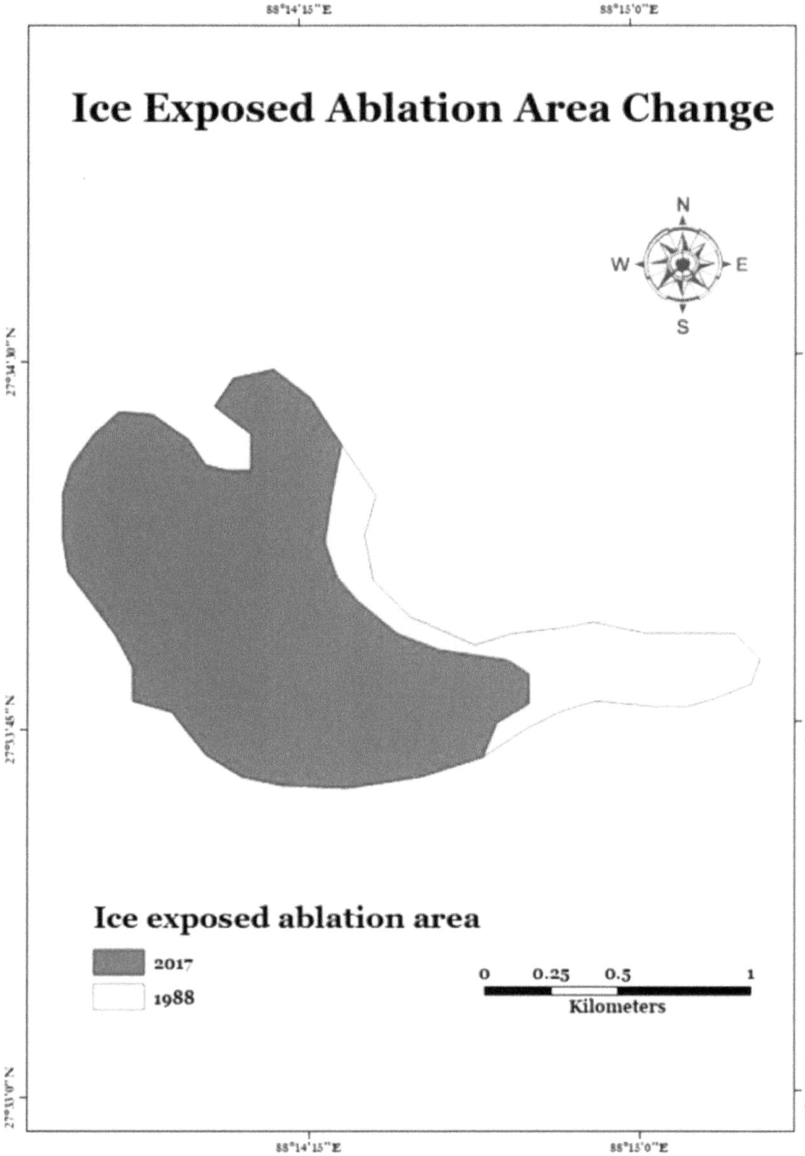

FIGURE 16.10 Map showing retreat in ice-exposed ablation area of glacier no. 23.

Debris-covered ablation area showed an increase as the accumulation and ice-exposed ablation area were converted in it. In 1988, the area was 156.57 km^2, which increased to 172.77 km^2 in 2017. Figure 16.11 represents the increase in the area of debris-covered ablation area, using glacier no. 16 as an example.

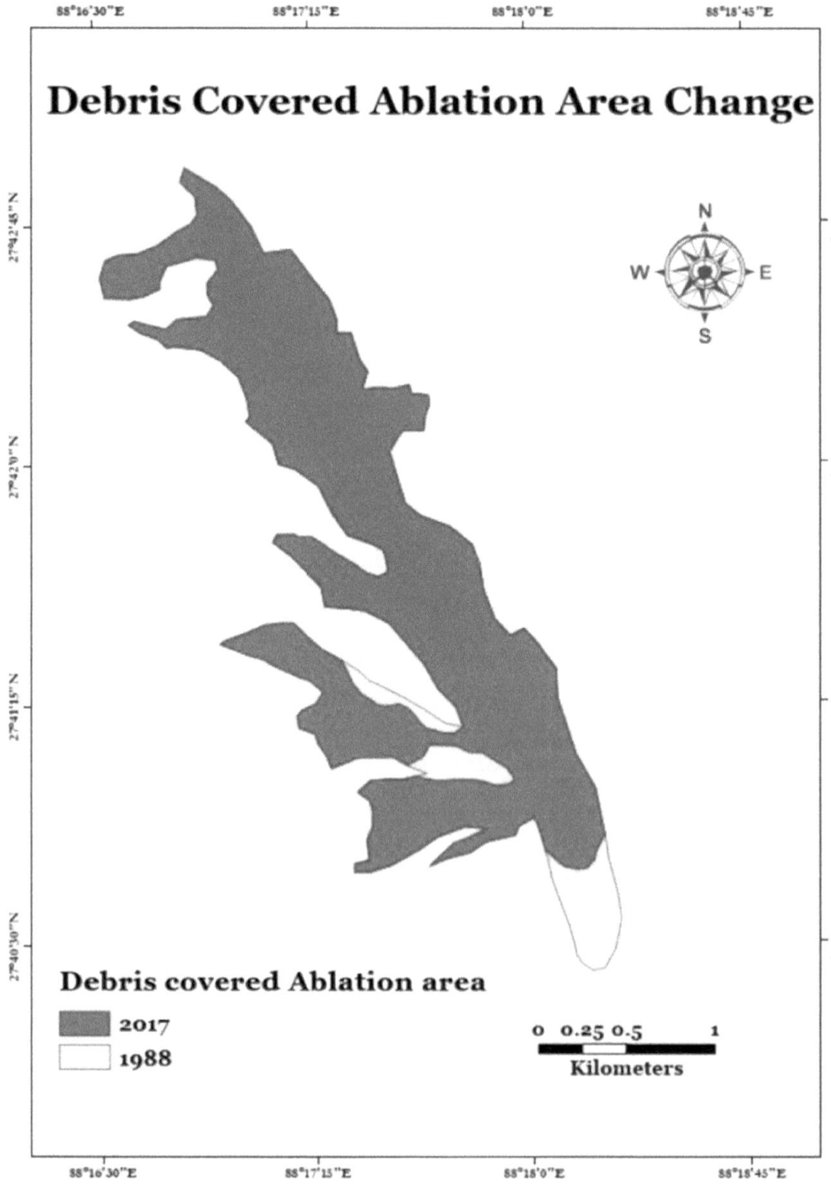

FIGURE 16.11 Map showing increase in debris covered ablation area of glacier no. 16.

16.4.7 CHANGE IN DEGLACIATED VALLEY

An increasing trend of deglaciated valley area was observed from 1988 to 2017. This is an indicator of glacier retreat, as the snout shifts backwards and the region is converted into deglaciated valley. As the snout shifts backwards and the stream generates, the debris covered ablation area converts into deglaciated valley. The area of deglaciated valley increased from 71.08 km^2 in 1988, to 85.98 km^2 in 2017. In Figure 16.12, the actual changes in the deglaciated valley over the years have been demonstrated using the map. Glacier no. 86 has been represented.

16.4.8 CHANGE IN SCARP FACE

Like the deglaciated valley, the increase in the area of scarp face is an indicator of glacier melting or retreat. An increasing trend of scarp face area was observed from 1988 to 2017. The accumulation area or the ice-exposed ablation area usually converts into scarp face due to melting of ice. The increase in the area of scarp face clearly proved glacier melting. The scarp face had an area of 202.29 km^2 in 1988, which increased to 225.85 km^2 in 2017. The change (increase) in scarp face is demonstrated using Figure 16.13 where glacier no. 6 is represented. The expansion of scarp face is clearly highlighted in this map.

16.4.9 CHANGE IN GLACIAL LAKES

An increasing trend in the area of glacial lakes has been observed. Increase in the area or formation of new lakes in the ablation zone of glacier is a sign of glacier retreat. New glacial lakes have formed behind loosely consolidated end-moraine dams in the Sikkim Himalayas due to the glacier thinning and retreat. It is nearly impossible to stop this kind of lake from draining catastrophically. As a result, these lakes can pose a threat to both people and property in the valleys below them (ICIMOD 2011).

For the past three decades, from 1988 to 2017, the temporal change in glacial lakes has been estimated. Since 1988, the area of glacial lakes has been steadily increasing. Between 1988 and 2017, the total area of lakes increased significantly, from 2.20 km^2 to 2.71 km^2. This indicates that glaciers in the Kangchenjunga region that are monsoon-fed as well as snow-fed are more vulnerable to rising temperatures. If the current trend of climate warming continues, glacial lakes in the Eastern Himalaya are expected to expand, increasing the risk of GLOF (glacial lake outburst floods) (Kumar et al., 2012). Figure 16.14 shows the gradual increase in the glacial lake area from the glacier no. 30.

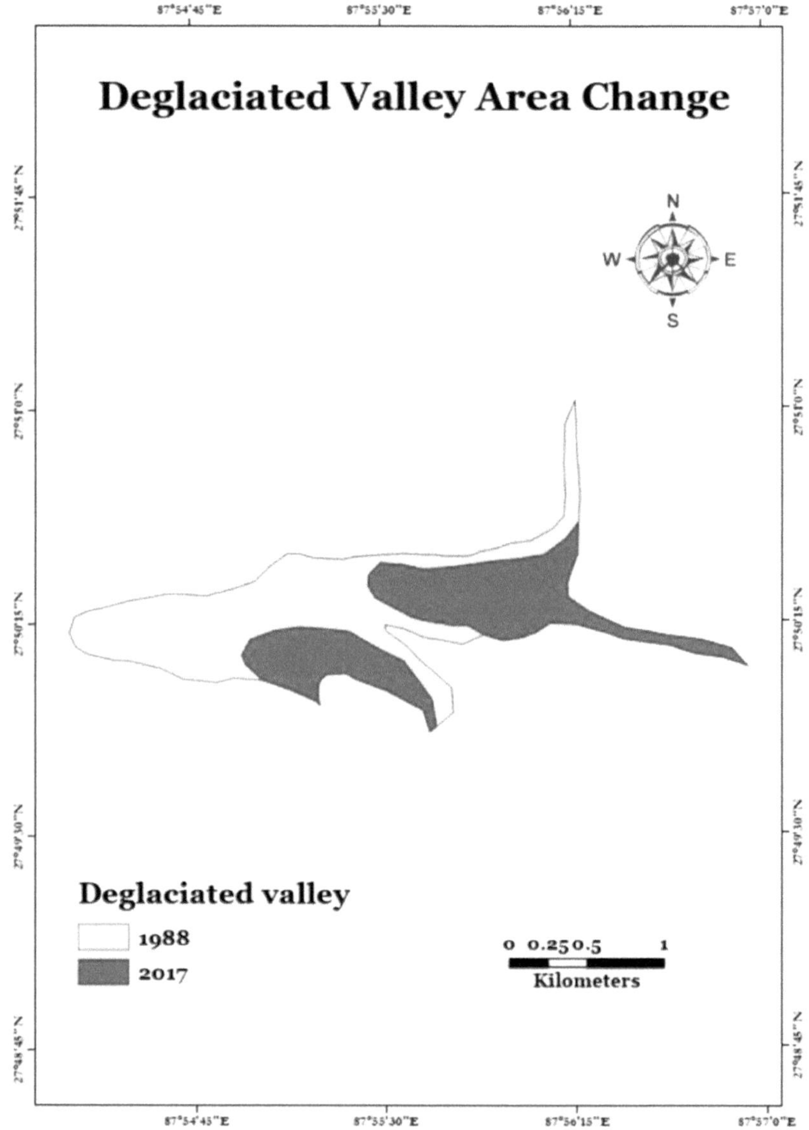

FIGURE 16.12 Map showing increase in deglaciated valley area of glacier no. 55 and no. 86.

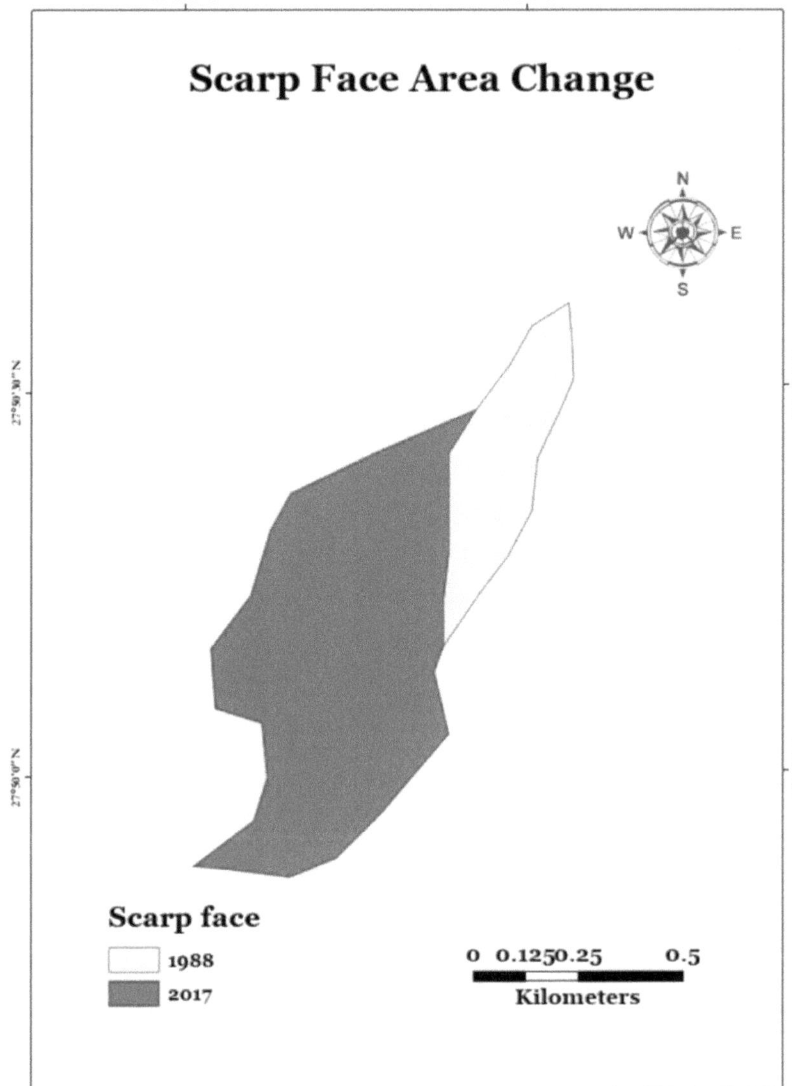

FIGURE 16.13 Map showing increase in scarp face area of glacier no. 6.

16.4.10 CHANGE IN SNOUT POSITION OF GLACIERS

As discussed earlier, the snout positions of glaciers were confirmed by locating the point of appearance of the stream from the glaciers because streams were a crucial factor that symbolized the change in snout position. The length of streams beyond the snout was 1898 km in the year 1988. This increased to 1937.96 km in 2017. It means that the monitoring of snout position revealed a retreat of 39.96 km and 1.33 km/per year from 1988 to 2017. Figure 16.15 depicts the change in the snout position of Glacier no. 55 over the 30 years. A remarkable shift in the snout position is easily identifiable.

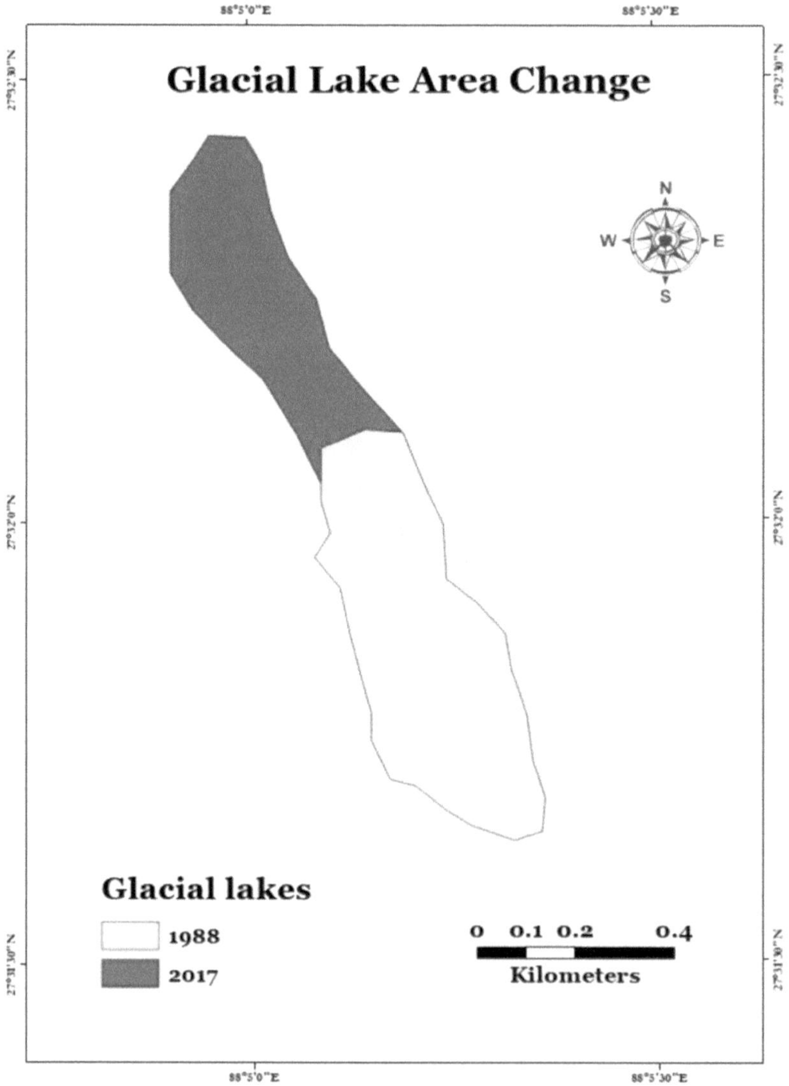

FIGURE 16.14 Map showing increase in the glacial lake area from glacier no. 30.

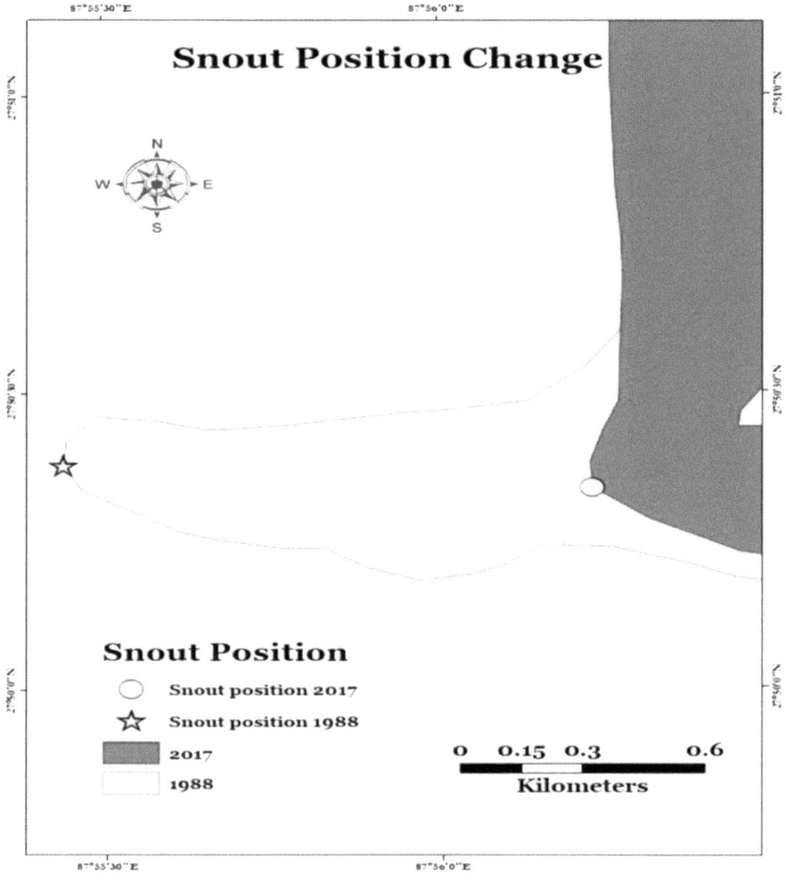

FIGURE 16.15 Map showing snout position change (retreat) in glacier no. 55.

16.5 CONCLUSIONS

In the study area, the accumulation and ice-exposed ablation area decrease over time; whereas debris-covered ablation, deglaciated, and glacial lakes area show increasing trend during 1988 to 2017 while glaciers snout position retreat back considerably. The glaciers in Kangchenjunga area are undoubtedly undergoing negative area change between 1988 and 2017 (30 years) with total glacial area decreases by 43.5 km^2. It also means that the rate of deglaciation noted in this study was ±1.45 km^2 per year accumulation area was 524.74 km^2 in 1988, which further decreased to 475.45 km^2 in 2017. It indicated that the total accumulation area decreases by 49.29 km^2 in between 1988 and 2017 and the rate of decrease in accumulation area was ±1.64 km^2 per year.

Ice-exposed ablation area also followed a decreasing trend with 246.77 km^2 in 1988 to 236.36 km^2 in 2017. It shows that ablation area which is ice exposed decreased by 10.41 km^2 between 1988 and 2017 and the rate of shrinking in Ice-exposed ablation area was ±0.35 km^2 per year. Debris-covered ablation area showed an increase as the accumulation and ice-exposed ablation area were converted in it. Ablation area in 1988, that was debris covered, accounted for 156.57 km^2, which increased to 172.77 km^2 in 2017. It indicates that the total debris-covered ablation area increases by 16.20 km^2 in between 1988 and 2017 and in these 30 years, the rate of growth in debris-covered ablation area was ±0.54 km^2 per year. The deglaciated and glacial lakes area also followed a decreasing trend by 14.90 and 0.51 km^2 respectively during 1988 to 2017 and the rate of shrinking was ±0.50 and ±0.017 km^2 per year respectively. The monitoring of snout position revealed a retreat of 39.96 km and 1.33 km/per year from 1988 to 2017. Thus, this study shows that Kangchenjunga region glaciers are monsoon-fed as well as snow-fed that are more particularly susceptible to temperature rise. It also confirms that rapid cryosphere thawing has the potential to have a negative impact on the hydrology of the study area.

ACKNOWLEDGMENTS

This work could not have been accomplished without the financial assistant of Savitribai Phule Pune University. The research work would not have been possible without the exceptional support of Giripremi's Kangchenjunga Eco Expedition 2019 and Guardian Giripremi Institute of Mountaineering (GGIM). Both the institutions organized and encouraged this "Eco expedition" to collect the different types of samples during the Mt. Kangchenjunga mountaineering expedition.

KEYWORDS

- **glacier retreat**
- **temporal assessment**
- **Kangchenjunga**
- **Himalayan glaciers**
- **glacial area**

REFERENCES

Bahuguna, I. M.; Rathore, B. P.; Brahmbhatt, R.; Sharma, M.; Dhar, S.; Randhawa, S. S.; Kumar, K.; Romshoo, S.; Shah, R. D.; Ganjoo, R. K.; Ajai. Are the Himalayan Glaciers Retreating? *Curr. Sci.* **2014,** 1008–1013.

Bajracharya, S. R.; Mool, P. K.; Shrestha, B. R. *Impact of Climate Change on Himalayan Glaciers and Glacial Lakes: Case Studies on GLOF and Associated Hazards in Nepal and Bhutan*; International Centre for Integrated Mountain Development (ICIMOD), 2007.

Benn, D. I.; Owen, L. A. The Role of the Indian Summer Monsoon and the Mid-Latitude Westerlies in Himalayan Glaciation: Review and Speculative Discussion. *J. Geol. Soc.* **1998,** *155* (2), 353–363.

Bhambri, R.; Bolch, T. Glacier Mapping: A Review with Special Reference to the Indian Himalayas. *Progr. Phys. Geog.* **2009,** *33* (5), 672–704.

Bhatt, B. C.; Nakamura, K. Characteristics of Monsoon Rainfall around the Himalayas Revealed by TRMM Precipitation Radar. *Monthly Weather Rev.* **2005,** *133* (1), 149–165.

Bolch, T.; Kulkarni, A.; Kääb, A.; Huggel, C.; Paul, F.; Cogley, J. G.; ...; Stoffel, M. The State and Fate of Himalayan Glaciers. *Science* **2012,** *336* (6079), 310–314.

Bolch, T.; Menounos, B.; Wheate, R. Landsat-Based Inventory of Glaciers in Western Canada, 1985–2005. *Remote Sens. Environ.* **2010,** *114* (1), 127–137.

Brahmbhatt, R. M.; Bahuguna, I. M.; Rathore, B. P.; Kulkarni, A. V.; Nainwal, H. C.; Shah, R. D.; Ajai. A Comparative Study of Deglaciation in Two Neighbouring Basins (Warwan and Bhut) of Western Himalaya. *Curr. Sci* **2012,** 298–304.

Gärtner-Roer, I.; Nussbaumer, S. U.; Hüsler, F.; Zemp, M. Worldwide Assessment of National Glacier Monitoring and Future Perspectives. *Mountain Res. Dev.* **2019,** *39* (2).

Gujree, I. et al. Evaluating the Variability and Trends in Extreme Climate Events in the Kashmir Valley Using PRECIS RCM Simulations. *Modeling Earth Systems and Environment* **2017,** *3* (4), 1647–1662.

ICIMOD. *Glacial Lakes and Glacial Lake Outburst Floods in Nepal*; ICIMOD: Kathmandu, Nepal, 2011; p 99.

Kääb, A.; Berthier, E.; Nuth, C.; Gardelle, J.; Arnaud, Y. Contrasting Patterns of Early Twenty-First-Century Glacier Mass Change in the Himalayas. *Nature* **2012,** *488* (7412), 495.

Kamp, U.; Byrne, M.; Bolch, T. Glacier Fluctuations between 1975 and 2008 in the Greater Himalaya Range of Zanskar, Southern Ladakh. *J. Mountain Sci.* **2011,** *8* (3), 374–389.

Kaushik, S.; Joshi, P. K.; Singh, T. Development of Glacier Mapping in Indian Himalaya: A Review of Approaches. *Int. J. Remote Sens.* **2019,** *40* (17), 6607–6634.

Krishna, A. P. Snow and Glacier Cover Assessment in the High Mountains of Sikkim Himalaya. *Hydrol. Process. Int. J.* **2005,** *19* (12), 2375–2383.

Kulkarni, A. V. Effect of Global Warming on the Himalayan Cryosphere. *Jalvigyan Sameeksha* **2007,** *22*, 93–108.

Kulkarni, A. V.; Rathore, B. P.; Alex, S. Is Global Warming Changing Snow Ablation Pattern and Reducing Glacial Extent of Baspa Basin in the Himalaya? *Int. Arch. Photogrammetry Remote Sens. Spatial Info. Sci.* **2003,** *34* (7/B), 1265–1269.

Kulkarni, A. V.; Rathore, B. P.; Mahajan, S.; Mathur, P. Alarming Retreat of Parbati Glacier, Beas Basin, Himachal Pradesh. *Curr. Sci.* **2005,** 1844–1850.

Kulkarni, A. V.; Rathore, B. P.; Singh, S. K.; Bahuguna, I. M. Understanding Changes in the Himalayan Cryosphere Using Remote Sensing Techniques. *Int. J. Remote Sens.* **2011,** *32* (3), 601–615.

Kumar, B.; Murugesh Prabhu, T. S. Impacts of Climate Change: Glacial Lake Outburst Floods (GLOFs). *Climate Change in Sikkim Patterns, Impacts and Initiatives. Information and Public Relations Department*; Government of Sikkim, Gangtok, 2012.

Kumar, V.; Venkataramana, G.; Høgda, K. A. Glacier Surface Velocity Estimation Using SAR Interferometry Technique Applying Ascending and Descending Passes in Himalayas. *Int. J. Appl. Earth Observ. Geoinf.* **2011,** *13* (4), 545–551.

Lama, S.; Devkota, B. Vulnerability of Mountain Communities to Climate Change and Adaptation Strategies. *J. Agric. Environ.* **2009,** *10*, 76–83.

Mason, L. E. *Abode of Snow: A History of Himalayan Exploration and Mountaineering*; Dutton, New York City, USA, 1954; p 372.

Miller, K. G.; Mountain, G. S.; Wright, J. D.; Browning, J. V. A 180-Million-Year Record of Sea Level and Ice Volume Variations from Continental Margin and Deep-Sea Isotopic Records. *Oceanography* **2011,** *24* (2), 40–53.

Mool, P. K.; Bajracharya, S. R.; Joshi, S. P.; Sakya, K.; Baidya, A. *Inventory of Glaciers, Glacial Lakes and Glacial Lake Outburst Floods Monitoring and Early Warning Systems in the Hindu-Kush Himalayan region, Nepal*; International Center for Integrated Mountain Development, Nepal, 2002.

Pall, I. A.; Meraj, G.; Romshoo, S. A. Applying Integrated Remote Sensing and Field-Based Approach to Map Glacial Landform Features of the Machoi Glacier Valley, NW Himalaya. *SN Appl. Sci.* **2019,** *1* (5), 488.

Rankl, M.; Braun, M. Glacier Elevation and Mass Changes over the Central Karakoram Region Estimated from TanDEM-X and SRTM/X-SAR Digital Elevation Models. *Ann. Glaciol.* **2016,** *57* (71), 273–281.

Robson, B. A.; Nuth, C.; Dahl, S. O.; Hölbling, D.; Strozzi, T.; Nielsen, P. R. Automated Classification of Debris-Covered Glaciers Combining Optical, SAR and Topographic Data in an Object-Based Environment. *Remote Sens. Environ.* **2015,** *170*, 372–387.

Romshoo, S. A. et al. Satellite-Observed Glacier Recession in the Kashmir Himalaya, India, from 1980 to 2018. *Environmental Monitoring and Assessment* **2020,** *192* (9), 1–17.

Satyabala, S. P. Spatiotemporal Variations in Surface Velocity of the Gangotri Glacier, Garhwal Himalaya, India: Study Using Synthetic Aperture Radar Data. *Remote Sens. Environ.* **2016,** *181*, 151–161.

Shrestha, B. B.; Nakagawa, H.; Kawaike, K.; Baba, Y.; Zhang, H. Glacial Lake Outburst Due to Moraine Dam Failure by Seepage and Overtopping with Impact of Climate Change, 2010.

Shukla, A.; Garg, P. K.; Srivastava, S. Evolution of Glacial and High-Altitude Lakes in the Sikkim, Eastern Himalaya Over the Past Four Decades (1975–2017). *Front. Environ. Sci.* **2018,** *6*, Article 81.

Singh, D.; Gupta, R. D.; Jain, S. K. Assessment of Impact of Climate Change on Water Resources in a Hilly River Basin. *Arab. J. Geosci.* **2015,** *8* (12), 10625–10646.

Space Applications Center (SAC). *Snow and the Glaciers of Himalayas*; Technical Report, 2011.

Zemp, M.; Hoelzle, M.; Haeberli, W. Six Decades of Glacier Mass-Balance Observations: A Review of the Worldwide Monitoring Network. *Ann. Glaciol.* **2009,** *50* (50), 101–111.

PART III

A Review of the Role of GIScience for Assessing Earth System Processes

CHAPTER 17

Application of GIScience for Soil and Water Conservation: A Review

ABIY GEBREMICHAEL*

Southern Agricultural Research Institute, Bonga Agricultural Research Center

**E-mail: gebremichaelaby@gmail.com*

ABSTRACT

The use of satellite data and geographic information systems (GIS) in water management science is discussed in this chapter. Satellite remote sensing and GIS have played a vital role in managing the earth's resources. Various satellite devices, such as the digital elevation model (DEM) and multispectral sensors, provide precise information on the state of the atmosphere at a specific location. Using GIS and modeling, all sorts of efficient strategies can be put into practice that have the best efficiency. In this chapter, the basics of GIScience are reviewed in light of the recent research in this scientific field.

17.1 INTRODUCTION

The success of planning and implementation of soil and conservation activities depends on the available data on soils, topography, climate, vegetation, land use, and other socioeconomic aspects (Pandey et al., 2010a). In this regard, application of Geographical data system (i.e., GIS) in the domain of water resources management, as well as remote sensing (RS) plays a critical role. One of the benefits of using Satellite Imagery

GIScience for the Sustainable Management of Water Resources. Gowhar Meraj, Shruti Kanga, Majid Farooq, Suraj Kumar Singh, & Sudhanshu (Eds.)
© 2023 Apple Academic Press, Inc. Co-published with CRC Press (Taylor & Francis)

data for natural resource management is the ability to obtain knowledge in both the spatial and temporal domains (Wang and Wang, 2005), which is critical for model simulation, prediction, and validation (Alejandra, 2008). It is well recognized that GIS and RS play an important role in the study of data from land image object recognition, as well as the interpretation of findings (Godert et al., 2001). In general, the author concluded which GIS as well as RS could be used to address the following issues: (1) To serve as a physiographic database for assessments of various types of soil degradation and management, (2) to incorporate input data for different feature units, (3) to calculate region, (4) to connect spatial information with non-spatial but far more accurate data elements, (5) to map (spatial-temporal changes in) land use land cover (LULC) use and (6) to classify areas of degradation. According to (Alejandra, 2008) traditional data collection methods produce point-based information, while the RS technique allows hydrologic parameters to be measured on spatial scales, and GIS incorporates spatial analytical features for spatially data sources. The simplicity of data update, information management, and statistical analysis in ways best suited to user needs are all advantages of GIS (Godert et al., 2001). It also enables the aggregation of large volumes of data from various sources on various topics. RS including multi-temporal satellite imagery has become a powerful method for tracking aspects such as land depletion, forest cover, urban development, and most forms of land use and land cover change (Mundia and Aniya, 2005; Gidey et al., 2017). As a result, major progress throughout the GIS application for seeking a solution to one of themes, water and soil conservation, has occurred in recent decades (Shi et al., 2004). The goal of this article was to provide a brief overview of the use of GIS as well as RS in soil conservation.

17.2 LITERATURE REVIEW

17.2.1 GEOGRAPHIC INFORMATION SYSTEMS

Although Geographical Information System (GIS) is defined based on different aspects like database used, structure, organizational set up; it is summarized as GIS is an integrated suite of computer-based tools that accommodate input data, its processing, display, and output of location based data. The definition adapted from (Burrough, 1986; Eastman, 2001; Burrough and McDonnel, 1998) shows two main points:

- GIS is a computer-based information management system for storing, evaluating, combining, and conditionally extracting information from maps and associated attribute data
- It is also defined as a computer-supported system for data storing, processing, integrating, manipulating, analyzing, checking, and displaying those location-based information on the earth's surface.

RS is defined by (Oettter et al., 2000) as technique to catch up the earth surface from remote by using satellites (space borne) or using aircrafts from air. It is a technique to observe the earth surface or the atmosphere from out of space using satellites (space borne) or from the air using aircrafts/airborne (Woodcock et al., 2001)]. The image of land cover and land use remotely accessed can be for a watershed management by utilizing a satellite image that covers a large scene. Landsat Thematic Mapper can be used at the regional level with its 30 m spatial resolution (Jensen, 2005). The point of classification with RS is to categorize every pixel in the image into themes or classes based on the reference spectral response of a band (Oettter et al., 2000).

17.2.2 GIS AND RS APPLICATION IN SOIL AND WATER CONSERVATION

Soil conservation is characterized as a combination of effective land use and development practices that encourages the productivity and long-term use of soil while also reducing soil erosion and degradation (Mitiku et al., 2006). Prediction of soil erosion threat, prioritization of the high risk, planning soil conservation, analysis of the effects of SWC, and Land Degradation Inventory are all examples of GIS and RS applications for soil water conservation (SWC). GIS is used to store and analyze geo-referenced data, as well as to manage the interconnections between high dimensionality and their discrete attributes, according to Pandey et al. (2010a). Watershed restoration, planning, and management have all benefited from the use of RS and GIS techniques. As a result, the following discussion will concentrate on SWC's core activities. The GIS is used to evaluate and calculate the function/object under analysis, while RS provides data for the proof (Oettter et al., 2000).

17.2.2.1 SOIL EROSION HAZARD PREDICTION AND ESTIMATION

Soil erosion is a process of topsoil particles being detached, transported, and accumulated from their original location to a new depositional region.

Raindrops, overland flow, or wind from highland to lower areas to streams or reservoirs are all possible causes (Wang and Wang, 2005). The key human impacts that accelerate soil erosion in Ethiopia are rapid population development, which leads to agriculture on steep slopes, clearing of vegetation, and overgrazing (Reusing et al., 2000; Rashid et al., 2013; Singh et al., 2017a).

Information on soil erosion, such as estimate or quantifying of soil loss from a given plot of land or data for prioritizing sustainable activities at the watershed/subwatershed stage, are critical inputs for long-term agricultural land management that prioritizes soil conservation. By integrating physiography, soil, land use/land cover, terrain map layers, and data of agriculture and soil physico-chemical properties, RS and GIS techniques are effectively used for the preparing of soil erosion inventories (Wang and Wang, 2005). According to (Zotaj and Meçaj, 2006), Soil Erosion Assessment can be set up using GIS and RS for erosion tracking, forecasting, and rapid investigation. The goal of the GIS application in terms of watershed management is to collect soil erosion data, create a database, and create a model for land degradation evaluation and management. Sakthivel et al. (2017) conducted a case study to map erosion-prone areas in North India (Song catchment, Uttar Pradesh). He used the Usle Formula (USLE) on the GIS interface to assess soil loss using satellite (IRS-1B, LISS-II) analyzed soil, land use/land cover, drainage, DEM-induced slope and slope length, and auxiliary agro-met and land characteristic data.

17.2.2.2 REQUIREMENTS FOR SOIL EROSION ASSESSMENT BY GIS AND RS

17.2.2.2.1 Selection of Suitable Models That Estimate Soil Erosion and Link with GIS Interface

Erosion estimation models that understand and determine erosion rate or amount at a point are important requirement for application of GIS that provide spatial information. Models represent equations that govern the process of erosion but the manipulation process takes much time and resource so that GIS interface is required. For example, Pandey et al. (2007) identified the critically erosion prone areas of the Karso watershed, Hazaribagh, India using the Universal Soil Loss Equation (USLE model) based average annual sediment yield estimates. USLE is applied with the help of GIS interface; otherwise, it will be tiresome to estimate erosion manually using equation used by the model. Likewise many scholars (Pandey et al., 2009, 2010b,

2013; Laflen and Dennis, 2013; deVente et al., 2013; Singh and Pandey, 2014; Bhatt et al., 2017; Sharma and Kanga, 2020) used different models to estimate soil erosion rates for different locations. The goals, time and spatial scales involved, as well as the conceptual basis of these models, differ significantly (Lane et al., 1992). Some of the prominent empirical and conceptual models such as RUSLE (Revised universal soil loss equation), MUSLE (Modified universal soil loss equation), SLEMSA (Soil Loss Estimation for South Africa), SEDD (Sediment Delivery Distributed), AGNPS (AGricultural Non-Point Source Pollution) and LASCAM (Large Scale Catchment Model) and physically based models, for example, ANSWERS (Areal Nonpoint Source Watershed Environment Response Simulation), LISEM (Limburg Soil Erosion Model), CREAMS (Chemical Runoff and Erosion from Agricultural Management Systems); WEPP (Water Erosion Prediction Project); MUSLE, and SWAT (Soil Water Assessment Tool) and applicability for soil erosion, and sediment yield prediction were reviewed in detail by (deVente et al., 2013; Merritt et al., 2003). From these erosion prediction models, those that have linkage with GIS can be used for spatial assessment of erosion hazard. Therefore, from this it shows that spatial analysis of erosion hazard needs application of GIS.

17.2.2.2.2 Generation of Major Input Data for GIS

The primary step in GIS analysis is to input spatial data produced from various sources (Pandey et al., 2010a). For computing the input data, conventional means seemed to be too expensive and time consuming. Deriving geospatial data on input variables has become more convenient and cost-effective since the introduction of RS technology (Sakthivel et al., 2017). GIS-based erosion modeling techniques have become more comprehensive and widely accepted as a result of their strong spatial process technology and consistency with RS data. Soil type, slope, drainage, geology, and land cover are some of the variables that can help improve the process (Lufafa et al., 2002). Seasonal land use patterns can be better understood using multi-temporal satellite images. It can be used to investigate erosion features like gullies, runoff interception by plants, and the LULC factor (Habtamu and Amare, 2016).

Digital Elevation Model

The fundamental requirement of any hydrologic or geomorphologic studies led by means of models is digital elevation model (DEM). It's one in all the vital inputs required for erosion modeling that created by analysis of stereoscopic optical and microwave RS data. DEM enables to derive various topographic attribute like elevation, slope, and aspect, etc., which are essential to investigate watershed physical characteristics (Habtamu and Amare, 2016). DEM data can be extracted from the satellite images of terrain such as stereo optical imagery. For spatial assessment of erosion risk, most of the spatial data are generated from DEM and satellite imageries in combination with ground control points collected using GPS. ASTER (Advanced Space borne Thermal Emission and Reflection Radiometer) NASA source DEM with a spatial resolution of 30 m can be obtained (Darlene, 2013) from http://www.neng.usu.edu/dtarb/giswr. The spatial pattern of soil loss was calculated using the Modified Universal Soil Equation (RUSLE) formula and GIS, which included a 1:50,000 scale topographic chart, an Aster-DEM, a 1:250,000 scale soil map, 13 years of weather records from four stations, and a satellite image (TM) with a spatial resolution of 30 m (Habtamu and Amare, 2016).

Land Use Land Cover Data

Another requirement of GIS based spatial analysis of erosion risk is LULC of the required location (Oettter et al., 2000). The LULC dynamics is an important factor forcing soil erosion and any soil erosion estimation model uses it and gets its response to hydrologic characteristics of soil and water catchment. Using image processing techniques, RS technique was used to provide land use/cover information (Yuksel et al., 2008). Many studies (Yuksel et al., 2008; Rode and Frede, 1997; Millward and Mersey, 1999; Jong et al., 2008) have been conducted to model soil erosion using geospatial technologies. RS and GIS enable homogeneous information over large regions and can therefore greatly contribute to regional erosion assessment (Jong et al., 2008). Soil erosion prediction using integrated application of RS data and GIS model could estimate and show spatial distributions soil loss in the watershed. As an account for seasonal variability, multitemporal satellite images are useful to extract the good information associated to seasonal land use dynamics for mapping land use/land cover. It may be accustomed generate a canopy and management factor (C-factor) that is one in all the input requirements of abrasion modeling (Morgan, 2001; Kanga et al., 2017a, 2017b, 2021; Rather et al., 2018; Hassanin et al., 2020). To account for spatial differences in erodibility which serve as input data for erosion modeling, parameters such as soil properties, climate, vegetation, topography, and lithology could be potentially mapped with satellite RS image (Pandey et al., 2010a). Especially optical satellite imagery can be used for soil mapping, mainly through visual delineation of soil patterns (Godert et al., 2001).

Meteorological Data

Rainfall is linked to soil loss, partly due to the detaching force of raindrops hitting the topsoil and partly due to rain's contribution to runoff (Morgan, 2001). USLE model rainfall erosional index, which is used as a source for GIS for soil loss estimation, needs meteorological data of determination of soil erosion utilizing USLE and GIS (Morgan, 2001)].

Drainage network data

In addition, inputs such as drainage network, contour map, and spot heights are important and can be generated from the survey of toposheet on a given scale (Lufafa et al., 2002). The drainage is typically dendritic there by reflecting homogeneity of lithology (Pandey et al., 2010a). According to (Hudson, 1992), knowledge of drainage condition is important since soil erosion by detachment and transport is drainage dependent.

17.2.2.2.3 Erosion Mapping Using GIS Interface

After fulfilling the quantitative and qualitative data as well as images, mapping of site suitability for soil and water conservation structures can be developed in the GIS interface (El Jazouli et al., 2017). Subsequently, adopting a holistic approach and pointing out an appropriate site for existing soil and water conservation structures require parameters of slope, land use/cover, drainage, and soils that integrated with the GIS environment (Gebreyesus et al., 2014; Singh et al., 2017b; Kanga et al., 2020a, 2020b).

17.2.3 PRACTICAL ACHIEVEMENTS WITH APPLICATION OF GIS AND RS IN SOIL EROSION ASSESSMENT

A study was conducted by (Habtamu and Amare, 2016) to estimate and map the mean annual soil loss form Koga watershed, Northwestern Ethiopia by using GIS and RS techniques. The soil loss was estimated by using by (Wischmeier and Smith,1965) using Universal Soil Loss Equation (USLE) model which considers the following factors such as:

- Topographic map scale (physical land features data)
- DEM of 20 m spatial resolution (LS factors)
- Digital soil map (for K-factor)
- Rainfall records of selected stations (for R-factor) and
- Landsat imagery (TM) with high spatial resolution m (for C and P factor)

These USLE variables were used in the ArcGIS 10.1 setting to estimate and chart the annually soil loss of the study region using the raster calculator inside the geo-processing tools. According to the findings, the watershed's annual soil loss ranged between 265 t/ha/y in the relatively steep slope portion of the watershed, with an annual average soil erosion of 47 t/ha/y, and zero in the middle and lower parts (Habtamu and Amare, 2016). The average annual soil loss inside the drainage basin was 255,283 tones, with 181,801 (71%) tones covering approximately 6691 (24%) hectares. The main areas spatially located at the watershed's upper steepest slope section (inlet) are found to be strongly affected by soil erosion.

Zotaj and Meçaj (2006) used a GIS-based scalogram modelling approach to assess and map soil erosion hazard in the Dangellia highlands (Permeti district, Estonia) employing RS-based physiography soil and land type (for C and P-factor determination) as well as DEM-derived surface slope map (for LS-factor determination) and field data of soil characteristics (for K-factor estimation) and contextual information of soil properties (for K-factor determination). Individual maps of parameter, such as R, K, LS, C, and P, will be overlaid to create a comprehensive map of soil erosion severity based on emerging GIS functionality, and the erosion severity groups will be prepared from the intensity map to display the key aspects in order of erosion danger. By combining erosion mitigating soil bunds, slope, and climatic parameters, the production results indicate that the RS and GIS applications are indeed the best technique for assessing soil erosion risk hotspots.

Lufafa et al. (2006) used USLE and GIS to forecast soil loss in a micro-catchment, Lake Victoria Region (LVB). Annual cropland use (93 t/ha/y) was expected to lose the most soil, followed by rangeland (52 t/ha/y), banana–coffee (47 t/ha/y), banana (32 t/ha/y), and woodland and papyrus swamp. Water erosion in small watersheds can be assessed using GIS and the WEPP model (Cocbrane and Flanagan, 1999). GIS is used in the WEPP model to assist in the creation of needed input files and the display of a map of erosion spatial distribution.

It was argued that having hydro-physical parameters in a GIS map format makes it easier to run erosion models and understand the spatial distribution of ecological problems like soil degradation (Gebreyesus et al., 2014;

Farooq and Muslim, 2014; Nathawat et al., 2010; Kumar et al., 2018; Joy et al., 2019). The author used a GIS and the Morgan–Morgan–Finney (MMF) model to measure and describe the geographic variation of erosion risk areas in the Mai-negus catchment in northern Ethiopia. The rate of soil detachment ranged from <20 t/ha/y to >170 t/ha/y.

However, Sakthivel et al. (2017) applied only RS and GIS Spatial analyst tool to classify degradation prone areas in India with image centered data sources unlike use of any models for erosion calculation. The parameters of soil erosion is estimation such as rainfall, slope, soil, land use, and land cover that are obtained from (1) IRS 1D satellite image, (2) topographic map, (3) Soil Survey data, and (4) Monthly rainfall data. The soil erosion hazard zones have been prepared using the above parameters based and ArcGIS Spatial analyst software.

17.3 SUMMARY AND CONCLUSION

SWC works are significant to control and minimize soil and sediment loss by erosion from the cultivated lands and to increase sustainable use of soil and land resources. Successful planning and implementation of soil and water conservation works depends on the quality and quantity of erosion intensity area identification based on information (soils, topography, climate, vegetation, land use, and other socioeconomic aspects) analysis. In this regard, GIS and the data from RS play an important role in linking and analyzing of such data particularly spatial basis, detection of erosion intensity risk area, extrapolation and interpretation for future monitoring and evaluation. Estimation of soil erosion threat, prioritization of the risk area, soil conservation planning, analysis of SWC effect, and soil loss inventory are all examples of GIS and RS applications for SWC. The most critical benefit of RS data application for management of natural resource is its ability to produce spatial and temporal information crucial for effective model analysis, prediction, and validation. In this chapter, it is clearly presented that with the help of models for erosion quantification as well as use of RS and GIS as source of data and spatial analysis tool, respectively. It is possible to know erosion intensity in any catchment so that planning and implement soil and water conservation activities would be simplified. In general, if a GIS-spatial analysis method is used in conjunction to proper erosion assessment models, assessing, and delineating erosion-prone areas and prioritizing conservation work can be very easy.

KEYWORDS

- **satellite remote sensing**
- **GIS**
- **MODIS**
- **LandSat**
- **DEM**
- **USLE**
- **soil erosion**
- **remote sensing**
- **GIS**
- **soil and water conservation**

REFERENCES

Alejandra, M. Soil Erosion Calculation Using Remote Sensing and GIS in Río-Grande De Arecibo Watershed, Puerto Rico. ASPRS. Annual Conference, Portland, Oregon, 2008.

Bhatt, C. M. et al. Satellite-Based Assessment of the Catastrophic Jhelum Floods of September 2014, Jammu & Kashmir, India. *Geomat. Nat. Hazards Risk* **2017,** *8* (2), 309–327.

Burrough, P.; McDonnel, A. *Principle of Geographic Information System*; Oxford University Press: Oxford, 1998.

Burrough, P. A. *Principles of Geographical Information Systems for Land Resources Assessment*. Monograph on Soils and Resources Survey No. 12; Oxford, UK: Clarendon Press, 1986.

Cocbrane, T.; Flanagan, D. Assessing Water Erosion in Small Watersheds Using WEPP with GIS and Digital Elevation Models Soil and Water Conservation Society. *J. Soil Water Conserv.* **1999,** *54* (4), 678–685. www.swcs.org

Darlene, T. Working with Digital Elevation Models and Digital Terrain Models in ArcMap 10.1. Data from NRVIS, ESRI. Toronto, 2013.

deVente, J.; Poesen, J.; Verstraeten, G.; Govers, G.; Vanmaercke, M.; VanRompaey, A.; Arabkhedri, M.; Boix-Fayos, C. Predicting Soil Erosion and Sediment Yield at Regional Scales: Where Do We Stand? *Earth-Sci. Rev.* **2013,** *127*, 16–29.

Eastman, J. R. *IDRISI 32 Release 2: Guide to GIS and Image Processing*; Clark Labs, Clark University: Worcester, MA, USA, 2001; 151 pp.

El Jazouli, A.; Barakat, A.; Ghafri, A.; El Moutaki, S.; Ettaqy, A.; Khellouk. R. Soil Erosion Modeled with USLE, GIS, and Remote Sensing: A Case Study of Ikkour Watershed in Middle Atlas (Morocco). *Geosci. Lett.* **2017,** *4*, 25. DOI: 10.1186/s40562-017-0091-6.

Farooq, M.; Muslim, M. Dynamics and Forecasting of Population Growth and Urban Expansion in Srinagar City-A Geospatial Approach. *Int. Arch. Photogrammetry Remote Sens. Spatial Inf. Sci.* **2014,** *40* (8), 709.

Gebreyesus, B.; Lulseged, T.; Paul, L. Soil Erosion Prediction Using Morgan-Morgan-Finney Model in a GIS Environment in Northern Ethiopia Catchment. *Hindawi Publ. Corp. Appl. Environ. Soil Sci.* **2014,** *15*, Article ID 468751.

Gidey, E.; Dikinya, O.; Sebego, R.; Segosebe, E.; Zenebe, A. Modeling the Spatio-Temporal Dynamics and Evolution of Land Use and Land Cover (1984–2015) Using Remote Sensing and GIS in Raya, Northern Ethiopia. *Model. Earth Syst. Environ.* **2017,** *3*, 1285–1301. https://doi.org/10.1007/s40808-017-0375-z

Godert, W. J.; van, L.; Stephan, M. The Role of GIS and Remote Sensing in Land Degradation Assessment and Conservation Mapping: Some User Experiences and Expectations. *Int. Soil Ref. Inf. Centre (ISRIC). JAG l* **2001,** *3* (1).

Habtamu, S.; Amare, S. Soil Loss Estimation Using GIS and Remote Sensing Techniques: A Case of Koga Watershed, Northwestern Ethiopia. *Int. Soil Water Conserv. Res.* **2016,** *4*, 126–136.

Hassanin, M.; Kanga, S.; Farooq, M.; Singh, S. K. Mapping of Trees Outside Forest (ToF) From Sentinel-2 MSI Satellite Data using Object-Based Image Analysis. *Gujarat Agric. Univ. Res. J.* **2020,** 207.

Hudson, N. *Land Husbandry*; BT Batsford Limited: London, 1992; pp 172–190.

Jensen, J. *Introductory Digital Image Processing: A Remote Sensing Perspective*, 3rd ed; Pearson Prentice Hall: Upper Saddle River, 2005.

Jong, S.; Paracchini, M.; Bertolo, F.; Folving, S.; Megier, J. Regional Assessment of Soil Erosion Using the Distributed Model SEMMED and Remotely Sensed Data. 2008, *37*, 291–308.

Joy, J.; Kanga, S.; Singh, S. K. Kerala Flood 2018: Flood Mapping by Participatory GIS Approach, Meloor Panchayat. *Int. J. Emerg. Technol.* **2019,** *10* (1), 197–205.

Kanga, S.; Singh S. K.; Sudhanshu. Delineation of Urban Built-Up and Change Detection Analysis Using Multi-Temporal Satellite Images. *Int. J. Recent Res. Aspects* **2017a,** *4* (3), 1–9.

Kanga, S.; Kumar, S.; Singh, S. K. Climate Induced Variation in Forest Fire Using Remote Sensing and GIS in Bilaspur District of Himachal Pradesh. *Int. J. Eng. Comput. Sci.* **2017b,** *6* (6), 21695–21702.

Kanga, S. et al. Modeling the Spatial Pattern of Sediment Flow in Lower Hugli Estuary, West Bengal, India by Quantifying Suspended Sediment Concentration (SSC) and Depth Conditions using Geoinformatics. *Appl. Comput. Geosci.* **2020a,** 100043.

Kanga, S.; Sheikh, A. J.; Godara, U. GIscience for Groundwater Prospect Zonation. *J. Crit. Rev.* **2020b,** *7* (18), 697–709.

Kanga, S.; Rather, M. A.; Farooq, M.; Singh, S. K. GIS Based Forest Fire Vulnerability Assessment and Its Validation Using Field and MODIS Data: A Case Study of Bhaderwah Forest Division, Jammu and Kashmir (India). *Indian Forester* **2021,** *147* (2), 120–136.

Kumar, S. et al. Delineation of Shoreline Change along Chilika Lagoon (Odisha), East Coast of India Using Geospatial Technique, 2018.

Laflen, J.; Dennis, C. The Development of U.SL. Soil Erosion Prediction and Modeling. *Int. Soil Water Conserv. Res.* **2013,** *1*, (2), 11.

Lane, L.; Renard, K.; Foster, G.; Laflen, J. Development and Application of Modern Soil Erosion Prediction Technology—the USDA Experience. *Aust. J. Soil Res.* **1992,** *30*, 893–912.

Lufafa, A.; Tenywa, M.; Isabirye, M.; Majaliwa, G.; Woomer, P. Prediction of Soil Erosion in a Lake Victoria Basin Catchment Using a GIS-Based Universal Soil Loss Model. *Agric. Syst.* **2002,** *76*, 883–894.

Merritt, W.; Letcher, R.; Jakeman, A. Review of Erosion and Sediment Transport Models. *Environ. Model. Softw.* **2003,** *18*, 761–799.

Millward, A.; Mersey, J. Adapting the RUSLE to Model Soil Erosion Potential in a Mountainous Tropical Watershed. *Catena* **1999,** *38*, 109–129.

Mitiku, H.; Herweg, K.; Stillhardt, B. *Sustainable Land Management–A New Approach to Soil and Water Conservation in Ethiopia*; Land Resources Management and Environmental Protection Department, Mekelle University: Mekelle, Ethiopia, Bern, Switzerland: CDE, University of Bern, NCCR North-South, 2006; 269 pp.

Morgan, R. A Simple Approach to Soil Loss Prediction: A Revised Morgan-Morgan- Finney Model. *Catena* **2001,** 305–322.

Mundia, C. N.; Aniya, M. Analysis of Land Use/Cover Changes and Urban Expansion of Nairobi City Using Remote Sensing and GIS. *Int. J. Remote Sens.* **2005,** *26*, 2831–2849. https://doi.org/10.1080/01431160500117865

Nathawat, M. S. et al. Monitoring & Analysis of Wastelands and Its Dynamics Using Multiresolution and Temporal Satellite Data in Part of Indian State of Bihar. *Int. J. Geomat. Geosci.* **2010,** *1* (3), 297–307.

Oettter, D.; Cohen, W.; Berterretche, M.; Maiersperger, T.; Kennedy, R. Land Cover Mapping in an Agricultural Setting Using Multiseasonal Thematic Mapper Data. *Remote Sens. Environ.* **2000,** *76*, 139–155.

Pandey, A.; Chowdary, V.; Mal, B. Identification of Critical Erosion Prone Areas in the Small Agricultural Watershed Using USLE, GIS and Remote Sensing. *Water Resour. Manage.* **2007,** *21*, 729–746.

Pandey, A.; Chowdary, V.; Mal, B.; Billi, M. Application of the WEPP Model for Prioritization and Evaluation of Best Management Practices in an Indian Watershed. *Hydrol. Process.* **2009,** *23*, 2997–3005.

Pandey, A.; Chowdary, V.; Mal, B.; and Dabral, P. Remote Sensing and GIS for Identification of Suitable Sites for Soil and Water Conservation Structures. *Land Degrad. Dev.* **2010a,** *22*, 359–372. (wileyonlinelibrary.com).

Pandey, A. C.; Singh, S. K.; Nathawat, M. S. Waterlogging and Flood Hazards Vulnerability and Risk Assessment in Indo Gangetic Plain. *Nat. Hazards* **2010b,** *55* (2), 273–289.

Pandey, A. C.; Singh, S. K.; Nathawat, M. S.; Saha, D. Assessment of Surface and Subsurface Waterlogging, Water Level Fluctuations, and Lithological Variations for Evaluating Groundwater Resources in Ganga Plains. *Int. J. Digital Earth* **2013,** *6* (3), 276–296.

Rather, M. A. et al. Remote Sensing and GIS Based Forest Fire Vulnerability Assessment in Dachigam National Park, North Western Himalaya. *Asian J. Appl. Sci.* **2018,** *11* (2), 98–114.

Reusing, M.; Schneider, T.; Ammer, U. Modeling Soil Loss Rates in the Ethiopian Highlands by Integration of High Resolution MOMS-02/D2-Streao-Data in a GIS. *Int. J. Remote Sens.* **2000,** *21* (9)

Rode, M.; Frede, H. Modification of AGNPS for Agricultural Land and Climate Conditions in Central Germany. *J. Environ. Qual.* **1997,** *26*, 165–172.

Sakthivel, R.; Jawahar, R.; Pugazhendi, V.; Rajendran, S.; Alagappamoses, A. Remote Sensing and GIS for Soil Erosion Prone Areas Assessment: A Case Study from Kalrayan Hills, Part of Eastern Ghats, Tamil Nadu, India. *Appl. Sci. Res.* **2011,** *3* (6), 369–376.

Sharma, A.; Kanga, S. Surface Runoff Estimation of Sind River Basin Using SCS-CN Method and GIS Technology, 2020.

Shi, Z.; Cai, C.; Ding, S.; Wang, T.; Chowb, T. Soil Conservation Planning at the Small Watershed Level Using RUSLE with GIS: A Case Study in the Three Gorge Area of China Catena **2004,** *55*, 33–48.

Singh, S. K.; Mishra, S. K.; Kanga, S. Delineation of Groundwater Potential Zone Using Geospatial Techniques for Shimla city, Himachal Pradesh (India). *Int. J. Sci. Res. Dev.* **2017b,** *5* (4), 225–234.

Singh, S. K.; Pandey, A. C. Geomorphology and the Controls of Geohydrology on Waterlogging in Gangetic Plains, North Bihar, India. *Environ. Earth Sci.* **2014,** *71* (4), 1561–1579.

Wang, G. W.; Wang, Z. Sedimentation Problems and Management Strategies of Sanmenxia Reservoir, Yellow River, China; *Water Resour. Res.* **2005,** *41*, W09417. DOI: 10.1029/2004WR003919

Wischmeier, W. H.; Smith, D. D. *Predicting Rainfall Erosion Losses from Cropland East of the Rocky Mountains*; Handbook No. 282; US Department of Agriculture (USDA): Washington, DC, USA, 1965.

Woodcock, C.; Macombe, R. S.; Pax-Lenny, M.; Cohen, W. Monitoring Large Areas for Forest Change Using Landsat: Generalization across Space, Time and Land Sensors. *Remote Sens Environ* **2001,** *78*, 194–203.

Yuksel, A.; Akay, E.; Gundogan, R.; Using ASTER Imagery in Land Use/Cover Classification of Eastern Mediterranean Landscapes According to CORINE Land Cover Project. *Sensors* **2008,** *8*, 1237–1251.

Zotaj, A.; Meçaj, N. Application of Remote Sensing and GIS on Soil Erosion Risk Assessment and Mitigation in Mountain Areas at Dangellia Highlands, Permet, Albania. *Geophys. Res. Abstracts* **2006,** *8*, 0**0180,** SRef-ID: 1607-7962/gra/EGU06-A-00180.

CHAPTER 18

Environmental Quality Monitoring Using Environmental Quality Indices (EQI), Geographic Information System (GIS), and Remote Sensing: A Review

FAHEEM AHAMAD[1*], RAKESH BHUTIANI[2], and MUKESH RUHELA[3]

[1]*Department of Environmental Science, Keral Verma Subharti College of Sciences (KVSCOS), Swami Vivekanand Subharti University, Meerut 250005, Uttar Pradesh, India*

[2]*Limnology and Ecological Modelling Laboratory Department of Zoology and Environmental Science, Gurukula Kangri Vishwavidyalaya, Haridwar 249404, Uttarakhand, India*

[3]*Department of Environmental Engineering (SITE), Swami Vivekanand Subharti University, Meerut 250005, Uttar Pradesh, India*

[*]*Corresponding author. E-mail: faheem.ahamad170390@gmail.com*

ABSTRACT

The quality of the environment and the health status of mankind are directly or indirectly associated with one another. When we discuss about the quality of the environment (water, air, and soil), it depends on different parameters associated with them. Environmental quality parameters are often categorized into physical, chemical, biological, microbiological, and heavy metals. But knowledge of these parameters is restricted to scientific community only. Knowledge about these parameters is out of reach of most of the folk, they

GIScience for the Sustainable Management of Water Resources. Gowhar Meraj, Shruti Kanga, Majid Farooq, Suraj Kumar Singh, & Sudhanshu (Eds.)
© 2023 Apple Academic Press, Inc. Co-published with CRC Press (Taylor & Francis)

know only about some parameters, having insufficient knowledge to know about the pollution level in particular part of the environment. Thus, quality indices (QI) are applied to the data of pollution levels to form it easy for folk to know the status of the environment. The basic aim of EQI (air and water quality index) is to categorize the quality of a source based on one or the other system which distilled the list of constituents, and their concentrations present in a sample into a single value. In the present study, an effort has been made to review the various methodologies available for the calculation of EQI (WQI or AQI) and the use of EQI to make the pollution level of air and water (surface water such as river or lake and underground water) understandable to every human being exclusively in India. The precious information on the use of Water Quality Indices (WQI), Air Quality Indices (AQI), remote sensing and Geographic Information System (GIS) for water and air quality mapping is also discussed in the present study.

18.1 INTRODUCTION

Environment and life, these two terms are dependent on each other. We know that without an environment, life does not exist. Therefore, great concern is paid in the present era on the environmental pollution at the local, regional, and global scale (Wald et al, 1999; Lim et al., 2009; Duan et al., 2013a, 2013b). The environment is categorized into the lithosphere (a rocky portion of the environment), hydrosphere (a watery portion of the environment), atmosphere (a blanket of gases), and biosphere (part of the lithosphere, hydrosphere, and atmosphere where life is possible). All these three components of the environment are equally important for the life of a living being whether plants or animals.

18.1.1 TERMINOLOGY AND DEFINITIONS

Pollutants, source, and sink are the three basic terms used in the study of environmental pollution. Substances that cause any type of pollution are known as pollutants. The places from where the pollutants get originated are termed as sources of pollution and places or substances or processes from where the pollutants get diminished or absorbed or get converted from a harmful category to normal category is termed as a sink. Vegetation is considered as the main sink of each type of pollutants. Based on the origin and forms, air pollutants are categorized into primary (exist in the form in which it gets originated from the source, such as sulfur oxides, carbon monoxide, volatile organic compounds, nitrogen oxides, and particulate matter) and secondary (formed in the atmosphere due to reaction between primary pollutants such as ozone, smog, acid rain, etc.) type. Water bodies are categorized into two types, that is, lentic and lotic stagnant water bodies such as lakes and ponds are called lentic while all the running water bodies, such as rivers and streams are called lotic water bodies. It has been observed that lentic water bodies are more vulnerable to pollution in comparison to lotic water bodies. Water pollution sources are categorized into point and nonpoint sources. Point sources of pollution are those whose locations and identification are fixed, such as pipe, smokestack, ship, factories, ditch, and sewage treatment plants. Nonpoint sources of pollution are those sources which are diffused and do not meet with the definition of point sources, such as agricultural runoff, urban runoff, abandoned mines, and residential areas (Ritchie et al., 2003).

18.1.2 PROBLEM OF POLLUTION

Pollution of the environment is defined as "Any undesirable change in the physicochemical and biological characteristics of environment to an extent that causes harm to the life of human beings and other living beings, material, and industrial processes." Before industrialization and urbanization, all the spheres of the environment were in perfect condition and the compositional components of lithosphere, hydrosphere, and atmosphere were in prescribed limits of Bureau of Indian Standard (BIS), Indian Council of Medical Research (ICMR), World Health Organisation (WHO), and Central Pollution Control Board (CPCB). It was the post-industrialization and urbanization era, in which an increase in most of the components was observed due to different industrial and domestic activities. Nowadays, the condition of each sphere of the environment is in serious condition and the extent of pollution is continuously growing at a very fast speed. Government officials, policy makers and scientific communities are unable to control this growing extent of pollution without the cooperation of the common man. It is the common man who can follow and apply the pollution control strategies and will play a crucial role in controlling the pollution, but the units and terms used to define the pollution of any spheres of the environment are out of reach of most of the folk.

18.2 ENVIRONMENTAL QUALITY INDEX, REMOTE SENSING AND GEOGRAPHIC INFORMATION SYSTEM

18.2.1 ENVIRONMENTAL QUALITY INDEX

In this chapter, the term EQI is used to define two different quality indices used for the assessment of air and water. EQI is defined as a unitless quantity obtained after different mathematical calculations performed on complex quality parameters of the selected source. The index used for the assessment of air is called air quality index (AQI) or air pollution index (API) while the index used for the assessment of water quality is called water quality index (WQI). WQI for water and AQI or API for air reflect the status of air and water of a particular place or water body in a simpler and easier manner. Still, there is a lack of standard and globally accepted methodology for the calculation of WQI and AQI (Singh et al., 2013a; Tiwari et al., 2014; Rawat and Singh, 2018).

18.2.1.1 WATER QUALITY INDEX

Evaluation of physicochemical parameters of water with the help of water quality index (WQI) provides an understandable and real picture of water quality (Kumar et al., 2020; Cude and Oregon, 2001; Kannel et al., 2007). Although traditional methods also quantify the quality of water but WQI provides the overall water quality of a resource or an area (Debels et al., 2005). WQI is a unitless mathematical quantity and a ranking tool first described by Horton (1965) in the early 1970s (Kumar et al., 2020; Bhutiani et al., 2018; Abbasi and Abbasi, 2012). Different methodologies were used for the calculation of WQI by different scholars in different parts of the world (Bordalo et al., 2001; Pesce and Wunderlin, 2000; Said et al., 2004). In India, the pioneer work in the field of WQI was done by Bhargava (Bhargava, 1983; Sharma and Arun, 2011). BIS, ICMR, and WHO standards were used as a reference value for the calculation of WQI. WQI can be applied to physico-chemical and heavy metal parameters of groundwater, lentic, and lotic water bodies to translate the information from a complex to a simple form. Different methodologies available for the calculation of WQI are given in Table 18.1.

18.2.1.2 AIR QUALITY INDEX

Like WQI, AQI is also a unitless and mathematical quantity which simplifies the complex data of air quality into a single value for better understanding.

TABLE 18.1 Methodologies Adopted by Various Researchers for the Calculation of WQI and AQI.

Index	Formula	Parameters used
NSF-WQI Brown et al. (1970)	$\sum_{i=1}^{N} SI_i W_i$	DO (%), FC, MPN/100 mL, pH, BOD, nitrates, total phosphate, temperature, turbidity and total Solids
O-WQI Cude and Oregon, (2001) and Dunnette and Oregon (1980)	$\sqrt{\frac{N}{\sum_{i=2}^{N} \frac{1}{SI_i^2}}}$	DO (%), FC, pH, BOD, ammonia+nitrate, nitrogen, total phosphorous, temperature, and total solids
PW-WQI Pesce and Wunderlin (2000)	$\frac{\sum_{i=1}^{3} SI_i}{3}$	DO (%), turbidity and conductivity
CPCB -WQI Sarkar and Abbasi (2006)	$\sum_{i=1}^{N} SI_i W_i$	DO (%), FC (MPN/100 mL), pH, BOD
River pollution index (RPI) Liou et al. (2004)	$\mathrm{SI_{temp} SI_{pH} SI_{box}} \left[\sum_{i=1}^{3} SI_i W_i \times \sum_{j=1}^{2} SI_j W_j \times \sum_{k=1}^{1} SI_k \right]^{1/3}$	DO (%), FC (MPN/100), pH, BOD, ammonia-nitrogen, temperature, turbidity, toxicity and suspended solids
U-WQI Boyacioglu (2007)	N/A $\frac{\sum_{i=1}^{N} SI_i}{N}$	Cadmium, fluoride, cyanide, mercury, arsenic, selenium, pH, nitrate-nitrogen, DO, BOD5, total phosphorus, and total coliform
S-WQI Said et al. (2004)	$\log\left[\frac{DO^{15}}{(3.8)^{15}(\mathrm{Turb})^{15} 15^{\mathrm{FC}/100} + 0.14(Con)^{0.5}}\right]$	DO (%), FC, MPN/100 mL, turbidity, EC, and total phosphorus
ISQA	$100 - \left(\frac{\sqrt{F_1^2 + F_2^2 + F_3^2}}{1.732}\right)$	DO (%), temperature, EC, suspended solids, and TOC
Weighted Arithmetic Index Brown et al. (1970)	$\mathrm{WQI} = \frac{\Sigma \mathrm{QiWi}}{\Sigma \mathrm{Wi}}$	

Source: Adapted from Abbasi and Abbasi (2012) and various other literatures.

The parameters which play an important role in the air quality assessment and mostly considered for the calculation of AQI include PM10, PM2.5, sulfur dioxide (SO_2), nitrogen dioxide (NO_2), and ozone. The methodology adopted for the calculation of AQI by the different researchers is different from that of WQI. In the literature survey, it was observed that no standard and universally accepted methodology is available for the calculation of AQI. Due to a lack of guidelines, researchers select the parameters for the calculation of AQI based on their convenience, hence there is a need to promote the use of EQI in the monitoring of the environment and development of suitable methodologies and guidelines for the calculation of EQIs including WQI and AQI. This chapter reviews the applications of the environmental quality index (EQI), remote sensing, and Geographic Information System (GIS) in the monitoring of environmental quality especially air and water quality. An attempt has also been made to reflect the overall scenario of the use of EQI, remote sensing, and GIS for the assessment of the environment.

18.2.1.3 REMOTE SENSING AND GEOGRAPHIC INFORMATION SYSTEM

Environmental quality can be determined by manual sampling and analysis. Although, the in situ analysis provides better results, yet in situ analysis is time-consuming and laborious (Gholizadeha et al., 2016)]. Easy detection of spatial or temporal variations of the environmental quality is not possible with in situ conventional sampling methods (Ritchie et al., 2003). With the advancement of space technology and increased use of computer application in recent decades, the use of computational technology has increased to a large extent to analyze and predict the quality of air, water, and soil in an efficient and effective manner (Gholizadeha et al., 2016; Lim et al., 2009). Remote sensing and GIS are two such technologies by which a large area of environment can be monitored in a rapid and cost-effective manner (Hadjimitsis et al., 2007a, 2007b). GIS is a computerized program of information collection, manipulation, analysis, presentation, and prediction or forecasting based on collected information with the help of modeling program without directly in contacts with the earth (Jensen et al., 2001; Burrough and McDonnell, 1998; Thomas et al., 2004). The reflected energy from the surface of the sun (passive sensors) and earth or object (active sensors) is measured by sensors and the collected information becomes visible in the form of images. Passive sensors record data only during day time due to the availability of sunlight while active sensors work on their own power supply and record data throughout day and night (Lim et al., 2009; Lillesand and Kiefer, 1994).

Aerosol optical thickness is the most concerning parameter for solving radiative transfer (RT) equation of atmospheric correction algorithm for the removal of atmospheric effects from the imageries are obtained with the help of GIS and remote sensing (Gholizadeha et al., 2016; Hadjimitsis, 1999; Hadjimitsis and Clayton, 2004; Hadjimitsis et al., 2002; Forster, 1984; Kanaroglou et al., 2002; Kaufman et al., 1990). Several methods are available in the literature for the calculation of aerosol optical thickness, such as the ocean method, contrast-reduction method, brightness method, differential textural analysis method, dark vegetation method, and temperature attenuation procedure (Hadjimitsis et al., 2007a).

Based on spatial, radiometric, temporal, and spectral resolution images obtained from remote sensing differ in quality. Different commercial and free satellite images are available at different resolutions, namely Landsat 8, 7 ETM and 5 TM, GeoEye, Medium Resolution Imaging Spectrometer (MERIS), IKONOS, WorldView-2 and 3, HICO, Moderate Resolution Imaging Spectroradiometer (MODIS), Nimbus 7, NOAA AVHRR, SPOT HRV, Cartosat, Advanced Spaceborne Thermal Emission and Reflection Radiation (ASTER), QuickBird, SeaWiFS, and Indian Remote Sensing (IRS) LISS-3 and P6.

18.3 ASSESSMENT OF WATER QUALITY USING WQI, REMOTE SENSING, AND GIS

Water is the basic need and essential thing of life for every living being. Due to the effect of substances present in water in the form of pollutants, the characteristics of reflected light from the surface of particular water bodies changes and these changes can be detected with the help of remote sensor in the form of imageries (Jerlov, 1976; Kirk, 1983). Literature suggests that remote sensing is a useful tool for the monitoring of water quality simultaneously in a large area on a regular basis (Ritchie and Charles, 1996; Schalles et al., 1998; Dekker et al., 2002; Zhang et al., 2002).

An empirical model of the sediment estimation using remote sensing was developed by Ritchie and coauthors (Ritchie et al., 1976). The author used the following equation:

$$Y=A+BX \text{ or } Y=AB^X$$

where

Y = Measurement of reflectance, radiance, and energy.

X = Concerned Water quality parameters.

A and B= Empirical factors of the concerned parameters (determined by the statistical relationship between the in situ observed value and spectral reflectance value).

Based on the optical properties and water quality parameters, Schiebe and coauthors (Schiebe et al., 1992) developed a model. The basic principle of working of this model is the detection of the relationship between the spectral and physical characteristics of the studied water. This physically based model coupled with statistical coefficients (Bi and Si) was used for the estimation of suspended sediment concentrations successfully (Schiebe et al., 1992; Harrington et al., 1992). Their model has the form

$$Ri = Bi^{[1\_e(c/Si)]},$$

where

Ri= ceflectance in waveband i determined by Landsat or SPOT.

C= concentration of suspended sediment.

Bi= reflectance saturation level in waveband i.

Si= concentration of parameter when the reflectance is 63% of the saturation in waveband i.

Groundwater quality vulnerability can be assessed by the DRASTIC index with the help of GIS and geoprocessing methodology (Nero and SajuSimon, 2015). In the DRASTIC model, seven parameters, such as water depth, recharge rate, topography, aquifer composition, soil type, impact of the vadose zone, and the hydraulic conductivity of the aquifer were used. Studies have shown a significant relationship of the concentration of suspended solids with that of reflectance or radiance at 700–800 nm (Ritchie et al., 1976). Radiance of the water is affected by type, texture, and color of the sediment, sensor view, depth of water, and incidence angles of sun rays (Novo et al., 1989; Blanchard and Leamer, 1973; Ritchie et al., 1975). Scattering of light process is much better than absorption. Absorption of light was more by chlorophyll than by suspended matter. For the chlorophyll concentration estimation, the most sensitive indicsator was the 690 nm reflectance height above the baseline from 670 to 850 nm (Gitelson et al., 1995). With the help of chlorophyll estimation, eutrophication of different water bodies was also detected (Schalles et al., 1998). Researchers also report that discrimination between the chlorophyll and high concentration of sediments is difficult by Landsat and SPOT due to broad wavelength spectral data (Dekker and Peters, 1993; Ritchie et al., 1974). The problem is now resolved with help of airborne and spaceborne sensors, such as SeaWiFS, Modular Optical Scanner (MOS), and Ikonos (Ritchie et al., 2003). Color dissolved organic matter (CDOM) affects the primary productivity of an ecosystem due to the absorption of the short-wavelength fraction of the electromagnetic spectrum which is important for photosynthesis (Usali and

Ismail, 2010; Meraj et al., 2021). Table 18.2 shows the platform used for the estimation of selected parameters.

TABLE 18.2 Remote Sensing and GIS Platform Used by Different Authors for Air Water Quality Analysis.

Sl. N.	Parameters	Platform used	References
1	Suspended matter in water	SPOT, Landsat, IRS, TM, even color and color IR photographs Coastal Zone Color Scanner (CZCS), and Sea- viewing Wide Field of View Sensor (SeaWiFS), MSS data	Dekker et al. (2002), Ritchie et al. (1976) and Liu et al. (2003)
2	Phytoplankton	AVHRR, CZCS, SeaWiFS, synthetic aperture radars	Siegel and Gerth (2000), Ha'kanson and Moberg (1994) and Svejkovsky and Shandley (2001)
3	Turbidity	IKONOS, Landsat 7 Enhanced Thematic Mapper (ETM)	Brezonik et al. (2005) and Wang et al. (2006)
4	PM10 and PM2.5	Landsat Thematic Mapper (TM) 5 IKONOS, SPOT, and QUICKBIRD	Lim et al. (2009) and Hadjimitsis and Thermistocleous (2015)

Water quality of an aquatic system is affected by suspended sediments, the concentration of dissolved organic matter (DOM), chlorophyll, carotenoids, pathogens, aquatic vascular plants, pesticides, metals, nutrients, thermal releases, and oils (Ritchie et al., 2003; Meraj et al., 2018). Remote sensing techniques assist in the monitoring and management of water quality and GIS makes the possible linking of water quality monitoring in space and time along with data collection, processing, and decision making (Ramadas and Samantaray, 2018). Srinivasa Rao and Jugran (2003) used IRS ID (LISS III), ILWIS, and Landsat 5 TM for the preparation of groundwater quality maps on the basis of TDS, hardness, and magnesium concentrations. Some researchers studied the deterioration of water quality using (LULC) land use land cover(Asadi et al., 2007; Singh et al., 2011; Farooq and Muslim, 2014; Nathawat et al., 2010; Kumar et al., 2018; Joy et al 2019). Srivastava and coauthors (2013) utilize the IRS 1D LISS III satellite imagery for the mapping of groundwater quality and concluded that temporal changes in underground water quality can be predicted with the help of these imageries. Babiker and coauthors (2004) performed the analysis of groundwater using GIS and observed a correlation between nitrate concentrations and land use vegetable cultivation.

Nero and Saju Simon (2015) wrote a review on the role of GIS and remote sensing for water quality assessment and conclude that these tools provide extra information in comparison to other tools and help in the tracing of contamination in water sources. Shallow water bodies are more susceptible to anthropogenic pollution (Farag El-Gamal, 2011). A small change in such type of susceptible water system will result in the death of aquatic animals and plants (Gholizadeha et al., 2016). Therefore, there is a need for continuous and regular monitoring of such types of water bodies for the maintenance of a healthy ecosystem. With the help of remote sensing technology, all the water quality parameters cannot be quantified. The parameters which can be quantified with the help of remote sensing are turbidity, the concentration of chlorophyll, color dissolved organic matter, and suspended inorganic sediment (Doerffer et al., 1999; Singh et al., 2017b; Kanga et al. 2020a, 2020b). Farag and El-Gamal (2011) analyzed the pollution status of Lake Burullus with the help of remote sensing and an in situ monitoring program.

Remote sensing and GIS along with an in itu monitoring program and WQI are more effective and sustainable methods of water quality monitoring (Usali and Ismail, 2010; Rashid et al., 2013; Singh et al., 2017a), but a very less number of studies were observed on the combination of RS and GIS. The process of water quality monitoring with the help of remote sensing was observed in the early 1970s (Gholizadeha et al., 2016; Usali and Ismail, 2010; Alparslan et al., 2007; Brando and Dekker, 2003; El-Din et al., 2013; Giardino et al., 2014; Koponen et al., 2002; Maillard and Santos, 2008; Pozdnyakov et al., 2005; Wang and Ma, 2001). Bhutiani and coauthors (2018) assessed the quality of the Malin River using WQI and concluded that the river was highly polluted due to organic waste. Kumar and coauthors (2020) studied the water quality of the uranium mining area using WQI and GIS and concluded that about half (42%) of the water sample was in very bad quality while GIS studies reflect that distribution of elements in the area is uneven. Rawat and Singh (2018) studied the groundwater quality using WQI and GIS and concluded that spatial distribution maps prepared with the help of GIS using WQI data provide the overall view of the water quality in the study area, and the integration of GIS and WQI provides the reliable information for the decision makers and policy makers (Singh et al., 2013b). Verma and coauthors (2013) performed the assessment of groundwater quality with the help of GIS and WQI and confirm the correlation between water use and LULC.

ASSESSMENT OF AIR QUALITY USING AQI, REMOTE SENSING, AND GIS

In the past, air pollution was a big issue in western industrial nations but now, air pollution becomes a crucial problem in Asian countries. In India, air pollution in urban areas especially in metro cities is a big issue due to increasing transport sector, industrial sector, and the agricultural sector. Among all the parameters of air quality, the main focus is on PM2.5 and further smaller size particles. Due to their small size, they penetrate through the respiratory system causing damage to lungs and other body organs along with the respiratory system. The main problem in the monitoring of air pollution with the help of remote sensing and GIS is the atmospheric effects due to alteration by aerosols (Lim et al., 2009; Kaufman et al., 1990). There is a lack of facts and concepts in the literature regarding the optical properties of aerosols (Hadjimitsis et al., 2004). The remedial measures of atmospheric effects are given in the remote sensing and GIS section of this chapter.

Remote sensing and GIS provide a complete picture of the study area including sources of pollution and their distribution (Narashid et al., 2010). This information obtained with the help of remote sensing and GIS simultaneously helps in the planning stage for the reduction of pollution and reduces the time consumed, efforts in a smart way, and also reduces the monetary consumption. The use of GIS has increased to a large extent in the last decade for inventory, analysis, understanding, modeling, and management in combination with the statistical interpolation method of the natural environment GIS-based approach for the estimation of pollutants (Carletti et al., 2000; Collins et al., 1995; Denby et al., 2005; Tayanc, 2000).

Atmospheric air pollution study was carried out by Lim and coauthors (2009) using Landsat TM 5.0 and concluded that the technique was reliable and cost-effective for environmental studies. Bhabesh and coauthors (2007) perform air quality monitoring using AQI and GIS. Jensen and coauthors (2001) studied air quality management and health impact using the GIS tool and also assessed the traffic air pollution. Elbir and coauthors (2009) developed the GIS-based decision-making tool for air quality management using dispersion modeling CALMET/CALPUFF and concluded that the system provides easy access to the software in air quality determination. Kumar and coauthors (2016) analyzed the air quality of Mumbai using ArcGIS and observed the satisfactory performance of the software on comparing with in situ data. Verma and Singh (2014) performed the seasonal mapping of the air quality using AQI and concluded that the air quality of Lucknow varies between very clean to poor category. In the development of AQI, equal importance must be given to all the concerned parameters (Bhabesh et al., 2007).

18.4 CONCLUSION

The present study was carried out to explore the use of WQI, AQI, remote sensing, and GIS for the assessment of environmental quality with the help of a literature survey. It was observed that the use of these methodologies for the assessment of the quality of the environment was started in the early 1970s. Still there is a lack of suitable information in the literature regarding the use of all these platforms. A lot of literature was available showing the use of indexing for the assessment of air and water quality, but exact guidelines and methodologies are not available yet. The use of remote sensing and GIS provides precious information on air and water quality. With the help of quality indexing, remote sensing, and GIS important findings were shown by several researchers. The authors concluded that the use of remote sensing and GIS along with quality indexing are the most suitable methodologies to know the sources of pollution in air and water with reference to land use land cover pattern, but there is a need to explore these methodologies to a large extent to make them easy to understand and accessible for the use of common peoples.

KEYWORDS

- **environmental quality indices**
- **water quality index**
- **air quality index**
- **remote sensing**
- **geographic information system**
- **remote sensor**

REFERENCES

Abbasi, T.; Abbasi, S. A. *Water Quality Indices*; Elsevier Publication, 2012; p 4.

Alparslan, E.; Aydoner, C.; Tufekci, V.; Tufekci, H. Water Quality Assessment at Omerli Dam Using Remote Sensing Techniques. *Environ. Monitor. Assess.* **2007,** *135* (1–3), 391–398. DOI: 10.1007/s10661-007-9658-6.

Asadi, S. S.; Rajani, G.; Reddy, M. A. Analysis and Interpretation of Groundwater Contamination Using Remote Sensing and GIS: A Case Study. *Curr. World Environ.* **2007,** *2* (2), 115–126.

Babiker, I. S.; Mohamed, M. A.; Terao, H.; Kato, K.; Ohta, K. Assessment of Groundwater Contamination by Nitrate Leaching from Intensive Vegetable Cultivation Using Geographical Information System. *Environ. Int.* **2004,** *29* (8), 1009–1017.

Bhabesh, C. S.; Bashab, N. M.; Kalyan, S.; Paul, P. R.; Singh, G. Geo-Environmental Quality Assessment in Jharia Coalfield, India, Using Multivariate Statistics and Geographic Information System. *Environ. Geol.* **2007,** *51*, 1177–1196.

Bhargava, D. S. Use of Water Quality Index for River Classification and Zoning of Ganga River. *Environ. Pollut. Ser. B Chem. Phys.* **1983,** *6* (1), 51–67. DOI: 10.1016/0143-148X(83)90029-0

Bhutiani, R.; Ahamad, F.; Tyagi, V.; Ram, K. Evaluation of Water Quality of River Malin Using Water Quality Index (WQI). *Environ. Conserv. J.* **2018,** *19* (1&2), 191–201.

Blanchard, B. J.; Leamer, R. W. Spectral Reflectance of Water Containing Suspended Sediment. In *Remote Sensing and Water Resource Management*; Thomson, K. P. B., Lane, R. K., Csallay, S. C., Eds.; American Water Resources Association: Urbana, IL, 1973; pp 339–347.

Bordalo, A. A.; Nilsumranchit, W.; Chalermwat, K. Water Quality and Uses of the Bangpakonk River (Eastern Thailand). *Water Res.* **2001,** *35* (15), 3635–3642.

Boyacioglu, H. Development of a Water Quality Index Based on a European Classification Scheme. *Water SA* **2007,** *33* (1), 101–106.

Brando, V. E.; Dekker, A. G. Satellite Hyperspectral Remote Sensing for Estimating Estuarine and Coastal Water Quality. *IEEE Trans. Geosci. Remote Sens.* **2003,** *41* (6), 1378–1387. DOI: 10.1109/TGRS.2003.812907.

Brezonik, P.; Menken, K. D.; Bauer, M. Landsat-Based Remote Sensing of Lake Water Quality Characteristics, Including Chlorophyll and Colored Dissolved Organic Matter (CDOM). *Lake Reserv. Manage.* **2005,** *21* (4), 373–382. DOI: 10.1080/07438140509354442.

Brown, R. M.; McClelland, N. I.; Deininger, R. A.; Tozer, R. G. A Water Quality Index: Do We Dare? *Water Sewage Works* **1970,** *117*, 339–343.

Burrough, P.; McDonnell, R. *Principles of Geographical Information Systems*; Oxford University Press Inc.: New York, 1998.

Carletti, R., Picci, M.; Romano, D. Kriging and Bilinear Methods for Estimating Spatial Pattern of Atmospheric Pollutants. *Environ. Monitor. Assess.* **2000,** *63*, 341–359.

Collins, S.; Smallbone, K.; Briggs, D. A GIS Approach to Modelling Small Area Variations in Air Pollution within a Complex Urban Environment. In *Innovations in GIS 2*; Fisher, P., Ed.; Taylor & Francis: London, 1995; pp 245–253.

Cude, C. G. Oregon Water Quality Index a Tool for Evaluating Water Quality Management Effectiveness. *J. Am. Resour. Assoc.* **2001,** *37* (1), 125–137. DOI: 10.1111/j.1752-1688.2001.tb05480.x

Debels, P.; Figueroa, R.; Urrutia, R.; Barra, R.; Niell, X. Evaluation of Water Quality in the Chillaan River (Central Chile) Using Physicochemical Parameters and a Modified Water Quality Index. *Environ. Monitor. Assess.* **2005,** *110*, 301–322. DOI: 10.1007/s10661-005-8064-1

Dekker, A. G.; Peters, S. W. M. The Use of the Thematic Mapper for the Analysis of Eutrophic Lakes: A Case Study in the Netherlands. *Int. J. Remote Sens.* **1993,** *14* (5), 799–821.

Dekker, A. G.; Vos, R. J.; Peters, S. W. M. Analytical Algorithms for Lake Water TSM Estimation for Retrospective Analyses of TM and SPOT Sensor Data. *Int. J. Remote Sens.* **2002,** *23,* 15–35.

Denby, B.; Walker, S. E.; Horálek, O. J.; Eben, K.; Fiala, J. Interpolation and Assimilation Methods for European Scale Air Quality Assessment and Mapping Part I: Review and Recommendations. European Topic Centre on Air and Climate Change Technical Paper 7, 2005.

Doerffer, R.; Sorensen, K.; Aiken, J. MERIS Potential for Coastal Zone Applications. *Int. J. Remote Sens.* **1999,** *20,* 1809–1818.

Duan, W.; He, B.; Takara, K.; Luo, P.; Nover, D.; Sahu, N.; Yamashiki, Y. Spatiotemporal Evaluation of Water Quality Incidents in Japan between 1996 and 2007. *Chemosphere* **2013a,** *93* (6), 946–953. DOI: 10.1016/j.chemosphere.2013.05.060.

Duan, W.; Takara, K.; He, B.; Luo, P.; Nover, D.; Yamashiki, Y. Spatial and Temporal Trends in Estimates of Nutrient and Suspended Sediment Loads in the Ishikari River, Japan, 1985 to 2010. *Sci. Total Environ.* **2013b,** *461–462,* 499–508. DOI: 10.1016/j.scitotenv.2013.05.022.

Dunnette, D. A. *Oregon Water Quality Index Staff Manual*; Oregon Department of Environmental Quality: Portland, OR, 1980.

Elbir, T.; Mangir, N.; Kara, M.; Simsir, S.; Eren, T.; Ozdemir, S. Development of a GIS-Based Decision Support System for Urban Air Quality Management in the City of Istanbul. *Atmos. Environ.* **2009,** *44,* 441–454.

El-Din, M. S.; Gaber, A.; Koch, M.; Ahmed, R. S.; Bahgat, I. Remote Sensing Application for Water Quality Assessment in Lake Timsah, Suez Canal, Egypt. *J. Remote Sens. Technol.* **2013,** 3, 61.

Farag, H.; El-Gamal, A. Assessment of the Eutrophic Status of Lake Burullus (Egypt) Using Remote Sensing. *Int. J. Environ. Sci. Eng. (IJESE)* **2011,** *2,* 61–74.

Farooq, M.; Muslim, M. Dynamics and Forecasting of Population Growth and Urban Expansion in Srinagar City-A Geospatial Approach. *Int. Arch. Photogrammetry Remote Sens. Spatial Inf. Sci.* **2014,** *40* (8), 709.

Forster, B. C. Derivation of Atmospheric Correction Procedures for Landsat MSS with Particular Reference to Urban Data. *Int. J. Remote Sens* **1984,** *5* (5), 799–817.

Gholizadeha, M. H.; Melesseb, A. M.; Reddia, L. Spaceborne and Airborne Sensors in Water Quality Assessment. *Int. J. Remote Sens.* **2016,** *37* (14), 3143–3180.

Giardino, C.; Bresciani, M.; Cazzaniga, I.; Schenk, K.; Rieger, P.; Braga, F.; Matta, E.; Brando, V. E. Evaluation of Multi-Resolution Satellite Sensors for Assessing Water Quality and Bottom Depth of Lake Garda. *Sensors* **2014,** *14* (12), 24116–24131.

Gitelson, A.; Yacobi, Y.; Karniell, A.; Kress, N. Remote Estimation of Chlorophyll Concentration in Polluted Marine Waters in Haifa Bay, Southeastern Mediterranean. In *Proceedings of SPIE,* 1995; pp 44–54.

Ha'kanson, B. G.; Moberg, M. The Algal Bloom in the Baltic during July and August 1991, as Observed from NOAA Weather Satellites. *Int. J. Remote Sens.* **1994,** *15,* 963–965.

Hadjimitsis, D. G. The Application of Atmospheric Correction Algorithms in the Satellite Remote Sensing of Reservoirs; PhD Thesis, University of Surrey, School of Engineering in the Environment, Department of Civil Engineering, Guildford, UK, 1999.

Hadjimitsis, D. G.; Clayton, C. R. I. Determination of the Aerosol Optical Thickness and Assessment of Atmospheric Conditions Using Satellite Image-Based Processing Algorithm and Radiative Transfer Theory. In *Proceedings 7th Pan-hellenic (International) Conference of Meteorology, Climatology and Atmospheric Physics*; University of Cyprus, Nicosia-Cyprus, 2004.

Hadjimitsis, D. G.; Clayton, C. R. I.; Hope, V. S. An Assessment of the Effectiveness of Atmospheric Correction Algorithms through the Remote Sensing of Some Reservoirs. *Int. J. Remote Sens.* **2004,** *25* (18), 3651–3674.

Hadjimitsis, D. G.; Retalis A.; Clayton, C. R. I. The Assessment of Atmospheric Pollution Using Satellite Remote Sensing Technology in Large Cities in the Vicinity of Airports. *Water, Air Soil Pollut.* **2002,** *2* (5–6), 631–640.

Hadjimitsis, D. G.; Themistocleous, K. Applications of Satellite Remote Sensing and Sun-photometers to Urban Air-Quality Monitoring: Potential Solutions to Cyprus. *Conference Paper* 2015.

Hadjimitsis, D. G.; Themistocleous, K.; Vryonides, P.; Toulios, L.; Clayton, C. R. I. Applications of Satellite Remote Sensing & GIS to Urban Air-Quality Monitoring: Potential Solutions and Suggestions for the Cyprus Area. In *Proceedings of the 6th International Conference on Urban Air Quality*; Sokhi, R. S., Neophytou, M.; Eds.; Limassol, Cyprus, March 27–29, 2007a. CD-disk: ISBN 978-1-905313-46-4; University of Hertfordshire and University of Cyprus; p 144.

Hadjimitsis, D. G.; Themistocleous, K.; Vryonides, P.; Toulios, L.; Clayton, C. R. I. Satellite-Derived Determination of Aerosol Optical Thickness for Air Pollution Monitoring: A Simplified Algorithm. In *Proceedings of the 6th International Conference on Urban Air Quality*; Sokhi, R. S., Neophytou, M.; Eds.; Limassol, Cyprus, March 27–29, 2007b. CD-disk: ISBN 978-1-905313-46-4; University of Hertfordshire and University of Cyprus; p 143.

Harrington, J. A.; Schiebe, F. R.; Nix, J. F.; Remote Sensing of Lake Chicot, Arkansas: Monitoring Suspended Sediments, Turbidity and Secchi Depth with Landsat MSS. *Remote Sens. Environ.* **1992,** *39* (1), 15–27.

Horton, R. K.; An Index Number System for Rating Water Quality. *J. Water Pollut. Control Federation* **1965,** *37* (3), 300–306.

Jensen, S. S.; Berkowicz, R.; Hansen, H.; Hertel, O. A Danish Decision-Support GIS Tool for Management of Urban Air Quality and Human Exposures. *Trans. Res. Part D Trans. Environ.* **2001,** *6* (4), 229–241.

Jerlov, N. G.; *Marine Optics*; Elsevier: Amsterdam, The Netherlands, 1976; 194 p.

Joy, J.; Kanga, S.; Singh, S. K. Kerala Flood 2018: Flood Mapping by Participatory GIS Approach, Meloor Panchayat. *Int. J. Emerg. Techn* **2019,** *10* (1), 197–205.

Kanaroglou, P. S.; Soulakellis, N. A.; Sifakis, N. I. Improvement of Satellite Derived Pollution Maps with the Use of a Geostatistical Interpolation Method. *J. Geog. Syst.* **2002,** *4*, 193–208.

Kanga, S. et al. Modeling the Spatial Pattern of Sediment Flow in Lower Hugli Estuary, West Bengal, India by Quantifying Suspended Sediment Concentration (SSC) and Depth Conditions using Geoinformatics. *Appl. Comput. Geosci.* **2020a,** 100043.

Kanga, S.; Sheikh, A. J.; Godara, U. GIscience for Groundwater Prospect Zonation. *J. Crit. Rev.* **2020b,** *7* (18), 697–709.

Kannel, P. R.; Lee, S.; Lee, Y. S.; Kanel, S. R.; Khan, S. P. Application of Water Quality Indices and Dissolved Oxygen as Indicators for River Water Classification and Urban Impact Assessment. *Environ. Monitor. Assess.* **2007**, *132* (1–3), 93–110. DOI: 10.1007/s10661-006-9505-1

Kaufman, Y. J.; Fraser, R. S.; Ferrare, R. A. Satellite Measurements of Large-Scale Air Pollution Methods. *J. Geophys. Res.* **1990,** *95*, 9895–9909.

Ketata, M.; Gueddari, M.; Bouhlila, R. Use of Geographical Information System and Water Quality Index to Assess Groundwater Quality in El Khairat Deep Aquifer (Enfidha, Central East Tunisia). *Arab. J. Geosci.* **2011,** 1–12.

Kirk, J. T. O. *Light and Photosynthesis in Aquatic Ecosystems*; Cambridge University Press: Cambridge, UK, 1983; 509 p.

Koponen, S.; Pulliainen, J.; Kallio, K.; Hallikainen, M. Lake Water Quality Classification with Airborne Hyperspectral Spectrometer and Simulated MERIS Data. *Remote Sens. Environ.* **2002,** *79* (1), 51–59.

Kumar, A.; Gupta, I.; Brandt, J.; Kumar, R.; Dikshit, A. K.; Patil, R. S. Air Quality Mapping Using GIS and Economic Evaluation of Health Impact for Mumbai City, India. *J. Air Waste Manage. Assoc.* **2016,** *66* (5), 470–481.

Kumar, M. P.; Nagalakshmia, K.; Jayarajub, N.; Prasada T. L.; Lakshmanna, B. Deciphering Water Quality Using WQI and GIS in Tummalapalle Uranium Mining Area, Cuddapah Basin, India. *Water Sci.* **2020**, *34* (1), 65–74.

Kumar, S. et al. Delineation of Shoreline Change along Chilika Lagoon (Odisha), East Coast of India Using Geospatial Technique, 2018.

Lillesand, T. M.; Kiefer, R. W. *Remote Sensing and Image Interpretation*, 3rd ed.; John Wiley and Sons: New York, 1994.

Lim, H. S.; Matjafri, M. Z.; Abdullah, K.; Wong, C. J. Air Pollution Determination Using Remote Sensing Technique. *Adv. Geosci. Remote Sens.*, Jedlovec, G., Ed.. ISBN: 978-953-307- 005-6, **2009**, InTech, http://www.intechopen.com/books/advances-in-geoscience-and-remotesensing/ air-pollution-determination-using-remote-sensing-technique

Liou, S. M.; Lo, S. L.; Wang, S. H. A Generalized Water Quality Index for Taiwan. *Environ. Monitor. Assess.* **2004,** *96* (40603), 35–52.

Liu, Y.; Islam, M.; Gao, J. Quantification of Shallow Water Quality Parameters by Means of Remote Sensing. *Progress Phys. Geog.* **2003,** *27* (1), 24–43.

Maillard, P.; Santos, N. A. P. A Spatial-Statistical Approach for Modeling the Effect of Non-Point Source Pollution on Different Water Quality Parameters in the Velhas River Watershed-Brazil. *J. Environ. Manage.* **2008,** *86* (1), 158–170.

Meraj, G. et al. Geoinformatics Based Approach for Estimating the Sediment Yield of the Mountainous Watersheds in Kashmir Himalaya, India. *Geocarto Int.* **2018,** *33* (10), 1114–1138.

Meraj, G.; Singh, S. K.; Kanga, S.; Islam, M. N. Modeling on Comparison of Ecosystem Services Concepts, Tools, Methods and Their Ecological-Economic Implications: A Review. *Model. Earth Syst. Environ.* **2021,** *2021*, 1–20.

Narashid, R. H.; Wan, M. N. Air Quality Monitoring Using Remote Sensing and GIS Technologies. In *International Conference on Science and Social Research (CSSR 2010)*, December 5–7, 2010, Kuala Lumpur, Malaysia, 2010.

Nathawat, M. S. et al. Monitoring & Analysis of Wastelands and Its Dynamics Using Multiresolution and Temporal Satellite Data in Part of Indian State of Bihar. *Int. J. Geomat. Geosci.* **2010,** *1* (3), 297–307.

Nero, A.; SajuSimon S. G.; Application of Remote Sensing and GIS in Water Quality Assessment—Review Paper. *Int. J. Appl. Eng. Res.* **2015,** *10* (59), 60–67.

Novo, E. M. M.; Hansom, J. D.; Curran, P. J. The Effect of Sediment Type on the Relationship between Reflectance and Suspended Sediment Concentration. *Int. J. Remote Sens.* **1989,** *10* (7), 1283–1289.

Pesce, S. F.; Wunderlin, D. A. Use of Water Quality Indices to Verify the Impact of Cordoba City (Argentina) on Suquıa River. *Water Res.* **2000,** *34* (11), 2915–2926.

Pozdnyakov, D.; Shuchman, R.; Korosov, A.; Hatt, C. Operational Algorithm for the Retrieval of Water Quality in the Great Lakes. *Remote Sens. Environ.* **2005,** *97* (3), 352–370.

Ramadas, M.; Samantaray, A. K. Applications of Remote Sensing and GIS in Water Quality Monitoring and Remediation: A State-of-the-Art Review. In *Water Remediation, Energy, Environment, and Sustainability*; Bhattacharya, S. et al., Eds.; 2018; pp 225–244.

Rashid, I. et al. Assessing the Impact of Anthropogenic Activities on Manasbal Lake in Kashmir Himalayas. *Int. J. Environ. Sci.* **2013,** *3* (6), 2036–2047.

Rawat, K. S.; Singh, S. K. Water Quality Indices and GIS-Based Evaluation of a Decadal Groundwater Quality. *Geol. Ecol. Landscapes* **2018.** https://doi.org/10.1080/24749508.2018.1452462

Ritchie, J. C.; Charles, M. C. Comparison of Measured Suspended Sediment Concentration Estimated from Landsat MSS Data. *Int. J. Remote Sens.* **1996,** *9* (3), 379–387.

Ritchie, J. C.; McHenry, J. R.; Schiebe F. R.; Wilson, R. B. The Relationship of Reflected Solar Radiation and the Concentration of Sediment in the Surface Water of Reservoirs. In *Remote Sensing of Earth Resources Vol. III*; Shahrokhi, F., Ed.; The University of Tennessee Space Institute: Tullahoma, TN, 1974; pp 57–72.

Ritchie, J. C.; Schiebe, F. R.; McHenry, J. R.; Remote Sensing of Suspended Sediment in Surface Water. *Photographic Eng. Remote Sens.* **1976,** *42*, 1539–1545.

Ritchie, J. C.; Schiebe, F. R.; McHenry, J. R.; Wilson, R. B.; May, J. Sun Angle, Reflected Solar Radiation and Suspended Sediments in North Mississippi Reservoirs. In *Remote Sensing of Earth Resources Vol. IV*; Shahrokhi, F., Ed.; University of Tennessee Space Institute: Tullahoma, TN, 1975; pp 555–564.

Ritchie, J. C.; Zimba, P. V.; Everitt, J. H. Remote Sensing Techniques to Assess Water Quality. *Photogrammetric Eng. Remote Sens.* **2003,** *69* (6), 695–704.

Said, A.; Stevens, D. K.; Sehlke, G. An Innovative Index for Evaluating Water Quality in Streams. *Environ. Manage.* **2004,** *34* (3), 406–414.

Sarkar, C.; Abbasi, S. A. Qualidex—A New Software for Generating Water Quality Indice. *Environ. Monitor. Assess.* **2006,** *119*, 201–231.

Schalles, J. F.; Gitelson, A. A.; Yacobi, Y. Z.; Kroenke, A. E. Estimation of Chlorophyll a from Time Series Measurements of High Spectral Resolution Reflectance in an Eutrophic Lake. *J. Physiol.* **1998,** *34*, 383–390.

Schiebe, F. R.; Harrington, J. A.; Ritchie, J. C. Remote Sensing of Suspended Sediments: The Lake Chicot, Arkansas Project. *Int. J. Remote Sens.* **1992,** *13* (8), 1487–1509.

Sharma, D.; Arun, K. Water Quality Analysis of River Yamuna Using Water Quality Index in the National Capital Territory, India. *Appl. Water Sci.* **2011,** *1* (3–4), 147–157. DOI: 10.1007/s13201-011-0011-4

Siegel, H.; Gerth, M. Remote-Sensing Studies of the Exceptional Summer of 1997 in the Baltic Sea: The Warmest August of the Century, the Oder Flood, and Phytoplankton Blooms. In *Satellites, Oceanography and Society*; Halpern, D., Ed.; Elsevier Science, 2000; pp 239–255.

Singh, C. K.; Shashtri, S.; Mukherjee, S.; Kumari, R.; Avatar, R.; Singh, A.; Singh, R. P. Application of GWQI to Assess Effect of Land Use Change on Groundwater Quality in Lower Shiwaliks of Punjab: Remote Sensing and GIS Based Approach. *Water Resour. Manage.* **2011,** *25* (7), 1881–1898.

Singh, S. K.; Srivastava, P. K.; Pandey, A. C. Fluoride Contamination Mapping of Groundwater in Northern India Integrated with Geochemical Indicators and GIS. *Water Sci. Technol. Water Supply* **2013a,** *13* (6), 1513–1523.

Singh, S. K.; Srivastava, P. K.; Pandey, A. C.; Gautam, S. K. Integrated Assessment of Groundwater Influenced by a Confluence River System: Concurrence with Remote Sensing and Geochemical Modelling. *Water Resour. Manage.* **2013b,** *27* (12), 4291–4313. DOI: 10.1007/s11269-013-0408-y

Singh, S. K.; Kanga, S. Role of Geoinformatics in Site Suitability Analysis of Infrastructures using PRA Approach. *Am. Int. J. Res. Sci. Technol. Eng. Math.* **2017a,** *1* (18), 81–85.
Singh, S. K.; Mishra, S. K.; Kanga, S. Delineation of Groundwater Potential Zone Using Geospatial Techniques for Shimla City, Himachal Pradesh (India). *Int. J. Sci. Res. Dev.* **2017b,** *5* (4), 225–234.
Srinivasa Rao, Y.; Jugran, D. K. Delineation of Groundwater Potential Zones and Zones of Groundwater Quality Suitable for Domestic Purposes Using Remote Sensing and GIS. *Hydrol. Sci. J.* **2003,** *48* (5), 821–833.
Srivastava, P. K.; Singh, S. K.; Gupta, M.; Thakur, J. K.; Mukherjee, S. Modeling Impact of Land Use Change Trajectories on Groundwater Quality Using Remote Sensing and GIS. *Environ. Eng. Manage. J. (EEMJ)* **2013**, *12*, 12.
Svejkovsky, J.; Shandley, J. Detection of Offshore Plankton Blooms with AVHRR and SAR Imagery. *Int. J. Remote Sens.* **2001,** *22*, 471–485.
Tayanc, M. An Assessment of Spatial and Temporal Variation of Sulphur Dioxide Levels over Istanbul, Turkey. *Environ. Pollut.* **2000,** *107* (1), 61–69.
Thomas, M. L.; Ralph, W. K.; Jonathan, W. C. *Remote Sensing and Image Interpretation*, 5th ed.; United States of America, 2004; p 461.
Tiwari, A. K.; Singh, P. K.; Mahato, M. K. GIS Based Evaluation of Water Quality Index of Groundwater Resources in West Bokaro Coalfield, India. *Curr. World Environ.* **2014**, *9* (3), 843–850. DOI: 10.12944/CWE.9.3.35.
Usali, N.; Ismail, M. H. Use of Remote Sensing and GIS in Monitoring Water Quality. *J. Sustain. Dev.* **2010,** *3* (3), 228–238.
Verma, A.; Singh, S. N. Seasonal Dynamics in Air Quality Index and Its Role in Pollution Mapping. *J. Ecophysiol. Occupational Health* **2014,** *14* (1 & 2), 32–47.
Verma, A.; Thakur, B.; Katiyar, S.; Singh, D.; Rai, M. Evaluation of Ground Water Quality in Lucknow, Uttar Pradesh Using Remote Sensing and Geographic Information System (GIS). *Int. J. Water Resour. Environ. Eng.* **2013,** *5*, 67–76.
Wald, L.; Basly, L.; Balleynaud, J. M. Satellite Data for the Air Pollution Mapping. Proceedings of the 18th EARseL Symposium on Operational Sensing for Sustainable Development (Enschede, Netherlands, May 11–14, 1998), In *Operational Remote Sensing for Sustainable Development*; Nieeuwenhuis, G. J. A., Vaugham, R. A., Molenaar, M., Eds.; **1999**; pp133–139.
Wang, F.; Han, L.; Kung, H. T.; Van Arsdale, R. B. Applications of Landsat-5 TM Imagery in Assessing and Mapping Water Quality in Reelfoot Lake, Tennessee. *Int. J. Remote Sens.* **2006,** *27* (23), 5269–5283. DOI: 10.1080/01431160500191704.
Wang, X. J.; Ma, T. Application of Remote Sensing Techniques in Monitoring and Assessing the Water Quality of Taihu Lake. *Bull. Environ. Contamin. Toxicol.* **2001,** *67* (6), 863–870.
Zhang, Y.; Pulliainen, J.; Koponene, S.; Hallikainen, M. Application of Empirical Neural Network to Surface Water Quality Estimation in the Gulf of Finland Using Combined Optical Data and Microwave Data. *Remote Sens. Environ.* **2002,** *81* (2–3), 327–336.

CHAPTER 19

Groundwater Contamination Control for Sustainable Water Resource Management: A Short Review

S. KARTHIGAYINI[1*] and S. SAHANA[2]

[1]*Department of Agricultural Engineering and Soil Science, Rajarata University of Sri Lanka, Sri Lanka*

[2]*Department of Biosystems Technology, Eastern University of Sri Lanka, Sri Lanka*

[*]*Corresponding author. E-mail: ammusacthivel@gmail.com*

ABSTRACT

Water is a vital factor for all living beings in the world. It exists in different forms and locations, such as in the air, on the surface, below the ground, and in the oceans. Water that is available below the ground is known as the groundwater. About only 2.5% of the Earth's water is freshwater, and mostly is frozen in glaciers and ice sheets. Among liquid freshwater, about 96% present in the groundwater and small remaining fraction is on the surface and in the air. So, the protection of groundwater from contamination is a must for water sustainability. Therefore, this chapter aims to provide some understanding about the groundwater protection and contamination control. It includes the occurrence of groundwater, storage forms of it, characteristics of groundwater, influence of discharge on its quality. Its dynamic and equilibrium, pollution factors, contamination survey, and the pollution remediation as well. We can conclude from this chapter about the importance of groundwater maintenance by every human beings.

GIScience for the Sustainable Management of Water Resources. Gowhar Meraj, Shruti Kanga, Majid Farooq, Suraj Kumar Singh, & Sudhanshu (Eds.)
© 2023 Apple Academic Press, Inc. Co-published with CRC Press (Taylor & Francis)

19.1 INTRODUCTION

The water on the earth exists in the atmosphere, biosphere, hydrosphere and lithosphere of the earth in different material states. The water that falls to the ground in the atmosphere is usually called atmospheric precipitation. The water in rivers, lakes, and oceans on the surface is called surface water. The water in various states buried in the pores, cracks, and crevice of rocks below the surface is collectively called groundwater. Groundwater is one of the important freshwater sources according for 0.59% of worlds total water and 22% of worlds total freshwater.

19.1.1 OCCURRENCE OF GROUNDWATER

Groundwater exists in the voids between rock and soil. The three types of rocks within 10 km of the surface of the earth's crust have different degrees of underground voids. This provides necessary space conditions for the formation, storage, movement, and development of groundwater. Therefore, studying the distribution and characteristics of voids in the rock soil framework is an important basis for studying groundwater. Rock voids are storage places and movement channels for groundwater.

The number, size, shape, connectivity, and distribution rules of voids all have an important impact on the distribution and movement of groundwater. According to the different origin of rock and soil voids, voids can be divided into three categories.

1. The pores in loose rocks—pore water.
2. Cracks in hard rock—fissure water.
3. Crevices in soluble rocks–karst water (Figure 19.1).

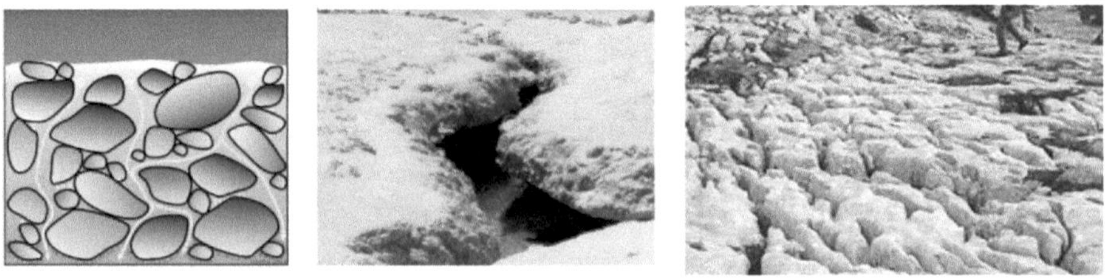

(a) (b) (c)

FIGURE 19.1 (a) Pore water (https://serc.carleton.edu/download/images/150635/diagram_illustrates_porosity_sedime.png). (b) Fissure water. (c) Karst water.

According to the physical state of water and the interaction between water and rock particles, the storage form of groundwater is divided into the following five types.

1. Gaseous water
2. Bound water
3. Gravity water
4. Capillary water
5. Solid water

According to its water permeability, rock formations can be divided into permeable layers and impermeable layers. If the permeable layer is saturated with groundwater, it is known as aquifer. If the permeable layer does not contain water or is not saturated with water, it is called a permeable nonaqueous layer. Therefore, the permeable layer can be an aquifer or a permeable nonaqueous layer. If there is an aquifer, there must be a water barrier. Water impermeable layer refers to aquiclude. Due to gravity under normal pressure conditions, a rock layer that cannot be penetrated and give a certain amount of water or the amount of water that is penetrated and given is insignificant. The impermeable layer hinders the movement of the groundwater. The part above the groundwater surface is called the vadose zone or the unsaturated zone. The part below the groundwater surface is called the saturated zone. Vadose zone is the necessary passage for the saturated zone to connect with the atmosphere and the surface hydrosphere. The liquid water in the saturated zone includes gravity water and bound water. Among them, gravity water is the main object of development, utilization, or elimination.

19.1.2 CHEMICAL CHARACTERISTICS OF GROUNDWATER

It includes various ions, gases, colloids, and organic matter. Almost all elements that exist in nature can be found in groundwater, but the content is different.

1. Main ion components: chloride ion, sulfate ion, hydrogen carbonate ion, sodium ion, potassium ion, calcium ion, and magnesium ion.
2. Gas components: carbon dioxide, oxygen, nitrogen, hydrogen sulfide and methane.
3. Colloidal components: ferric hydroxide, aluminum hydroxide and silica.
4. Organic matter; carbon, hydrogen, oxygen-based polymer compounds.

These mainly depend on the nature and solubility of surrounding rocks. The influence of different chemical components in groundwater on its chemical properties. Its chemical properties mainly include pH, salinity, and hardness of groundwater (Pereira et al., 2012).

pH

Among them, the degree of acidity and alkalinity of groundwater depends on the concentration of hydrogen ions in the water. It is usually expressed as a logarithm and the negative sign is used to obtain the pH value of the water body.

Groundwater can be divided into the following types according to its pH value:

1. Weak acid water
2. Strong acid water
3. Neutral water
4. Weak alkaline water
5. Strong alkaline water

TDS (total dissolved solids)

The total mineralization of groundwater is the total amount of ions, molecules, and compounds contained in the groundwater (Md. Bodrud-Doza et al., 2016). It represents the degree of mineralization of groundwater, and is one of the important indicators for evaluating water quality. According to the total salinity, groundwater can be divided into

1. Freshwater
2. Weak brackish water
3. Strong brackish water
4. Saltwater

Hardness

The hardness of groundwater refers to the total amount of dissolved calcium and magnesium ions in the water. The higher the amounts of calcium and magnesium ions per unit volume of groundwater, the greater the hardness of the water.

Groundwater with excessive hardness is very unfavorable for domestic and industrial use. According to the total hardness of the groundwater, the groundwater can be divided into the following types:

1. Extremely soft water
2. Soft water
3. Weakly hard water
4. Hard water
5. Extremely hard water

The chemical composition of groundwater is mainly affected by the recharge, runoff, and discharge processes of groundwater. These are the three basic links of the groundwater cycle. And are also the most important controlling factors for the formation of groundwater, volume, and quality.

Recharge

Groundwater can be recharged through

1. Atmospheric precipitation
2. Surface water
3. Condensed water
4. Recharge between aquifers
5. Other sources (Passarello et al., 2014)

Atmospheric Precipitation

Water that falls on the ground becomes surface runoff or is consumed by evaporation. Part of the infiltrated water is retained in the vadose zone and part of the water evaporates back to the atmosphere after rain. The rest is the amount of water to recharge the aquifer under normal circumstances, the amount of water infiltrated to recharge the aquifer only accounts for 20–50% of the precipitation.

The main factors affecting the recharge of atmospheric precipitation include

1. Precipitation characteristics
2. Evaporation intensity
3. Lithology and thickness of vadose zone
4. Topography
5. Vegetation coverage

19.1.3 PRECIPITATION CHARACTERISTICS

Among them, precipitation characteristics include rainfall, rainfall intensity, and duration. A considerable part of the precipitation will be used to supplement the water deficit caused by the evaporation of the unsaturated zone during the dry season. Therefore, when the annual precipitation is less than a certain value, there is actually no recharge effect on the groundwater. When the precipitation intensity exceeds the ground infiltration rate, ground runoff will be generated. Drizzle is very beneficial to the recharge of groundwater. Although the precipitation intensity is not large, it lasts for a long time and has a wide distribution area. So, the amount of groundwater recharge is also large.

Evaporation Intensity

The stronger the evaporation intensity, the less favorable it is for groundwater to receive atmospheric precipitation recharge.

19.1.4 LITHOLOGY AND THICKNESS OF VADOSE ZONE

The influence of this is reflected in the vadose zone and has good permeability and is good for absorbing precipitation. The greater the thickness of the aeration zone (diving buried depth), which is not conducive to recharge groundwater, the greater the amount of retained water. But, if the phreatic burial is too shallow and the capillary saturated zone is very close to the ground, the infiltration rate of precipitation will be reduced, and a large amount of it twill be converted into surface runoff, which is not conducive to recharge groundwater.

19.1.5 TOPOGRAPHY

The surface slope will flow away quickly, if the terrain has a large slope while the gentle terrain and local depressions will accumulate the slope increasing the infiltration share of precipitation.

19.1.6 VEGETATION COVERAGE

It is reflected in forests can store precipitation and reduce surface runoff. The soil under the forest is rich in organic matter, the root system is developed, and the crown and fallen leaves can protect the surface soil structure, which is conducive to the infiltration of precipitation. Forests can also increase the amount of precipitation in local areas and help groundwater to get more recharge.

All the above factors are not independent. All are mutually related and analyzed. For example, in mountainous areas with strong rock melting rock formations although the topography is steep and the groundwater level is hundreds of meters deep due to the high permeability of the rock formations, even continuous heavy rain can be fully absorbed, and the share of precipitation recharge groundwater can be as high as 70–90%.

19.1.7 SURFACE WATER

It includes rivers, lakes, reservoirs, and oceans, all of which can become groundwater recharge sources under certain conditions.

Relationship between river and groundwater recharge under normal circumstances, regional river valley deep cut and the river water is lower than the groundwater level all year around. At this time, river water plays a role of draining groundwater, instead of recharging groundwater.

Due to the accumulation of rivers, the riverbed is usually at a high level. In this case, the groundwater is always supplied by the river. In some parts of alluvial plains and basins, the relationship between river water and groundwater changes with seasons. In this case, in some alluvial plains, the riverbed forms the so-called "overground river" due to strong accumulation and groundwater is often recharged by river water.

When river water recharges groundwater, the amount of recharge depends on the length of the permeable riverbed and the perimeter of flooding, the permeability coefficient of the riverbed, the height difference between the river water and the groundwater level, and the water passing time of the riverbed.

The coordinated relationship between the address structure and topography of an area will have a great impact on the recharge of the confined aquifer. The aquifer that exists between the aquifers is a confined aquifer. When the aquifer is exposed on high terrain, it can only be recharged by atmospheric precipitation within the exposed range. When the aquifer is exposed on low terrain, the precipitation in the entire catchment area may collect and supplement the groundwater.

When a region develops a water-conducting fault that cuts through the waterproof of a confined aquifer under favorable topographic conditions, a large area of precipitation can also be introduced into the aquifer for recharge. The size of the catchment area also affects the recharge received by the diving aquifer.

19.1.8 CONDENSED WATER

Condensation is the process by which, as the temperature drops, the absolute humidity of the air exceeds the saturation humidity and the excess of the absolute humidity of the air condenses into liquid water.

In general, the amount of condensed water is quite limited. But for mountains, deserts, and areas with large temperature differences between day and night, condensation water is also one of the main sources of recharge groundwater.

19.1.9 RECHARGE BETWEEN AQUIFERS

It occurs when there is a water head difference between two water linked aquifers and water in the aquifer with a higher water head will be fed to the aquifer with a lower water head through the recharge channel. The supply channels just mentioned are mainly "skylights."

Water-conducting faults, boreholes, and half of water-resisting layer in river alluvial plain and coastal plain areas, the upper part is usually divided into fine sand, medium sand, or coarse sand aquifers and the middle part is separated by a certain thickness of clayey or clay which is not completely watertight but has weak permeability.

When the head difference between two aquifers reaches a certain value, the water in the aquifer with high head will permeate through the weak previous layer to the other aquifer. This process is called overcurrent.

Although the vertical permeability coefficient of the weak previous layer is quite small and the flow rate per unit area is not large, the overflow occurs in the entire distribution range of the weak previous layer and the water cross-section is quite large and the total overflow rate is also quite large when considering a long time scale.

In addition to the above sources, certain human activities may also lead to the recharge of groundwater. Groundwater recharge from surface water, such as reservoirs, ponds, pits, and canals as well as from irrigation, industrial and domestic wastewater discharge and artificial recharging after it is replenished with some chemical water.

Diving in arid areas often forms highly mineralized water due to long-term evaporation and concentration. In those areas where low mineralized water is often supplied, fresh water suitable for drinking can often be found. The recharge of groundwater with high salinity and polluted water will lead to the deterioration of aquifer water quality. Much of this happens under human influence (Ronayne et al., 2017). For example, unreasonable discharge of industrial wastewater and domestic sewage precipitation leaching waste and absorption of waste gas to recharge groundwater, excessive drainage of coastal areas or freshwater aquifers associated with the saltwater layer can also cause the deterioration of water quality by supplying seawater to the freshwater layer.

19.1.10 RUNOFF

It is the process of groundwater flowing from the recharge area to the discharge area. Except in some special circumstances, groundwater is often subject to constant runoff. It transfers the amount of groundwater and salt from the recharge area to the discharge area. Thus, affecting the spatial distribution of the amount of water and water quality in aquifers or water-bearing systems.

The main contents of groundwater runoff research include

1. Runoff direction
2. Runoff intensity
3. Runoff conditions
4. Runoff volume

Groundwater runoff system from the watershed in the interfluve region to the valleys on both sides of the direction of underground runoff has undergone a complex transformation process from top to near level. And then from bottom to top on the watershed of groundwater recharge area.

With the increase of depth, the groundwater head pressure decreases gradually, while in the valley area of groundwater discharge area, the groundwater head pressure shows an opposite trend.

The streamlines from the watershed to the valley became more and more dense, indicating the intensification of underground runoff and the increase of runoff. Even if the entire interfluve block is a homogeneous aquifer. However, in different sections and different depths of the aquifer, the groundwater head value also changes.

For deep wells dug in river valleys, there is no need for a water barrier and artesian or confined water wells can also be dug.

19.1.11 DISCHARGE

It can be carried out by

1. Spring
2. Discharge to the river
3. Evaporation

Spring

The formation of the springs is mainly due to the corrosion of the terrain exposing the aquifer to the surface or groundwater is blocked during the movement. The elevation of the spring can reflect the elevation of the local groundwater level. The distribution of springs reflects the distribution of aquifers or water channels and the location of recharge areas and drainage areas. The chemical composition and physical properties of the spring can reflect the characteristics of local groundwater quality and burial condition.

19.1.12 DISCHARGE TO THE RIVER

When rivers, lakes, oceans etc. erode into the aquifer, the groundwater is dispersed along the perimeter of the surface water circumference, which is called aerial drainage.

The discharge is not as concentrated as the spring. Therefore, usually select a section on the river and regularly measure the flow of the river (Figure 19.2).

- At point A, the flow process line rises. Start to draw a horizontal line intersect at point D of the preceding section.
- The portion below AD = amount of groundwater that is supplied to the river by the discharge.

The river that flows all the year-round, the flow of the river in the dry season is all provided by the groundwater discharge. Therefore, the above-mentioned flow difference method can be used to measure the groundwater discharge flow. However, in the flood season, the river flow is mainly formed by the accumulation of precipitation in the basin may also contain a part of the leakage.

Therefore, the river section method cannot determine the discharge flow during the flood season.

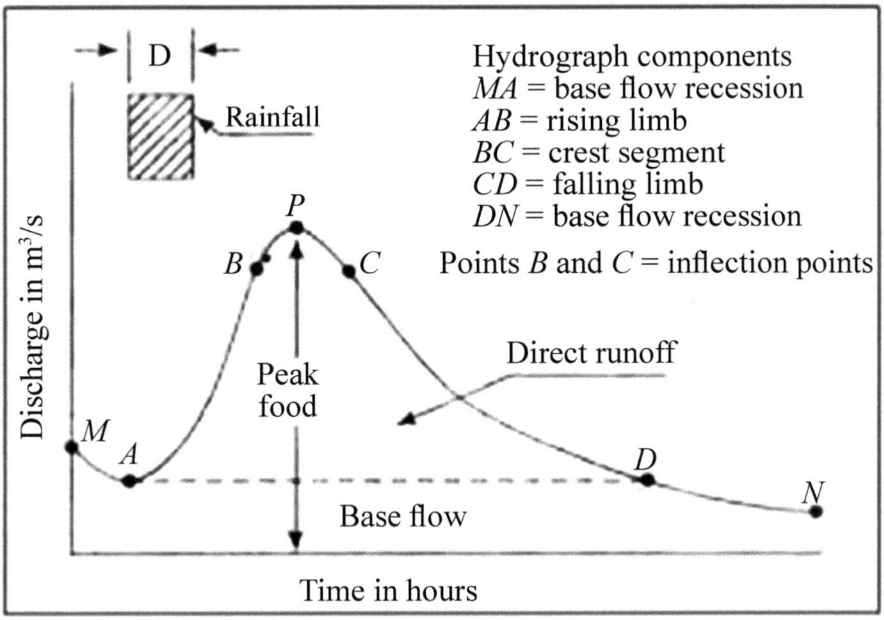

FIGURE 19.2 Flow curve with time.

Evaporation

Process of water dissipating from liquid to gas including soil evaporation, leaf evaporation, water evaporation, soil evaporation, plant evaporation, and groundwater evaporation.

The evaporation discharge of groundwater can be divided into two types:

1. Soil water evaporation—not directly related to the saturated zone.
2. Phreatic water evaporation.

Factors, such as climate, buried depth of phreatic water and lithology of vadose zone, thus determine the salinization degree of soil and groundwater.

19.1.13 INFLUENCE OF DISCHARGE ON GROUNDWATER QUALITY

The process of groundwater discharge also has an impact on the quality of groundwater (McCutcheon et al., 1993). Groundwater discharge can be divided into two categories.

1. Runoff discharge—salt is removed along with the water.
2. Evaporation—salt is not excreted with the amount of water.

According to the characteristics of recharge and discharge, groundwater circulation can be divided into two categories (Figure 19.3)

1. Infiltration through flow type.
2. Infiltration evaporation type.

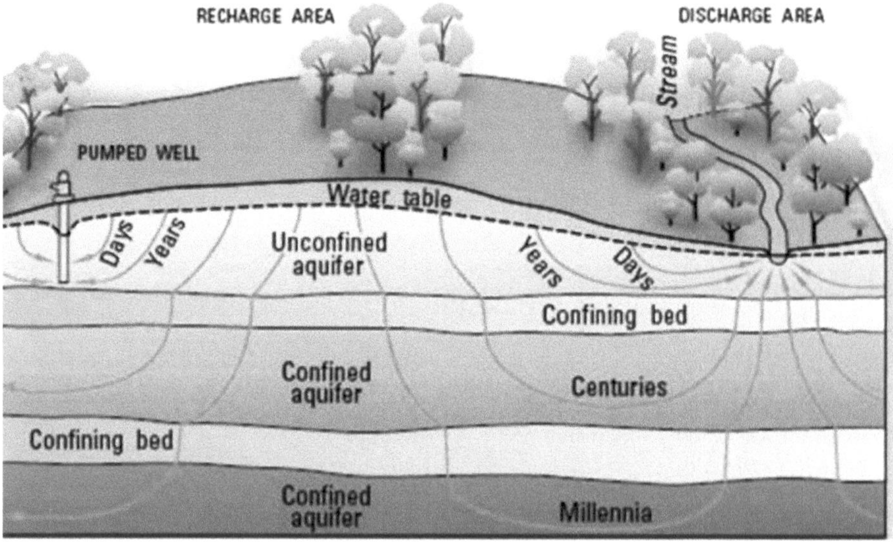

FIGURE 19.3 Aquifer.
Source: Environment and climate change Canada.

The results of the former long-term cycle, it is to make the rock and soil and the groundwater in it develop in the direction of leaching and desalination. The results of long-term circulation of the latter, are desalination and salination of rock soil and groundwater in recharge area. And salination of groundwater and soil in discharge area.

19.2 GROUNDWATER DYNAMICS AND EQUILIBRIUM

It refers to the change of groundwater level, water quantity, water quality, water temperature, and other dynamic factors over time. That is to say, the response of the output information (water quantity, quality, water temperature etc.) of the groundwater system to the input information.

The groundwater system often exchanges material and energy with the outside world, which results in the corresponding change of the states of the system itself. Under the action of external factors, the groundwater system has the function of self-regulation and dynamic balance.

When the external factors change, i will inevitably cause the change of the internal state of the system, which is the destruction of the original balance of the system. The groundwater system will make internal adjustments according to its own functions.

The natural factors affecting groundwater dynamics include

1. Meteorological factors
2. Hydrological factors
3. Geological factors

When the groundwater system is stimulated by input information, such as atmospheric precipitation, surface water, irrigation water or artificial extraction, and natural discharge as a response, the internal state of the groundwater system will change in response and it twill be reflected by output information.

The output information as the response is the groundwater dynamic elements (water level, water volume, water quality, water temperature etc.) and then, the input information as the incentive can be understood as the groundwater equilibrium factors.

When the balance of the groundwater system is destroyed by external action, and exceeds the self-regulation ability of the system, the groundwater system will produce abnormal groundwater level and groundwater pollution. This result is usually caused by natural disasters and human activities.

Then, when the groundwater is polluted, it is necessary to investigate and understand the situation of groundwater pollution before taking remediation measures.

19.3 GROUNDWATER POLLUTION

It means that groundwater is affected by human activities, which causes the quality of groundwater to deteriorate. So, it is no longer suitable for use.

Groundwater pollution on the one hand is greatly affected by human social and economic activities. On the other hand, its pollution poses a serious threat to the ecological environment especially the ecological safety including human life. Therefore, it is of great significance to strengthen research on groundwater pollution.

The substance that causes pollution is called groundwater pollutant. And its source is the pollution source (Bhuiyan et al., 2011).

According to the different causes of pollution sources, they are basically divided into

1. Waste discharge
2. Waste storage
3. Transport
4. Human activities

19.4 POLLUTION SOURCE THAT DISCHARGE WASTE

Such sources include

1.1 Septic tanks
1.2 Deepwater injection wells
1.3 Land use

Septic Tanks

It is used to store domestic sewage. Water in the septic tanks come from toilets, bathing, dishwashing, laundry water, etc. when the sewage enters the septic tank, it undergoes precipitation and anaerobic decomposition discharged into the soil through the drainage system, thus constituting groundwater pollution source.

Deepwater Injection Wells

It is to inject liquid wastes or other liquids into a certain area below the groundwater surface. The injected liquids generally include hazardous wastes, oil well brine, rainwater, municipal sewage, mine wastewater, etc. If the injected wastewater enters the drinking water aquifer, it will cause groundwater pollution.

Land Use

It is reflected in irrigating the land with treated or untreated municipal and industrial sewage or fertilizing the land with sludge or use of pesticides to kill insects will bring many organic pollutants and they become a source of groundwater pollution.

19.4.1 POLLUTION SOURCES FOR STORING OR PROCESSING WASTE

Mainly reflected in the areas, such as landfill, open dump, residential garbage dump, mining waste pit, solid material accumulation place, animal burial site, various explosion test sites, and radioactive waste disposal sites.

These places are long-term accumulation of pollutants and pollutants will pass through the shower, infiltrate into the groundwater system and cause groundwater pollution.

19.4.2 POLLUTION DURING TRANSPORTATION

It is caused by leakage of transportation pipelines or traffic accidents. The pollution sources mentioned bring organic or inorganic biological or nonbiological and radioactive pollutants into the underground environment.

Thereby, modeling chemical pollution, bacterial pollution, and radioactive pollution of groundwater.

19.4.3 POLLUTION SOURCES CAUSED BY HUMAN ACTIVITIES

These pollution sources come from various planned human activities including

1. Agricultural irrigation
2. Use of pesticides
3. Use of fertilizer
4. Use of salt on highways
5. Use of air pollutants
6. Infiltration and mine drainage

These are having a common feature, that is, they pollute groundwater through planned human activities. And also, these pollution sources are usually only indirect pollution sources of groundwater, unless there are direct recharge channels, such as boreholes, natural collapse columns, etc. which generally act on groundwater through vadose zones or surface waters. At the same time, human activities may also produce or aggravate natural pollution sources. Such pollution sources include the interaction of surface water and groundwater. For example, natural leaching, and seawater intrusion.

19.4.4 GROUNDWATER CONTAMINATION SURVEY

Groundwater pollution control is a complicated problem. Even if expensive measures are taken, it can only prevent pollutants from entering the aquifer and spreading in it. Once pollution occurs, it is far more complicated to restrict and restore measures than preventive measures. Therefore, strengthening the investigation and monitoring of groundwater pollution are important measures to prevent problems before they occur (Farnham et al., 2003).

The survey of groundwater pollution generally includes

1. Land use types investigation and research.
2. Sources and routes of pollution.
3. Lithology of vadose zone and groundwater transport characteristics.
4. Change of groundwater and pollution transport characteristics.

19.4.5 LAND USE TYPES

First find out the quantity and regional distribution of various types of land and evaluate the quality and development potential of the land. Then clarify the rationality of the land use structure, reveal the existing problems of the land use. It will provide a basis for making rational use of land resources.

19.4.6 SOURCES AND ROUTES OF POLLUTION

On the basis of land type surveys carrying out pollution source surveys, groundwater pollution sources are usually classified as industrial pollution sources, agricultural pollution sources, domestic pollution sources, and surface polluted water bodies.

According to the pollution source from, it can be divided into the following:

Point pollution source.
Surface pollution source.
Diffuse pollution source.

The way of water pollution is the way that the polluted solution enters the groundwater from the pollution source.

19.5 LITHOLOGY OF VADOSE ZONE AND GROUNDWATER TRANSPORT CHARACTERISTICS

Study the permeability characteristics of the geological structure of the strata and the migration, degradation, and removal process of pollutants in the vadose zone. Find out the lithological composition, thickness, and regional distribution characteristics of the vadose zone, find out the soil type in the key zone of the lithology, thickness, structure, and distribution characteristics of the vadose zone. This is a major factor to detect groundwater dynamics and forecast the development trend of groundwater pollution.

19.6 CHANGE OF GROUNDWATER AND POLLUTION TRANSPORT CHARACTERISTICS

Investigate the impact of change in precipitation, water conservancy projects, land use, etc. on groundwater recharge, investigate the main ways of groundwater discharge and changes in discharge volume, and analyze groundwater flow-field changes and their causes. The regional survey should carry out the unified measurement of groundwater level and conduct a balanced dynamic analysis of the regional groundwater level runoff and discharge conditions. The survey of key areas should be based on a series of dynamic data through mathematical statistics or the establishment of numerical model simulation analysis of groundwater flow-field changes.

Finally, on the basis of the above pollution investigation, develop measures to prevent and eliminate groundwater pollution, comprehensive antiseepage measures, and other engineering measures.

19.7 GROUNDWATER POLLUTION REMEDIATION

Broadly speaking, groundwater pollution remediation, groundwater pollution control, and groundwater pollution prevention are three ways humans deal with pollution.

Groundwater pollution prevention means the pollution treatment methods that use ideal facilities or environmentally friendly materials that do not cause pollution from the source.

Groundwater pollution control refers to the control of pollution caused by human beings in production and life before the pollutants are released into the environment. The pollutants are captured or changed to achieve the purpose of purification.

Groundwater pollution remediation is the pollution caused by humans in production and life, then implement the purification method of pollution treatment. Remediation of groundwater pollution is extremely expensive (Foster, 1995). So when considering about the methods for groundwater restoration, groundwater adsorption technologies can be divided into three major categories.

1. Natural repair method or passive repair method—simplest and cheapest.
2. Application technology is to dig out the soil and transport it to a suitable place for treatment.
3. Groundwater extraction—treatment systems and soil vapor extraction systems.

19.7.1 NATURAL REPAIR OR PASSIVE REPAIR METHOD

It relies on natural process effects including biodegradation, volatilization, and adsorption. The mechanism of natural repair is complex and important chemical and physical characteristics are extremely variable. This makes it difficult to improve groundwater quality through natural evolution. It can be proved that the natural restoration protect the environment.

19.7.2 APPLICATION TECHNOLOGY

Application type technology was mainly developed in the mid-1980s. Due to factors, such as limited waste sites, high processing costs, and long-term debt operations, although this technology is simple, it has been abandoned by people. This technology is to transfer the problem elsewhere, not as encouraged by the latest regulations, that is, the pollutants are treated or controlled in situ.

19.7.3 GROUNDWATER EXTRACTION—TREATMENT SYSTEMS AND SOIL VAPOR EXTRACTION SYSTEMS

In addition, there are some in situ remediation technologies, such as soil infiltration remediation systems such as:

- Treatment and recharged sewage by the soil leakage system.
- Control and defense of groundwater pollution.

KEYWORDS

- **contamination**
- **control**
- **groundwater**
- **pollution**

REFERENCES

Bhuiyan et al. Investigation of the Possible Sources of Heavy Metal Contamination in Lagoon and Canal Water in the Tannery Industrial Area in Dhaka, Bangladesh. *Environ. Monitor. Assess.* **2011,** *175*, 633–649.

Farnham, I. M.; Johannesson, K. H.; Singh, A. K.; Hodge, V. F.; Stetzenbach, K. J. Factor Analytical Approaches for Evaluating Groundwater Trace Element Chemistry Data. *Analy. Chim. Acta* **2003,** *490*, 123–138.

Foster, S. S. D. Groundwater for Development – Overview of Quality Constraints. In Nash, H., McCall, G. J. H., Eds. Groundwater Quality. *Water Resour. Bull.* **1995,** *28* (3), 525–533.

https://serc.carleton.edu/download/images/150635/diagram_illustrates_porosity_sedime.png

McCutcheon, S. C.; Martin, J. L.; Barnwell, T. O. Water Quality. In *Handbook of Hydrology*; Maidment, D. R., Ed.; McGraw-Hill, Inc.: New York, 1993; pp 11.1–11.73.

Md. Bodrud-Doza, A.R.M.; Islam, T.; Ahmed, F.; Das, S.; Saha, N.; Rahman, M. S. Characterization of Groundwater Quality Using Water Evaluation Indices, Multivariate Statistics and Geostatistics in Central Bangladesh. *Water Sci.* **2016,** *30* (1), 19–40.

Passarello, M. C.; Pierce, S. A.; Sharp, J. M. Uncertainty and Urban Water Recharge for Managing Groundwater Availability Using Decision Support. *Water Sci. Technol.*, **2014,** *70*, 1888–1896.

Pereira, M. G. de; de Andrade, M. V. A. S.; Ornelas, V. C.; de Almeida, R. A. N.; Ribeiro, J. N. et al. Assessment of Physical-Chemical Characteristics of Water and Sediments from a Brazilian Tropical Estuary: Status and Environmental Implications. *Sci. World J.* 2012.

Ronayne, M. J.; Roudebush, J. A.; Stednick, J. D. Analysis of Managed Aquifer Recharge for Retiming Streamflow in an Alluvial River. *J. Hydrol.* **2017,** *544*, 373–382.

Sedki, A.; Ouazar, D. Simulation-Optimization Modeling for Sustainable Groundwater Development: A Moroccan Coastal Aquifer Case Study. *Water Resour. Manage.* **2011,** *25*, 2855–2875.

Siebert, S.; Burke, J.; Faure's, J. M.; Frenken, K.; Hoogeveen, J.; Döll, P.; Portmann, F. T. Groundwater Use for Irrigation–A Global Inventory. *Hydrol. Earth Syst. Sci.* **2010,** *14*, 1863–1880.

CHAPTER 20

Assessing the Impact of Urbanization on Deepor Beel: A Review

PRIYANKA SENGUPTA* and SNEHA RANI DEB

Department of Geography, Gauhati University, Guwahati, Assam, India

**Corresponding author. E-mail: senguptapriyanka741@gmail.com*

ABSTRACT

The tempestuous Brahmaputra gave rise to innumerable wetlands during its senescent course; one such instance is the Deepor Beel—a wetland of international significance as recognized by the Ramsar Convention in 2002. It acts as a local carbon sink because phytoplanktons growing therein are great carbon sequesters. It may well be labeled as the lungs of Guwahati city. The Beel recharges groundwater tables and "acts like a sponge to control floods." A part of it has been designated as a bird sanctuary, courtesy of the myriad avifauna that finds their abode in the landscape. A large chunk of the wetland has been infested by water hyacinth, an invasive species, leading to disappearance of native floral species. The hit and miss development spree that the Guwahati city has embarked on has caused a shear stress on the wetland causing the system's endemic processes to distort. Transport arteries are the major trails of human encroachment in the region. Untreated sewage from the city is discharged into the Beel through the Bahini and Bharalu rivers. The contaminated ground water from the peripheral landfill site of Guwahati Municipal Corporation (GMC) infiltrates into the Beel. The forests in the southern precincts have been subjected to logging and this has aggravated erosion and consequent siltation in the region. Studies on the physicochemical properties of water reveal that the water is contaminated, explaining the high

GIScience for the Sustainable Management of Water Resources. Gowhar Meraj, Shruti Kanga, Majid Farooq, Suraj Kumar Singh, & Sudhanshu (Eds.)
© 2023 Apple Academic Press, Inc. Co-published with CRC Press (Taylor & Francis)

turbidity. Water Quality Index (WQI) in most parts of Deepor Beel is quite high, rendering it unfit for drinking. Sizeable concentrations of heavy and trace metals have been found in the wetland in recent research endeavors. The increase of fishing practices along with the frequent disturbance of lake water by boats causes extensive churning of organic matter, leading to sprawl of aquatic flora and reduction in fish population. The wetland ecosystem is actually shrinking and the aquatic vegetation is increasing rapidly. The wetland ranges from mesotrophic to eutrophic in different times of the year. The increasing eutrophication of the lake has led to levels of dissolved oxygen plummeting as low as to be able to create a hypoxic condition for aquatic fauna. It is causing gradual conversion of the wetland to grassland. Pesticides used in the adjacent agricultural fields find their way into the lake ecosystem by means of runoff, causing biomagnification. The absorption capacity of the Beel during flood events has now decreased leading to artificial water-logging in the city. Self-purifying capacity of the wetland is also compromised because of pollution. The dumping site needs to be relocated to an alternative area away from the wetland. Ecotourism can be used as leverage to conserve the wetland through wise planning for sustainable management of the ecology along with revenue gains and improved well-being of local communities through "responsible tourism" in a natural landscape. This paper is an attempt at reviewing few of the noted research works done on the ecology of Deepor Beel, its vulnerability to urban encroachments, and solutions posited for the same.

20.1 INTRODUCTION

The landscape wherein water dictates the physicobiochemistry of the varied life forms and the numerous relationships and interdependences of the biotic and abiotic factors of an area can be claimed as a wetland. Geographically quoting, any spatial expanse that remains flooded partially or fully throughout the year, with abundance of hydric soil, fall under the terminology of "wetlands" (Crandell, 2020). The Ramsar Convention defines wetlands in its Article 1.1 as *"areas of marsh, fen, peatland or water, whether natural or artificial, permanent or temporary, with water that is static or flowing, fresh, brackish or salt, including areas of marine water the depth of which at low tide does not exceed six meters"* (*An Introduction to the Convention on Wetlands, 2016)***.** Contrary to nearby lands, the plumb of groundwater is much higher in the wetlands. The role of wetlands is much like that of table salt—whose presence is hardly acknowledged, but the absence is pronounced pronto. Among the most high-yielding ecological systems that prevail on the earth, wetlands top the list. They are often prescribed as "the kidneys of the landscape" because of their inherent filtering capacity. Some scholars also label them as "biological supermarkets" due to the complex food webs they sustain (Barbier et al., 2003; Brander and Schuyt, 2010). S.R. Arya and Elizabeth K. Syriac referred to wetlands as living waters in their paper "Wetlands: The living waters—A review" (Arya and Syriac, 2018). The initiation of the formation of wetlands is attributed to several mechanisms of climatic and geological phenomena, notably glaciations, sea level rise, de-glaciations followed by postglacial rebound or simply isostatic rebound. The manifold wetlands found throughout the globe owe its antecedent to the geography and geology of the realm. Wetlands are features most pronounced in the deltaic part of a river, near kettle lakes, in the close vicinity of large lakes, and sea portions which intruded inland. But wetlands do form when icy-cold water get locked up due to sea level drop. Many plateaus are known to host wetlands within their depressions. Whatever is the causation process and wherever be the site of the wetlands, the essentiality of wetlands cannot be denied (Crandell, 2020).

Most wetlands harbor biocolanders (specific biological communities) to filter pollutants and accumulate hazardous wastes within themselves, thus sieving off toxins. Wetlands are sponges, absorbing the heat emanating from the area and regulating temperature fluctuations and moisture flow. Most species of aquatic flora within wetlands act as effective hyperaccumulators. Carbon sequestration by phytoplankton is quite expeditious compared with terrestrial flora. "Wetlands are, thus, also considered as local carbon sinks" (Pandit, 2016).

The Paleozoic era is known to host one of the most important events, that is, the GOBE (Great Ordovician Biodiversification Event) of which the wetlands are key witnesses. Wetlands served as safe havens for species that phased habitats from water to land while progressing atop evolutionary wheels. Apart from the ecological services that the wetlands purvey today, they in the past acted as "masts" of an ecosystem, thus furnishing the basic shelter for the new genus of flora and fauna (Rashid et al., 2013; Singh et al., 2017a). The viaduct of terrestrialization of watery substrate craving species was the wetlands which proffered sufficient water for sustenance, successful reproduction, and dispersal and germination of various faunal species. From a paleontological standpoint, it can be inferred that the improved modes of sexual reproduction in flora and fauna, followed by high survival rate of tetrapods after terrestrialization of life forms in Devonian period was due to the presence of ample wetlands amidst the "cycle of Super Continental

Drifts" (From the Cambrain Explosion to The Great Dying, 2018; The Whole Saga of Supercontinents, 2018).

Wetlands are dynamic in nature. Over the course of years, some wetlands are wiped out and some new ones are born. However, the rate of alteration of wetlands has been quite significantly affected by anthropogenic pursuits. Flawed consideration of wetlands as wastelands has given way to urban encroachments into these sensitive ecosystems and their rapid conversion into land use categories which are able to yield more direct and tangible benefits to human populations (Barbier et al., 1997; Meraj et al., 2021). According to a study conducted by Global Wetland Outlook report, "wetlands are vanishing 3 times faster than forests." Presently, the global coverage of wetlands is around 12 million km^2 with an ever accelerating trend of areal decrease since the last century (Tooth and Waal, 2020). This chapter looks forward into systematic accumulation of studies done about an inland wetland, namely, Deepor Beel in the heart of the city of Guwahati, Assam, India.

The study area comprises a wetland named Deepor Beel which is located in the southwestern fringe of Guwahati city in Kamrup (M) district in the state of Assam, India (Figure 20.1). It is one of the 21 national wetlands which have been declared so far (Mondal and Sharma, 2011). It is considered as one of the largest wetlands of the Brahmaputra Valley of Lower Assam. Though locally called as Beel, it gained the title of wetland of international importance under the Ramsar Convention in 2002.

Deepor Beel spans for 40.14 km^2, of which 30.8 km^2 is subject to encroachment. The average elevation of the wetland is 53 m above MSL (WWF (India)). The depth of the wetland seasonally fluctuates within bounds of 6 and 1.5 m (Pandit, 2016). "A part of the wetland (4.14 km^2) was declared as a Bird Sanctuary by Assam Govt. in 1989, and managed by the State Forest Department" (Mahanta, 2018).

The depth of the Beel varies from 1 to 4 m based on flooding intensity. It helps in mitigating extreme flood events during monsoons. It receives most of its water load from its parent rivers, namely, Basistha and Kalmani and some from excess monsoon downpour. The Khonajan channel acts as the gateway for the Beel to drain into the Brahmaputra (Mondal and Sharma, 2011). "The Beel forms the 'lung space' that constantly replenishes the increasingly choked metropolitan air" (Mahanta, 2018).

The wetland is sandwiched between the National Highway No. 37 and the Northeast Frontier Railway track. Toward the eastern side of the Beel is the vast maw of an ever-urbanizing Guwahati. To the south, there lies a stretch of vegetated land classified as Rani-Garbhanga Reserve Forest. The western part of the Beel is skirted by the Lokpriya Gopinath Bordoloi International

Airport. Also, the railway network runs zigzag from the northern–western part of the Beel toward the southern side then paralleling along the southern edge, thereafter forwarding in the east (Figure 20.1).

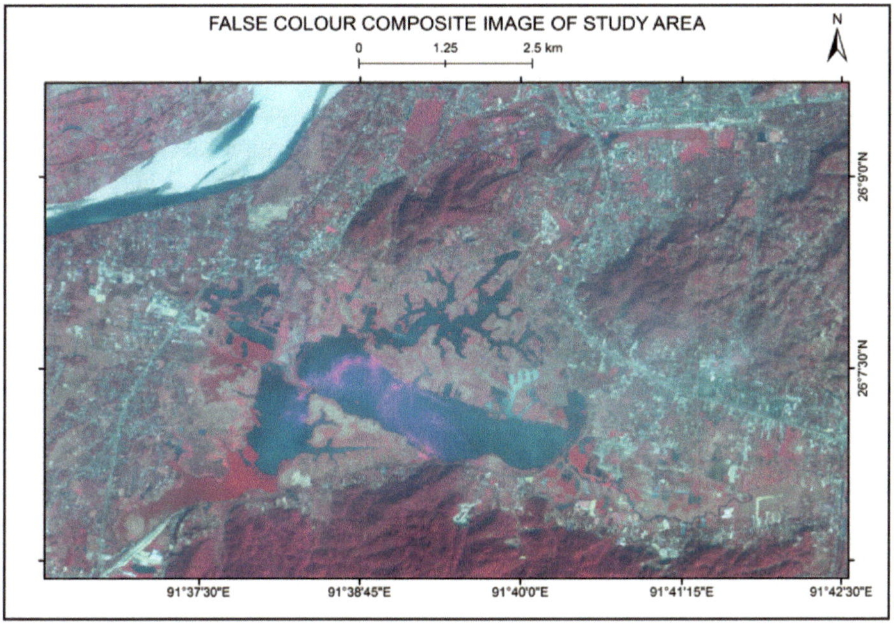

Source: www.earthexplorer.usgs.gov
Satellite: Sentinel 2A
ID: L1C_T46RCP_A023868_20200117T043313
Acquisition date: 17-01-2020

FIGURE 20.1 False color composite image of the study area.

20.2 THREATS TO THE WETLAND

The learned wise folks of Assam have knitted eyebrows toward the present dilemma of increasing anthropogenic pressure on the Beel and its immediate surroundings. In an interview, Professor Abani Kumar Bhagawati of Department of Geography, Gauhati University vocalized that all the rivulets, streams, catchment of Deepor Beel, and other wet lands disbanded throughout Guwahati are all interconnected and are alleys of the same major system (Bhagawati, 2017). The institutionalized encroachment in Deepor Beel and its adjacent connected landmasses have proven perilous to the ecosystem.

The sanctity of an ecosystem and the fragile bonds it shares with avifauna, mega fauna and cultural life of humans is irrefutable.

The wetland is shrinking rapidly, owing to large-scale encroachment by myriad groups, each seeking personal gains. Mondal and Sharma in their paper “Identification of Significant Environmental Aspects and Factors Affecting Wetland Dynamics and Ecological Characters of Deepor Beel Wetland Using Geoinformatic Techniques” delved into significant environmental aspects and factors affecting the Deepor Beel dynamics and ecological character. For assessing changes at the site, they performed supervised classification applying maximum likelihood classifier for the years 1990, 1997, and 2007. The area statistics were also derived to get an idea of the relative change in the wetland area and other land use categories.The relative change in wetland ecosystem from 1990 to 2007 was found to be a whopping −21.1% (Mondal and Sharma, 2011).

Similar patterns of wetland shrinkage were recorded by Pandit in her paper ‘Transformation of Urban Wetlands As An Effect of Urban Development: An Analysis of Deepor Beel in Guwahati, Assam.”The wetland was found to shrink drastically from 1991 to 2010 (Pandit, 2016).

Changes around the site or at the catchment were deemed by Mondal and Sharma to be caused by some principal factors which are summarized as follows:

a. Deforestation —Forests adorning the southern precincts of the wetland are subjected to extensive logging for supplying timber to sawmills that have sparingly popped up in recent years. This renders the top soil loose and fragile, and consequent erosion follows. Eroded soil from these erstwhile forested lands leads to aggravated siltation in the wetland (Mondal and Sharma, 2011).
b. Disturbance from transport arteries—Transport arteries are the major means of encroachment to the wetland ecosystem of the Beel (Figure 20.2). The Rangia-Guwahati broad gauge line touches the Borhola Beel, lying to the northeast of Deepor Beel and being a part of the latter itself (Mondal and Sharma, 2011). The Northeast Frontier Railway track constructed in the 1990s runs right through the natural Elephant Corridor near the Rani-Garbhanga Reserve Forest and is the primary cause of many unfortunate deaths of these mammals. The natural response of the elephants was to destroy human habitats that have encroached into their otherwise peaceful existence (Pandit, 2016). It has also obstructed the course of natural runoff into the Beel (Mondal and Sharma, 2011). This railway track divided the Beel into

two parts and the smaller southeastern section has dried up drastically. This segregation of the wetland has led to creation of smaller and more sensitive subecosystems within it (Pandit, 2016). The Beel is skirted by National Highway No. 37 on the east and northeast, a PWD road on the south, Dharapur–Kahikuchi section of NH 37 on the west and Engineering College Road on the north (Mondal and Sharma, 2011; WWF (India)). Deepor Beel once had a linkage to the Brahmaputra through the Borsola Beel but the stampede of transport arteries have attenuated the links and even terminated them in certain areas (Pandit, 2016).

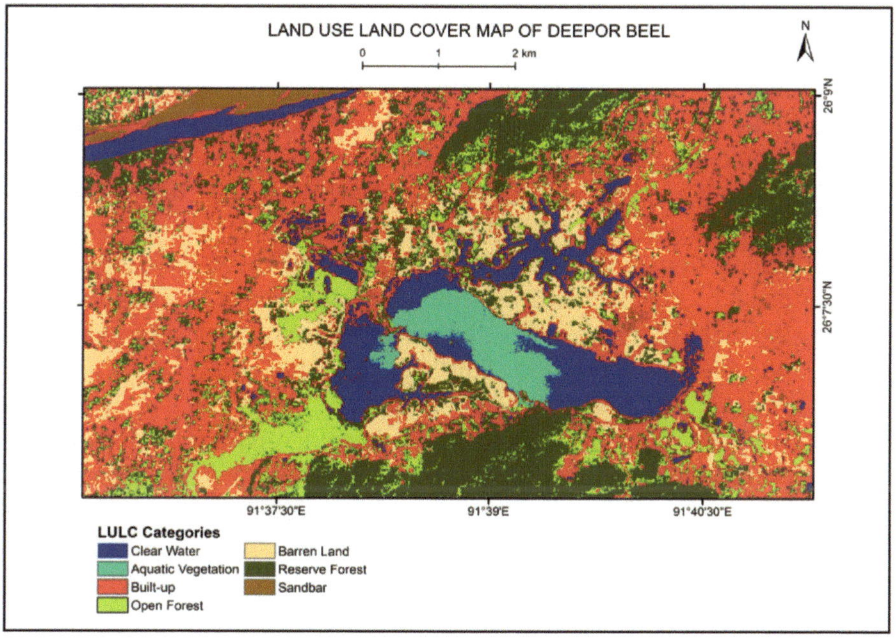

Source of Base Map: www.earthexplorer.usgs.gov
Satellite: Sentinel 2A
ID: L1C_T46RCP_A023868_20200117T043313
Acquisition date: 17-01-2020
Prepared using Maximum Likelihood Classification Algorithm

FIGURE 20.2 Land use land cover map of Deepor Beel.

c. Encroachment and settlement expansion—The extent of built-up land within the precincts of the Beel has been found to spike up quite rapidly. The area expanse of the built-up land was 287.11, 424.05

and 533 hectares, respectively in 1987, 1997, and 2007. All of the land around the Beel except the Garbhanga Reserve Forest is under private ownership. The government as well as private stakeholders are encroaching into the wetland and expanding settlements (Figure 20.2). Villages that are located within the Beel precincts include Tetelia, Pamohi, and Mikirpara. The crop grown by the people is boro paddy. The farmers use considerable doses of fertilizers and pesticides in hope of better yields. These chemicals find their way to the inner circles of the Beel through runoff and boost eutrophication. Assam Engineering College, Assam Ayurveda College, Government Sanskrit College and a part of Gauhati University, Tetelia High School and a few primary schools, Mother Teresa Hospital and temporary farm houses are certain institutions responsible for encroachment into the Beel (Mondal and Sharma, 2011; WWF (India)).

d. Industrial pursuits—An industrial complex that goes by the name Artfed Industrial Complex has come up in the periphery of the wetland and it poses a serious threat to the ecosystem of the Beel (Mondal and Sharma, 2011).
e. Excavations and brick kilns—Soil is dug out from the peripheral areas of the Beel and used in producing bricks in kilns located in the vicinity. These excavated areas are left on their own as pits with no efforts at restoration. This aggravates the problem of soil erosion (Mondal and Sharma, 2011).
f. Landfill site—The Guwahati Municipal Corporation (GMC) dumps all the sewage and solid waste of Guwahati city in the Pamohi landfill site located quite close to the wetland. This pollutes the surface water as well as the groundwater (through seepage), and this polluted water finds its way to the Beel, thus causing inhabitable conditions for aquatic fauna (Mondal and Sharma, 2011).
g. Illegal poaching—Many of the native as well as migratory species are illegally hunted down by poachers. Species count has reached unimaginable lows in recent years (Mondal and Sharma, 2011).
h. Unplanned and unregulated fishing—Excessive fishing with no regulations over mesh size and water pumps has led to dwindling fish population in the wetland (Mondal and Sharma, 2011). Also, ghost gear (parts of fishing nets left behind by fisher folk) causes many aquatic fauna to get trapped and eventually die.

Kapil and Bhattacharyya, in their paper entitled "Spatial, Temporal and Depth Profiles of Trace Metals in an Urban Wetland System: A Case Study with Respect to the Deepor Beel, Ramsar Site 1207, India," the team inferred that most of the contamination was caused by elements, such as arsenic, cadmium, lead, mercury, and nickel (Kapil and Bhattacharyya, 2013). Anthropocentric drivers have allowed for accumulation of trace elements in the Beel waters. Most of these lead to slow poisoning as arsenic, chromium, nickel and cause allergic reactions and are carcinogenic in nature (Awofolu et al., 2005). These elements are the root cause of fish deaths and unhealthy aquatic life. Once inside a living body, cadmium causes kidney stones, cell injury by interfering with calcium regulation (Woodworth and Pascoe, 1982). Mercury is a lethal element as it is both an enzyme and protein inhibitor and it magnifies in living body with time. Such baleful elements, upon intruding into the Beel ecosystem, penetrate into the existing food web and sublimate to consequent trophic levels through the process of biomagnification. Metals are soaked up by the mucilage in many algae present in the wetland. As soon as the algae enter the food chain, they magnify into the subsequent trophic levels. Most of the metals also return into the water bed later (Pandit, 2016).

A study conducted on the Beel posited that the water quality and the diversity of aquatic insects were directly influenced by the waste dumps. The water samples collected from the vicinity of the dump recorded low dissolved oxygen, but high electrical conductance, dissolved solids, alkalinity, and free CO_2. All of these impede the diversity of aquatic insects. The Shannon values were higher at the dump site indicating less genetic and species diversity while other sites deep within the Beel had low Shannon values. The diversity of aquatic insects is an important of the health of any water body, as they are the preliminary food of higher order fauna (Choudhury and Gupta, 2017).

Baruah, in her paper "Potential of Urban Wetlands for Ecotourism Development: A Case of Deepor Beel, Guwahati," calculated the Water Quality Index (WQI) of Deepor Beel wetland using the weighted arithmetic WQI. WQI in most parts of the Deepor Beel is quite high, rendering it unfit for drinking. In few areas, it is even more than the permissible limit of WHO standard of 100 (Baruah, 2020).

A paper entitled "Ecosystem evaluation (1989–2012) of Ramsar wetland Deepor Beel using satellite-derived indices"' states "The landscape development index (LDI) indicated a rapid increase in the impact of the surrounding land use on the wetland from 1989 to 2012" (Mozumder et al, 2014). It shows that the pressure of planned encroachment into the wetlands is ever increasing. The absorption capacity of the Beel during flood events has now decreased leading to artificial water-logging in the city (Pandit, 2016).

The paper "Valuing Fishing Activity of the Deepor Beel" advocates that the Beel is the reservoir of fish and other necessities for livelihood, yet, the resources go undervalued. This is due to the fact that most services cannot be marketed as these are the part of culture. Thispaper calculated the overall value of fishing to be INR 116,469,375 (Dutta and Sharma, 2000).

The erosion of nearby hills and siltation into the Beel has led to an awful ballad, where the mineral load has increased several folds causing distortions of continuous chains of life within the Beel. The flimsy interactions and interdependence within the system is stormed due to high levels of siltation causing sudden increase of minerals which favor parasites and pathogen growth. Eutrophication is a menace within the lake with water hyacinth being the most notorious species. The decline in the native fish and microfauna species is pushing the wetland toward a slow death and no restoration can help the pristine Beel to rebound into its original shape.

Presently, Deepor Beel is at the intersection of a logical debate: 'Ecology versus Economy'

When ecology commanded the clearance of humans from the very immediate surroundings of the wetland in recent times, stories written in blood smeared the newspapers and magazines. The low wage earners who resided in the nearby hills and wetland zone were asked to evacuate their houses for the sake of ecology. A decade later, studies reveal that a major portion of urbanized Guwahati stand on the ribs of trimmed hills and filled water bodies. Likewise, the tug of war between ecology and economy stresses on natural ecosystems and niches with strain being suffered by endemic flora and fauna, migratory fauna, and lastly the tribal communities who nurture the ecology.

In the paper entitled "Urban Stream and Wetland Restoration in the Global South—A DPSIR Analysis," the team of researchers wrote that water bodies in Global South endure "Urban Stream Syndrome." To tackle and manage the urban hydrosystems, the acceptance and commitment of the local, often poor people should be made aware of the River Culture and the fragile bond shared between urban water system and urban dwellers, which runs beyond ecological justification. The DPSIR approach used in this paper answers some very basic yet deep-rooted questions. The approach promises to suggest answers to few never ending debates. It reflects how politics, market economy and strategies, and fissured planning of wetland resources cause distress to the harmony of Beel waters. It sheds light onto how the urge to urbanize the rural, semirural and non-urbanized lands sicken the pristineness of urban water systems. This approach does not refrain itself from knowing the present status of urban wetlands. It also acts as a mediator between the health of wetlands and urban policy makers (Wantzen, 2019).

Eutrophication is one of the most perilous effects on water systems; natural eutrophication in wetlands takes several decades but after the onset of the decades of industrialization during 1960s and 1970s, cultural eutrophication has fastened the process (Kanga et al., 2017a, 2017b, 2021; Rather et al., 2018; Hassanin et al., 2020). The harmful consequences of eutrophication are well established in the scientific world, yet, in Deepor Beel, the haven of innumerable organisms, the effects are unfathomable. The creation of dark algal blooms thwarts the penetration of light thus reducing the rate of photosynthesis. A reduced photosynthetic rate disturbs the energy flow and reduces biomass production. The ultimate result is the massive die-offs of flora causing the water to turn turbid. The turbidity of the water distorts the prey–predator mechanism (Lehtiniemi et al., 2005). The detritus results in noisome intermix with the chemical composition of water and thus species depending on the chemosensory mechanisms sketch a fallacious surrounding, in turn, hampering their cognition of the niche they are in (Turner and Chislock, 2010; Farooq and Muslim, 2014; Nathawat et al., 2010; Kumar et al., 2018).

20.3 RESTORATION OF THE WETLAND

The present deplorable condition of the wetland calls for immediate efforts at restoration. The ecological principles that light the way to wetland restoration are the context and position of the landscape, types of natural habitats, specific hydrological regime, different paces of ecosystem attributes, nutrient supply rates, specific disturbance regimes, seed banks and dispersal, environmental conditions and life history traits, and succession theory and genotypes (Zedler, 2000). Another strategy to restore wetlands is the creation or maintenance of buffers. Buffers are meant to reduce the potential of damage to a wetland caused by human advances. These buffers may span to cover reserved forests forming a protective ring around the sensitive ecology of the wetland. Buffer widths depend on the valuation of the wetland in question. For low-value wetlands, smaller buffers (5–10 m) will suffice. But for higher value wetlands, larger buffers (15–30 m) are necessary so as to effectively provide a barrier to all possible threats to the wetland and to best preserve the aquatic resources (Castelle et al., 1994).

Ecotourism is one way of restoring the wetland to its pristine state. It can aid in achieving a balance between tourism and ecology, as opposed to conventional tourism which leaves ecology out of the equation. It can be beneficial to both the protected areas and the inhabitants as it fosters biodiversity conservation simultaneously with social and economic development. An ecotourist always goes for raw experiences of the nature and culture of a place (Deori and Das, 2013). Ecotourism can be used as leverage to conserve the wetland through wise planning for sustainable management of the ecology along with revenue gains and improved well-being of local communities through "responsible tourism" in a natural landscape (Mahanta, 2018).

Some environmental conservation and restoration strategies for the wetland forwarded by Baruah include the following:

a. Maximizing ecotourism circuits might help in generating greater revenues for the local residents. The proposed ecotourism circuits include cultural ecotourism circuit, leisure ecotourism circuit, education ecotourism circuit, and one day transit tour (Baruah, 2020).
b. Conservation zoning of the wetland has been proposed by Baruah (2020) to better conserve and preserve the natural habitat. The four zones include the core area aquatic ecosystem (Deepor Beel lake), the ecological restoration zone (Retention Zone), the riparian buffer zone (Native Green Belt) and the eco-sensitive zone (NATE-based Recreation) (Baruah, 2020).
c. Creation of floating islands with indigenous plant species can be done to generate a congenial habitat for local as well as migratory water birds. This initiative can be worth a shot to revive the pristine biodiversity that the wetland once boasted. These floating islands can serve as ideal nesting grounds and safe havens for many species. These islands are deemed to be located in the extreme interior of the lake (Baruah, 2020).
d. Interceptor sewerage treatment plant can be taken up to solve the crisis around the sewage inflow into the Deepor Beel ecosystem from Dispur subbasin, which has led to deteriorating water quality of the wetland (Baruah, 2020).

20.4 CONCLUSION

The preservation, conservation, management and restoration of the Deepor Beel, though easy on pen and paper is yet a strenuous task. The Beel is, in itself, a system where each individual species, factor and linkages differentiate and integrate the pristine beauty. Neither the individual diverse species, nor the waters alone characterize the Deepor Beel. Instead, the assemblage and network of grasses, the grasshoppers, the frogs, the snakes, the birds, the people, etc. woven into an intricate eco-economical fabric represents the Beel. It would be judicious if further distortion of the Beel is avoided and it is allowed to have rights of its own.

KEYWORDS

- **wetland**
- **sewage**
- **eutrophication**
- **biomagnification**
- **ecotourism**

REFERENCES

An Introduction to the Convention on Wetlands (previously The Ramsar Convention Manual). Gland, Switzerland: Ramsar Convention Secretariat, 2016.

Arya, S. R.; Syriac, E. K. Wetlands: The Living Waters-A Review. *Agric. Rev.* **2018,** *39* (2), 122–129.

Awofolu, O. R.; Mbolekwa, Z.; Mtshemla, V.; Fatoki, O. S. *Levels of Trace Metals in Water and Sediment from Tyume River and Its Effects on an Irrigated Farmland*; Water, SA, 2005.

Barbier, E. B.; Acreman, M.; Knowler, D. *Economic Valuation of Wetlands: A Guide for Policy Makers and Planners*; Ramsar Convention Bureau: Gland, Switzerland, 1997.

Baruah, P. Potential of Urban Wetlands for Ecotourism Development—A Case of Deepor Beel, Guwahati. *Nat. Environ. Pollut. Technol.* **2020,** *19* (2), 611–625.

Bhagawati, A. K. Geography and Landscape of Deepor Beel. Interview with Author, 2017, Nov 1.

Brander, L.; Schuyt, K. The-Economic-Value-of-the-Worlds-Wetlands, 2010. http://teebweb.org/wp-content/uploads/2013/01/The-economic-value-of-the-worlds-wetlands.pdf

Castelle, A. J.; Johnson, A. W.; Conolly, C. Wetland and Stream Buffer Size Requirements—A Review. *J. Environ. Qual.* **1994,** *23*, 878–882.

Choudhury, D.; Gupta, S. Impact of Waste Dump on Surface Water Quality and Aquatic Insect Diversity of Deepor Beel (Ramsar Site), Assam, North-east India. *Environ. Monitor. Assess.* **2017**.

Crandell, C. J. Science/Wetland. **2020,** Mar 05. https://www.britannica.com; https://www.britannica.com/science/wetland (accessed Dec 20, 2020).

Deori, S.; Das, N. Ecotourism in Wetland Environment: A Case Study of Deepar Beel Wildlife Sanctuary (Ramsar Site) of Assam (India). *Revista Brasileira de Ecoturismo* **2013,** *6* (2), 354–365.

Dutta, J.; Sharma, A. *Valuing Fishing Activity of the Deepor Beel*; Space and Culture, India, 2000; pp 122–132.
From the Cambrian Explosion to the Great Dying, 2018. [Motion Picture].
Goswami, M.; Nautiyal, S. Transitional Peri-Urban Landscape and Use of Natural Resource for Livelihoods. In *Socio-Economic and Eco-Biological Dimensions in Resource Use and Conservation*; 2020; pp. 435–457.
Hassanin, M.; Kanga, S.; Farooq, M.; Singh, S. K. Mapping of Trees Outside Forest (ToF) From Sentinel-2 MSI Satellite Data Using Object-Based Image Analysis. *Gujarat Agric. Univ. Res. J.* **2020,** *207*.
Kanga, S.; Kumar, S.; Singh, S. K. Climate Induced Variation in Forest Fire Using Remote Sensing and GIS in Bilaspur District of Himachal Pradesh. *Int. J. Eng. Comput. Sci.* **2017b,** *6* (6), 21695–21702.
Kanga, S.; Rather, M. A.; Farooq, M.; Singh, S. K. GIS Based Forest Fire Vulnerability Assessment and Its Validation Using Field and MODIS Data: A Case Study of Bhaderwah Forest Division, Jammu and Kashmir (India). *Indian Forester* **2021,** *147* (2), 120–136.
Kanga, S.; Singh, S. K.; Sudhanshu. Delineation of Urban Built-Up and Change Detection Analysis using Multi-Temporal Satellite Images. *Int. J. Recent Res. Aspects* **2017a,** *4* (3), 1–9.
Kapil, N.; Bhattacharyya, K. G. Spatial, Temporal and Depth Profiles of Trace Metals in an Urban. *Environ. Pollut.* **2013,** 51–72.
Lehtiniemi, M.; Engström-Öst, J.; Viitasalo, M. Turbidity Decreases Anti-Predator Behaviour in Pike Larvae, Esox lucius. *Environ. Biol. Fishes* **2005,** 1–8.
Mahanta, N. R. Urban Wetlands and the Built Environment; Eco-Tourism as a Strategy for Sustainable Co-existence and Management of Deepor Beel—A Ramsar Wetland in India. M. Phil Thesis, Indian Institute of Technology, Guwahati, Centre for the Environment, Guwahati, 2018.
Meraj, G.; Singh, S. K.; Kanga, S.; Islam, M. N. Modeling on Comparison of Ecosystem Services Concepts, Tools, Methods and Their Ecological-Economic Implications: A Review. *Model. Earth Syst. Environ.* **2021,** 1–20.
Mondal, M. S.; Sharma, P. Identification of Significant Environmental Aspects and Factors Affecting Wetland Dynamics and Ecological Characters of Deepor Beel Wetland Using Geoinformatic Techniques. *J. Indian Water Resour. Soc.* **2011,** *31* (3–4), 33–40.
Mozumder, C.; Tripathi, N.; Tipdecho, T. Ecosystem Evaluation (1989–2012) of Ramsar Wetland Deepor Beel Using Satellite-Derived Indices. *Environ. Monitor. Assess.* **2014,** 7909–7927.
Pandit, S. Transformation of Urban Wetlands as an Effect of Urban Development: An Analysis of Deepor Beel in Guwahati, Assam. *Creative Space (CS)* **2016,** *4* (1), 103–114.
Rashid, I. et al. Assessing the Impact of Anthropogenic Activities on Manasbal Lake in Kashmir Himalayas. *Int. J. Environ. Sci.* **2013,** *3* (6), 2036–2047.
Rather, M. A. et al. Remote Sensing and GIS Based Forest Fire Vulnerability Assessment in Dachigam National Park, North Western Himalaya. *Asian J. Appl. Sci.* **2018,** *11* (2), 98–114.
Singh, S. K.; Kanga, S. Role of Geoinformatics in Site Suitability Analysis of Infrastructures using PRA Approach. *Am. Int. J. Res. Sci. Technol. Eng. Math.* **2017,** *1* (18), 81–85.
The Whole Saga of Supercontinents, 2018. [Motion Picture].
Tooth, S.; Waal, B. v. d. The 'Global Wetland Outlook' Report. *Geog. Rhodes Univ.***2020,** *104* (3), 154–159.

Turner, A. M.; Chislock, M. F. Blinded by the Stink: Nutrient Enrichment Impairs the Perception of Predation Risk by Freshwater Snails. *Ecol. App.*, 2010.
Wantzen, K. M. Urban Stream and Wetland Restoration in the Global. Sustainability—A DPSIR Analysis, 2019.
Wetland Inventory—Ministry of Environment and Forests (n.d.). https://doczz.net/doc/1560740/wetland-inventory---ministry-of-environment-and-forests
Wetland of Global Importance Is Used for Garbage Dumping. **2013,** December 18. https://wildflowerassam.wordpress.com/2013/12/18/wetland-of-global-importance-is-used-for-garbage-dumping/
Woodworth, J.; Pascoe, D. Cadmium Toxicity to Rainbow Trout, Salmo Gairdneri Richardson: A Study of Eggs and Alevins. *J. Fish Biol.* **1982,** 47–57.
WWF (India) (n.d.). Information Sheet on Ramsar Wetlands (RIS). https://rsis.ramsar.org: https://rsis.ramsar.org/RISapp/files/RISrep/IN1207RIS.pdf
Zedler, J. B. Progress in Wetland Restoration Ecology. *TREE* **2000,** *15*, 402–407.

CHAPTER 21

Atmospheric Boundary Layer and Geographic Information System (GIS): A Review

C. NAVIN KUMAR*, N. THAVAPRAKAASH, S. PANNEERSELVAM, and S. ARUL PRASAD

Tamil Nadu Agricultural University, Coimbatore 641003, Tamil Nadu, India

**Corresponding author. E-mail: navindpi@gmail.com*

ABSTRACT

Based on the time scales that prevail in the atmosphere, the atmospheric boundary layer (BL) is divided into microscale and partly to meso- and gamma-scales. Scaling is the basic processes for measuring the atmospheric process. This scaling of atmosphere mainly helps in atmospheric prediction. This entire layer had a relationship with micrometeorology and carbon fluxes prevailing inside and outside the crop growth and canopy. The factors that affect the microclimate are landform, vegetation, water bodies, soil, and artificial structures like buildings, trees etc. Crops in relationship with the BLs include surface and internal BL in micrometeorology. The parameters which would favor crop production are relative humidity, canopy temperature, soil temperature, photosynthetically active radiation, and leaf gas exchange parameters. Each parameter has specific instrument for measuring apart from using the sensors. Every single factor has a wider role, for example, a root growth is only concerned with soil temperature and shoots growth fretful by canopy temperature and relative humidity. Satellite-based crop monitoring for

GIScience for the Sustainable Management of Water Resources. Gowhar Meraj, Shruti Kanga, Majid Farooq, Suraj Kumar Singh, & Sudhanshu (Eds.)
© 2023 Apple Academic Press, Inc. Co-published with CRC Press (Taylor & Francis)

pest and diseases uses methane assessment that is done by change in carbon and nitrogen flux in inside and outside the plant canopy. These applications are helpful in crop monitoring. Each factor of BL in micrometeorology had a greater impact in satellite and land-based observations of methane.

21.1 ATMOSPHERIC SCALES

Long et al. (2014) found this classification system in 1975. This atmospheric scale was based on time and spaces. For measuring atmospheric process, this scaling of atmosphere is the basic principle and this system was broadly classified into three types:

1. Macroscale
2. Mesoscale
3. Microscale

Macroscale

It deals with larger scale atmospheric motions, that is, more than 100 km the atmospheric composition of gases may vary at specific layers.

Mesoscale

It lies between the macroscale and mesoscale. It doesn't have a role in favoring or maintaining a microclimate.

Microscale

As micrometeorology is concerned, microscale plays a role in microclimate.

All the three scales in micrometeorology are interpreted with alpha, beta, and gamma. The weather occurrences in the atmosphere are classified according to time scales prevailing (Table 21.1) and similarly, the weather and climate prediction models are also classified in the similar manner. For example, large-scale global circulation models are linked to the range of macro-β type. The weather forecast was associated to the meso-α range. On the other hand, higher range resolution models are in relation with the meso-β,γ scale. Micrometeorology is the relation between the microscales in the atmosphere and partly related to the meso-γ scale. High resolution (small scale) measurements are necessary to measure the spatial extension under a low-pressure system of the atmosphere. These high-resolution modules are used in micrometeorological studies. The frequency of these measuring atmospheric systems is directly related to the velocity of the pressure system. This is effective for all the time scales and these occurrences can be only observed if the measurements are made on time scales within a smaller space. Microclimatology describes the long time average of micrometeorological processes concerned with the wing and temperature fluctuations (Glickman, 2000).

TABLE 21.1 Scale of the Layer with Height (Arya and Wyngaard, 1975).

Scale roughness layer	Boundary layer height (m)
Roughness layer planetary	0.0–0.1
Turbulent surface layer	0–10
Boundary layer	0–1000
Troposphere	$0–1 \times 10^4$
Stratosphere	$> 1 \times 10^4$

21.2 DIFFERENT ATMOSPHERIC SCALES IN THE LOWER LAYER

1. Surface boundary layer (BL)
2. Internal BL
3. Planetary BL
4. Laminar BL

21.2.1 SURFACE BOUNDARY LAYER

This layer is immediately above the surface so it is in contact with the atmosphere. Larger energy fluctuation occurs in this layer as there is interaction of atmosphere gases with this layer. Because of the roughness of the soil surface layer, the energy obtained from the light is reradiated onto the Earth's surface. This layer has significant temperature, humidity, and wind vertical gradients. In a time scale of an hour or less, the atmosphere reacts to surface forcing. The direction of the wind does not shift when you get higher (Wan et al., 2015). Surface BL depth is around 10% of planetary BL depth. These layers interact not only with the Earth but also with the atmospheric BL.

21.2.2 INTERNAL BOUNDARY LAYER

This layer is in touch with the soil since it is directly above it. Constant flux exists, that is, heat, momentum, and mass transfer are all invariant with height. The layer develops in depth as a result of surface roughness and distance from the tip. According to a study titled "Seasonal change of average aerosol scale height, Trivandrum" (Jung et al., 2009), there is a connection between aerosol depositions and the internal BL. His coworkers confirmed that there is almost 1.3 km more aerosol accumulation in the inner BL mostly during month of August. The height of the aerosol in the atmosphere is closely related to the height of the thermal internal boundary layer (TIBL). The mixing field between aerosol as well as the atmosphere constitutes nearly 10–30% to the aerosol optical depth (AOD) at 0.5 m wavelength. Between 1989 and 1997, in the mixing area of AOD, a long-term rise in aerosol pattern of 2.8% per year was observed, and aerosol quantity continues to increase year by year in Trivandrum, Kerala (Figure 21.1).

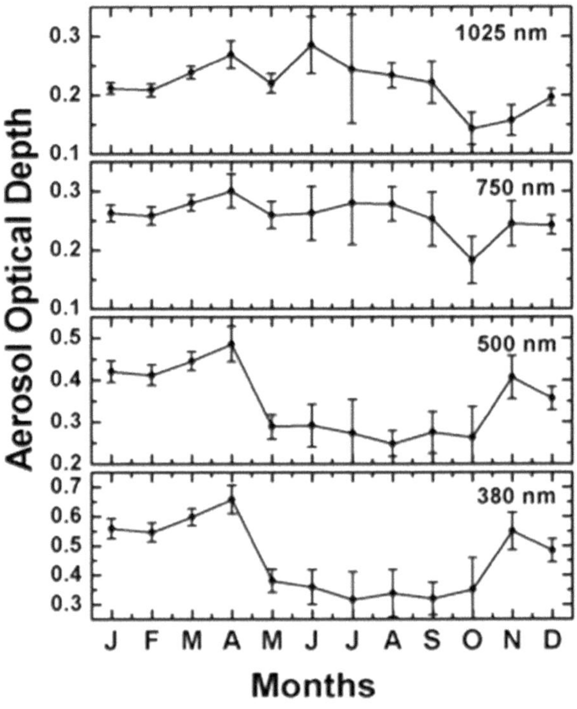

FIGURE 21.1 Long-term seasonal variation of mean aerosol with height of internal boundary layer in Trivandrum, Kerala.

Source: Reprinted from Moorthy et al., 2007. https://creativecommons.org/licenses/by/3.0/

21.2.3 PLANETARY BOUNDARY LAYER (PBL)

It is also known as the atmospheric BL and weather phenomenon occurs in this layer.

Two types of layers present in PBL are:

1. Lowest 10% of the layer is surface layer or Prandtl layer.
2. Layer above atmospheric BL is Ekman layer

The variation in the wind direction takes place in this layer. The structure is due to turbulence and most of the living organisms present in this layer are strongly influenced directly by interaction with the Earth's surface (Fröhlich-Nowoisky et al., 2016). Thereby causing a large velocity gradient near the Earth's surface, which leads to turbulent eddies. It is a constant flux layer, that is, the change and exchange of heat over a specific layer is constant and due to the constant flux behavior two types of stratification occurs:

1. 20–50 m in unstable stratification (above PBL)
2. Few meters for stable stratification (within PBL).

In relation to plants, the carbon flux occurs and the carbon is transported into the atmosphere and causes warming up of the atmosphere contributing plants as a minor component in carbon dioxide transport.

The PBL height depends on:

1. Amount of heating or cooling of the surface layer.
2. Wind strength.
3. The topographical and roughness characteristics of the surface.
4. Greater scale in vertical motions of the atmosphere (cm/s).
5. Heat and moisture prevailed by horizontal advections.

During daytime: PBL height on the order of 1 km (0.2–5 km).
During nighttime: PBL height on the order of 100 m (20–500 m).

21.2.4 LAMINAR BOUNDARY LAYER

The flow streamlines become almost identical to the object at very small scales (mm). Flow is not unstable, and molecular diffusion transports across the BL (Saad et al., 2014).

21.3 EARTH'S SURFACE ENERGY BALANCE

The Earth surface is warmed by the incoming shortwave radiation from the sun (K↓), and only a part of this incoming radiation is reflected back (K↑) into the atmosphere. Moreover, the surface engrosses incoming longwave radiation due to longwave emitted by the clouds, smaller atmospheric particles and gases (I↓). The energy absorbed is partially emitted into the atmosphere as longwave outgoing solar radiation (I↑). The black body absorbs more radiant energy whereas the white body absorbs less compared with black body. In total, the surface of the Earth receives more radiation energy, that is, the net radiation is positive at the ground surface (Khosa et al., 2010).

The excess amount of energy supplied is carried back to the atmosphere by changes in the turbulent energy fluxes.

1. The sensible heat flux (QH)
2. Heating the atmosphere from the surface to 100 m, except for days with strong convergence, is a result of sensible heat flux.
3. The latent heat flux (QE)
   The energy is transferred into the soil because of the ground heat flux (QG) and will be put in storage in the plants and buildings (ΔQS).

According to the law of energy conservation, the energy balance at the Earth's surface is $Q^* = Q_H + Q_E - Q_G + \Delta Q_s$

Q^* = total radiation upward at surface
Q_H = sensible upward heat flux
Q_E = latent upward heat flux
Q_G = heat flux upward molecular into the bottom
ΔQ_S = storage/residual

21.4 TURBULENT FLUXES

Turbulence is generally the mixing of air. The interchange of latent and sensible heat among land and the atmosphere occurs due to the turbulent mixing of air. Turbulence is created within the crop canopy due the larger variation of wind and temperature at the microscale range. The air moves in form of horizontal as well as the vertical direction. As air parcel moves, it carries heat, momentum, and water vapor. It creates eddies that mix air from above downward and from below upward and transports some amount of heat and water vapor. During this transport, the water is carried from the warmer to the cooler surface. Near the surface, the turbulence are vertical motions; vertical turbulence heated due to the buoyancy and heating up the surface.

21.5 FACTORS INFLUENCING THE MICROCLIMATE

21.5.1 TOPOGRAPHY

Topography is the shape of the land that includes large features like mountains or small features like little hills, buildings, etc. Here, the temperature varies with altitude. For each 0.1 km upsurge in elevation, cooling proportion is about 0.65°C. On mountain slopes, an air speed increases or decreases on windward or leeward side. On leeward side, difference in air speed causes the low pressure.

21.5.2 VEGETATION

It is the efficient absorber of solar radiation. Most of the radiant energy is absorbed during photosynthesis and synthesizes carbohydrate, starch, and glucose. It controls the microclimate. Shading is more significant in hot and dry climate. It cools the environment by absorbing solar radiation. Cooling effect of hot and dry climate is due to vegetative factor.

21.5.3 WATER BODIES

All the water bodies are heated up during a day. A thermal mass that can act as a thermal buffer/heat sink. Evaporation of water increases humidity. It useful in maintaining a hot and dry climate. Land heated up during day so low pressure created on land which causes the wind to move from land to sea breeze. Land breeze is vice versa of sea breeze. Land breeze occurs only at night.

21.5.4 SLOPE

The sloping land can improve the solar angle and allows the solar radiation to penetrate inside the crop canopy. The leaf angle can modify microclimate on reflecting and absorbing solar radiation the Earth surface.

21.5.5 ALTITUDE

For every 120 m increase in elevation causes delay of flowering by average of 4 days in unprotected crops (Horton et al., 2015). Elevation is a good example for microclimate modification. In mountain areas, the temperature decreases with 1°degree for every 330 ft increases in summer and every 400 ft increases during the winter.

21.5.6 WIND BREAK DENSITY

Density is the ratio of solid portion to the total area of the barrier. By adjusting the windbreak density, different wind flow pattern could be identified. By combining low growing shrubs with medium and tall deciduous trees, dense plant material (Brandle et al., 2009) is provided at three levels (low, middle, and upper) of the windbreak during the growing season (Figure 21.2).

	Open wind speed 20mph Decidious 25-30% density			Open wind speed 20mph Decidious 40-60% density			Open wind speed 20mph Multi row 60-80% density			Open wind speed 20mph Solid fence 100% density		
Distance from wind break	15H	20H	30H	15H	20H	30H	15H	20H	30H	15H	20H	30H
Miles per hour	16	17	20	12	15	19	13	17	19	18	19	20
% of open wind speed	80%	85%	100%	60%	75%	95%	65%	85%	95%	90%	95%	100%

FIGURE 21.2 Wind speed reduction with different densities.

Source: Reprinted with permission from Brandle et al. © 2009 John Wiley and Sons.

21.5.7 LEAF ARCHITECTURE

Some of the leaves present in the plants with serrated margins. This also decides the photosynthetic factor. Canopies with straight leaves can achieve photosynthetic rates of 70–100% inordinate than those whose leaves are horizontally arranged. Crop canopy photosynthesis is affected by the spatial pattern of plant (Orlanski, 1975).

21.5.8 WATER RELATIONS

When plants are exposed to cold and humid air and sufficient soil moisture, the diurnal pattern of canopy carbon dioxide exchange rates reaches maximum in near midday. Reduction in canopy photosynthesis during periods of soil moisture deficit condition leads to the closure of the stomatal.

21.5.9 EVAPORATIVE COOLING

Water is sprayed at a high pressure (30–70 kg/m^2) and it is sprayed as a fine mist that cools the air. The mist evaporates before reaching the plant canopy and reduces the air temperature by 5–10°C.

21.5.10 GIS RELATIONSHIP WITH ATMOSPHERIC BOUNDARY LAYER

21.5.10.1 COLLECTION OF AIR SAMPLES FOR METHANE ESTIMATION

Jiménez et al. (2009) described that 10 min before the sample collection, a chamber should be placed on the surface of the soil with 4–5 cm insert to lessen the disruption of the sampling site. To avoid sample damage, extra caution is required. After replacing the plants with both the chamber with a syringe and one-way valve pump, four air specimens were taken in Tedlar bags, starting at zero time and increasing by 15 min. The air samples have to be introduced into the analyzer in the morning (between the hours of 9–10 am) and in the evening (between the hours of 2–3 pm). The total amount of the morning and evening fluxes are used as the value for the day. According to the time programmed analyzer, which activated the valve and back-flushed the samples, the samples were flushed in the programmed amount of time. As the principal standard curve is linear over other concentration ranges used for sample analysis, the gas chromatography must be calibrated before and after each set of sample measurements using 1ppm, 2.3ppm, and 5ppm different standards. CH_4 flux expresses as mg/m^2/hr^1 using the equation given by Stull (2012). The obtained CH_4 concentrations determine by peaked area and change in the flux to be measured based the formula proposed by Duchemin et al. (1999) to evaluate methane emission.

$$f = (V/A)\ (\Delta C/\Delta t)$$

where

f is the emission rate of greenhouse gas (mg m^{-2} h^{-1}) in the field,
V, volume of above the soil chamber (m^3),
A, chamber cross-section (m^2),
ΔC difference between zero and t times concentration (mg cm^{-3}), and
Δt two sampling periods equal to time duration (h).

21.5.10.2 ESTIMATION OF METHANE EMISSION USING SATELLITE DATA

21.5.10.2.1 Land Surface Temperature (LST)

LST is one of the most significant criteria for studying the land surface and methane emissions. Several researchers have proposed LST determination methods such as the spectral ratio (Stevenson et al., 2020), reference channel technique, emissivity normalization method, alpha residue method, and Split window method (Song et al., 2013). The emissivity, on the other hand, is one of the major issues that arise when estimating the kinetic temperature of numerous land cover types. Split window technique (SWM) corrects for atmospheric effects based on differential absorption in adjoining infrared bands but each type of feature requires the exact emissivity. Temperature-dependent factor (T factor) must be calculated using MODIS LST data. T factor is being used to design the transformation in methanogenic activity as a function of the temperature. The large percentage of methanogens prefer temperatures between 30 and 40°C (Wan et al., 2002). The following equation describes methane emission from wetlands:

$$\text{ECH4} = E_{obs}\ \text{Ft}\ A$$

E_{obs}, methane flux observed from different categories, Ft, T-factor, and A, area.

Using ArcGIS 10.1, the model was created by applying the measured methane flux, T factor, and area map to any and every pixel of rice area maps in various temporal resolutions in order to obtain min and max emissions at various times from the start of next season (SoS).

21.5.10.2.2 Tropospheric Methane Sink

The primary sink for methane is oxidation by the hydroxyl radical present in the troposphere for the period of 9 years constrained in the upper atmosphere by methyl chloroform (MCF) (Ostler et al., 2016). OH is produced by photochemical process and its concentration present is operated by intricate chemistry that is not well characterized in the models (Krishna et al., 2007). However, the methane loss is suitably slow so that variation in the OH concentrations affects concentrations of the methane on seasonal, interannual, and interhemispheric scales. It doesn't affect the regional-scale gradients relevant to reverse analyses of the satellite data.

The methane sink is typically calculated using defined global 3D fields of OH concentration level from an independent modeling of tropospheric oxidant chemistry in global reverse analyses. Cressot et al. (2014) combined methane and MCF carbon emissions in their inversion, allowing for OH concentrations to be adjusted. For limited-domain reversals with spatial boundary conditions, specifying OH quantities is not a problem. Since the modeling domain is represented on a timescale that is significantly shorter than the methane's 9-year life span. In that case, the methane sink and the boundary conditions are effectively combined.

21.5.10.2.3 Methane Accounting in the Stratosphere

The presence of stratospheric methane inversions in satellite methane data necessitates proper stratospheric secretarial. In the tropics, the stratosphere accounts for about 5% of total methane gas emissions, and in higher latitudes, it accounts for up to 25%. Methane is carried to higher elevations for about 5 years after it reaches the stratospheric interface in the tropics. OH, O(1D), and Cl atoms photochemically oxidize it over time, causing a variability throughout the column mean mole fraction XCH4 that is out of phase of tropospheric methane (Naik et al., 2013). In model projections, these stratospheric methane transports tend to be too fast, resulting in overestimation of stratospheric methane concentrations at high latitudes. Inversions that do not account for this cause effect result in a 5% overestimation of methane production at northern midlatitudes and a 40% overestimation at the Arctic pole.

The emission levels from boreal wetlands are severely balanced when accounting for them. To assess the stratospheric methane simulation, the highest number of observation-based data sets is available. This included balloons, TCCON stratospheric retrievals (Naik et al., 2013), and solar occultation and limb satellite observations. Jacob et al. (2016) did a thorough evaluation of their CTM using balloon observational data as a precursor to inversion of SCIAMACHY data, which led people to restrict their inversion to the 50°S–50°N latitudinal range, where model bias was low. Another option is to use a latitudinal bias adjustment to compensate for the difference between both the CTM and the satellite imagery. Mauder et al. (2020) proposed a method to rectify for stratospheric methane prejudice in CTMs by incorporating some of the restrictions derived from perpendicular sulphur hexafluoride profiles on the age of air samples in the stratosphere (SF6).

21.5.10.3 EVAPOTRANSPIRATION ESTIMATION USING REMOTE SENSING

According to the water balance equation,

$$\mathrm{ET} = P - R - \Delta \mathrm{SW}$$

where

- P is precipitation,
- R is runoff, and
- ΔSW is the change in water storage.

Recent advancements in remote sensing technology, particularly satellite-based remote sensing, have made it possible to map and monitor precipitation and changes in water storage. Resources, on either hand, are accessible via land-based networks. Because remotely sensed SW data is only available from the GRACE satellites, the water balance method for evapotranspiration estimation is commonly used over month-to-month or even higher temporal resolution. For example, Zeng et al. (2012) used GRACE data as well as in situ runoff measurement techniques to estimate evapotranspiration over 59 of the world's major river basins, and then built a spatial regression analysis of annual E by incorporating P, temperature, and NDVI data. Long et al. (2014) used the water balance method to calculate monthly E in several South Central United States basins, and Wan et al. (2015) used the water balance method to calculate monthly E for 592 subbasins from across continental United States (CONUS) from 2002 to 2014 by downscaling of GRACE data using land surface models.

21.5.10.4 EMPIRICAL MODELS

Based on the physically based methods, there are also a diversity of empirical methods used for RS-based E estimation and mapping using a satellite data. More or less all of the empirical models use statistical methods, extending from simple single-parameter or multivariate methods to composite machine learning approaches, linking E to various illustrative variables that are more easily observed or obtained from ground and satellite RS. For example, Wan et al. (2002) developed a simple empirical regression model that expresses E is a function of temperature of the surface (Ts) and global radiation (Rn) and it is accounted for the impacts of relative humidity. Jung et al. (2009) designed a machine learning method called the MTE to assess monthly E using temperatures, P, fAPAR, and potential radiation as explanatory variables. Jiménez et al. (2009) built a nonlinear statistical model that predicts E using remotely-sensed visible and near infrared spectral reflectances, thermal infrared (TIR) skin temperature, active microwave.

21.6 CONCLUSION

Observing atmospheric methane from space and the need for inverse analyses to correct the recognized factor of methane emissions. Shortwave infrared (SWIR) observations are the most important for calculating emissions. They are receptive to the entire atmospheric column all the way down to the ground. Reclamations that combine the SWIR as well as the TIR will indeed separate the lower troposphere, which leads to methane emissions, and thus reduce uncertainty in measuring the free tropospheric backdrop and stratosphere. The GOSAT satellite's current SWIR observations are of excellent quality but limited. On a 100–1000-km resolution scale, inverse analyses and annual averaging can account for emissions from various regions. The TROPOMI instrument, which is set to launch in 2017, will be aims to measure methane emissions on a daily scale as well as detect and quantify massive point sources. As a result, the significance of satellite data for climate strategy will be significantly enhanced. If it meets its 1–5% precision specification, the GHGSat instrument, which was launched in 2016 and has a 50×50 m^2 pixel resolution over 12×12 km^2 viewing domains, will be able to detect methane fact sources efficiently. Methane sources in the Arctic are difficult to observe from space. Because of the scarcity of solar radiation and the difficulty of measuring stratospheric methane, MERLIN's upcoming Lidar observations will provide an exclusive asset for observing the Arctic in the dark.

KEYWORDS

- **microscale**
- **atmospheric boundary layer**
- **micrometeorology**
- **carbon flux**

REFERENCES

Arya, S. P. S.; Wyngaard, J. C. Effect of Baroclinicity on Wind Profiles and the Geostrophic Drag Law for the Convective Planetary Boundary Layer. *J. Atmos. Sci.* **1975,** *32*(4), 767–778.

Brandle, J. R.; Hodges, L.; Tyndall, J.; Sudmeyer, R. A. Windbreak Practices. In *North American Agroforestry: An Integrated Science and Practice*, 2009; pp 75–104.

Cressot, C.; Chevallier, F.; Bousquet, P.; Crevoisier, C.; Dlugokencky, E. J.; Fortems-Cheiney, A.; Montzka, S. A. On the Consistency Between Global and Regional Methane Emissions Inferred from SCIAMACHY, TANSO-FTS, IASI and Surface Measurements. *Atmos. Chem. Phys.* **2014,** *14*(2), 577–592.

Duchemin, E.; Lucotte, M.; Canuel, R. Comparison of Static Chamber and Thin Boundary Layer Equation Methods for Measuring Greenhouse Gas Emissions from Large Water Bodies. *Environ. Sci. Technol.* **1999,** *33*(2), 350–357.

Fröhlich-Nowoisky, J., Kampf, C. J; Weber, B.; Huffman, J. A.; Pöhlker, C.; Andreae, M. O.; Su, H. Bioaerosols in the Earth System: Climate, Health, and Ecosystem Interactions. *Atmos. Res.* **2016,** *182*, 346–376.

Glickman, T. S. *Glossary of Meteorology*; American Meteorological Society: Boston, MA, 2000; p 855.

Horton, D. E.; Johnson, N. C.; Singh, D.; Swain, D. L.; Rajaratnam, B.; Diffenbaugh, N. S. Contribution of Changes in Atmospheric Circulation Patterns to Extreme Temperature Trends. *Nature* **2015,** *522*(7557), 465–469.

Jacob, D. J.; Turner, A. J.; Maasakkers, J. D.; Sheng, J.; Sun, K.; Liu, X.; Frankenberg, C. Satellite Observations of Atmospheric Methane and Their Value for Quantifying Methane Emissions. *Atmos. Chem. Phys.* **2016**.

Jayadeva, H. M.; Prabhakara Setty, T. K.; Gowda, R. C.; Devendra, R.; Mallikarjun, G. B.; Bandi, A. G. Methane Emission as Influenced by Different Crop Establishment Techniques and Organic Manures. *Agric. Sci. Digest.* **2009,** *29*(4), 241–245.

Jiménez, C.; Prigent, C.; Aires, F. Toward an Estimation of Global Land Surface Heat Fluxes from Multisatellite Observations. *J. Geophys Res. Atmos.* **2009,** *114*.

Jung, M.; Reichstein, M.; Bondeau, A. Towards Global Empirical Upscaling of FLUXNET Eddy Covariance Observations: Validation of a Model Tree Ensemble Approach Using a Biosphere Model. *Biogeosciences* **2009,** *6*, 2001–2013.

Khosa, M. K.; Sidhu, B. S.; Benbi, D. K. Effect of Organic Materials and Rice Cultivars on Methane Emission from Rice Field. *J. Environ. Biol.* **2010,** *31*, 281–285.

Krishna Moorthy, K.; Suresh Babu, S.; Satheesh S. K. Temporal Heterogeneity in Aerosol Characteristics and the Resulting Radiative Impact at a Tropical Coastal Station? Part 1: Microphysical and Optical Properties. *Ann. Geophys. Eur. Geosci. Un.* **2007,** *25*(11), 2293–2308.

Long, D.; Longuevergne, L.; Scanlon, B. R. Uncertainty in Evapotranspiration from Land Surface Modeling, Remote Sensing, and GRACE Satellites. *Water Resour. Res.* **2014**, *50*, 1131–1151.

Mauder, M.; Foken, T.; Cuxart, J.; Surface-Energy-Balance Closure Over Land: A Review. *Bound. Layer Meteorol.* **2020**. https://doi. org/10.1007/s10546-020-00529-6.

Naik, V.; Voulgarakis, A.; Fiore, A. M.; Horowitz, L. W.; Lamarque, J. F.; Lin, M.; Cionni, I. Preindustrial to Present-Day Changes in Tropospheric Hydroxyl Radical and Methane Lifetime from the Atmospheric Chemistry and Climate Model Intercomparison Project (ACCMIP). *Atmos. Chem. Phys.* **2013,** *13*(10), 5277–5298.

Orlanski, I A. Rational Subdivision of Scales for Atmospheric Processes. *Bullet. Am. Meteorol. Soc.* **1975,** 527–530.

Ostler, A.; Sussmann, R.; Patra, P. K.; Houweling, S.; De Bruine, M.; Stiller, G. P.; Saunois, M. Evaluation of Column-Averaged Methane in Models and TCCON with a Focus on the Stratosphere. **2016**.

Saad, K. M.; Wunch, D.; Toon, G. C.; Bernath, P.; Boone, C.; Connor, B.; Roehl, C. Derivation of Tropospheric Methane from TCCON CH 4 and HF Total Column Observations. *Atmos. Measurement Tech.* **2014**, *7*(9), 2907–2918.

Song, Q.; Zhang, G.; Zhu, X. G. Optimal Crop Canopy Architecture to Maximise Canopy Photosynthetic CO2 Uptake Under Elevated CO2–a Theoretical Study Using a Mechanistic Model of Canopy Photosynthesis. *Funct. Plant Biol.* **2013**, *40*(2), 108–124.

Stevenson, D. S.; Zhao, A.; Naik, V.; O'Connor, F. M.; Tilmes, S.; Zeng, G.; Horowitz, L. W. Trends in Global Tropospheric Hydroxyl Radical and Methane Lifetime Since 1850 from AerChemMIP. *Atmos. Chem. Phys.* **2020**, *20*(21), 12905–12920.

Stull, R. B. *An Introduction to Boundary Layer Meteorology*, Springer Science Business Media, 2012; p 13.

Wan, Z.; Zhang, K.; Xue, X. W.; Hong, Z.; Hong, Y.; Gourley, J. J. Water Balance Based Actual Evapotranspiration Reconstruction from Ground and Satellite Observations Over the Conterminous United States. *Water Resour. Res.* **2015,** *51*, 6485–6499. DOI: 10.1002/2015 WR017311.

Wan, Z.; Zhang, Y.; Zhang, Q.; Li, Z. L. Validation of the Land-Surface Temperature Products Retrieved from Terra Moderate Resolution Imaging Spectroradiometer Data. *Remote Sens. Environ.* **2002,** *83*(1–2), 163–180.

Watson, P. L. Methane Exchange Reactions of Lanthanide and Early-Transition-Metal Methyl Complexes. *J. Am. Chem. Soc.* **1983,** *105*(21), 6491–6493.

Wilson, D.; Alm, J.; Laine, J.; Byrne, K. A.; Farrell, E. P.; Tuittila, E. S. Rewetting of Cutaway Peatlands: Are we Re-Creating Hot Spots of Methane Emissions?. *Restor. Ecol.* **2009,** *17*(6), 796–806.

Zeng, Z. Z.; Piao, S. L.; Lin, X.; Yin, G. D.; Peng, S. S.; Ciais, P.; Myneni, R. B. Global Evapotranspiration Over the Past Three Decades: Estimation Based on the Water Balance

Equation Combined with Empirical Models. *Environ. Res. Lett.* **2012,** *7*, 014026. DOI: 10.1088/1748-9326/7/ 1/014026.

Zhang, G.; Murakami, H.; Yang, X.; Findell, K. L.; Wittenberg, A. T.; Jia, L. Dynamical Seasonal Predictions of Tropical Cyclone Activity: Roles of Sea Surface Temperature Errors and Atmosphere/Land Initialization. *J. Climate* **2020,** *1*, 1–61.

Zinder, S. H.; Anguish, T.; Cardwell, S. C. Effects of Temperature on Methanogenesis in a Thermophilic (58°C) Anaerobic Digestor. *Appl. Environ. Microbiol.* **1984,** *47*(4), 808–813.

Index

D

E

F

G

T

U

V

W